T0399690

Evolution of Neurosensory Cells and Systems

Evolutionary Cell Biology

Series Editors

Brian K. Hall – Dalhousie University, Halifax, Nova Scotia, Canada
Sally A. Moody – George Washington University, Washington DC, USA

Editorial Board

Michael Hadfield – University of Hawaii, Honolulu, USA
Kim Cooper – University of California, San Diego, USA
Mark Martindale – University of Florida, Gainesville, USA
David M. Gardiner – University of California, Irvine, USA
Shigeru Kuratani – Kobe University, Japan
Nori Satoh – Okinawa Institute of Science and Technology, Japan
Sally Leys – University of Alberta, Canada

Science publisher

Charles R. Crumly – CRC Press/Taylor & Francis Group

Published Titles

Evolving Neural Crest Cells
Edited by Daniel Meulemans Medeiros, Brian Frank Eames, Igor Adameyko

Development of Sensory and Neurosecretory Cell Types: Vertebrate Cranial Placodes, volume 1
By Gerhard Schlosser

Evolutionary Origin of Sensory and Neurosecretory Cell Types: Vertebrate Cranial Placodes, volume 2
By Gerhard Schlosser

Evolutionary Cell Processes in Primates: Bones, Brains, and Muscle, Volume I
Edited by M. Kathleen Pitirri and Joan T. Richtsmeier

Evolutionary Cell Processes in Primates: Genes, Skin, Energetics, Breathing, and Feeding, Volume II
Edited by M. Kathleen Pitirri and Joan T. Richtsmeier

The Notochord: Development, Evolution and contributions to the vertebral column
Eckhard P. Witten and Brian K. Hall

Evolution of Neurosensory Cells and Systems: Gene regulation and cellular networks and processes
Edited by Bernd Fritzsch and Karen L. Elliott

For more information about this series, please visit:
www.crcpress.com/Evolutionary-Cell-Biology/book-series/CRCEVOCELBIO

Evolution of Neurosensory Cells and Systems

Gene Regulation, Cellular Networks and Processes

Edited by

Bernd Fritzsch, Karen L. Elliott

CRC Press
Taylor & Francis Group
Boca Raton London New York

CRC Press is an imprint of the
Taylor & Francis Group, an **informa** business

First edition published 2022
by CRC Press
6000 Broken Sound Parkway NW, Suite 300, Boca Raton, FL 33487-2742

and by CRC Press
2 Park Square, Milton Park, Abingdon, Oxon, OX14 4RN

ISBN: 978-0-367-55211-4 (hbk)
ISBN: 978-0-367-55287-9 (pbk)
ISBN: 978-1-003-09281-0 (ebk)

DOI: 10.1201/9781003092810

Typeset in Times LT Std
by KnowledgeWorks Global Ltd.

Contents

Contributors

Jeremy S. Duncan
Department of Biological Sciences
Western Michigan University
Western Michigan School of Medicine
Kalamazoo, Michigan

Karen L. Elliott
Department of Biology
The University of Iowa
Iowa City, Iowa

Jacob Engelmann
AG Active Sensing
Faculty of Biology
Bielefeld University
Bielefeld, Germany

Bernd Fritzsch
Department of Biology, CLAS
Department of Otolaryngology
The University of Iowa
Iowa City, Iowa

Takeshi Imai
Graduate School of Medical
    Sciences
Kyushu University 3-1-1, Maidashi,
    Higashi-ku
Fukuoka, Japan

Sarah Nicola Jung
AG Active Sensing
Faculty of Biology
Bielefeld University
Bielefeld, Germany

Robin F. Krimm
Anatomical Sciences and
    Neurobiology
Louisville, Kentucky

Valerie Lucks
AG Active Sensing
Faculty of Biology
Bielefeld University
Bielefeld, Germany

Paul R. Martin
Save Sight Institute & School of
    Medical Sciences & ARC Centre for
    Integrative Brain Function
University of Sydney
Australia

Stephen D. Roper
Department of Physiology & Biophysics
    and Otolaryngology
Miller School of Medicine
University of Miami
Miami, Florida

Sydney N. Sheltz-Kempf
Department of Biology
Western Michigan School of
    Medicine
Kalamazoo, Michigan

Bernd Sokolowski
University of South Florida
Morsani College of Medicine
Department of Otolaryngology-HNS
Department of Molecular
    Pharmacology & Physiology
Tampa, Florida

Hans Straka
Department of Biology II
Ludwig Maximilians University of
    Munich
Planegg, Germany

Ebenezer N. Yamoah
Department of Physiology and Cell
    Biology
School of Medicine
University of Nevada
Reno, Nevada

Preface

The various vertebrate sensory systems share a common theme: the presence of specialized peripheral sensory cells which detect specific stimuli, and the ability to encode and transmit such stimuli into the nervous system. While the former feature is common to all senses, the latter can be distinguished by either direct connections from the sensory cell itself, or through intermediate sensory neurons. Evolution has assembled these morphophysiological characteristics through optimization of functional cell types, driven largely by selective pressure acting upon the components of the various systems. This selection pressure predominately shaped the fundamental gene regulatory networks and associated cellular mechanisms which underly the development of such cell types and their functional properties. Information gained from gene expression profiles have guided our understanding of the evolution and development of these various sensory components. While sensory systems have embraced unique developmental gene expression profiles in an effort to accommodate environmental demands of their individual sensory modalities, they have nevertheless maintained a similar foundation of gene expression across neurosensory development. For instance, the dominant expression of *Eya1*, which drives *Pax2/3/6* expression, subsequently leading to *Sox2* upregulation.

Gene regulation strategies are the primary determinant of cellular networks and are mainly understood in the context of establishing specific cellular and molecular characteristics. Each of the vertebrate senses (olfaction, vision, trigeminal, taste, vestibular, auditory, lateral line, and electroreception) have neurosensory cells at the periphery that depend upon a small set of conserved genes. Sensory cells depend on *Atoh1* (5 different sensory receptors), *Neurod1* (retina sensory cells), *Sox2/Shh* (taste receptors), and a few unique genes for olfaction. In addition, the Merkel cells of the trigeminal systems depend on Piezo and sensory hair cells in the vestibular, auditory, lateral line and depend on *TMC1/2* for mechanotransduction. The encoding and transmission of sensory information, accomplished largely by differentiated sensory neurons, is heavily influenced by the role of *Neurog1/2* in nearly all sensory systems, with the notable exception of the visual system, which depends on *Atoh7*.

In the brain, central nucleus neurons depend on specific genes based predominately on their location in the brain. For example, Atoh1 is expressed in the most dorsal portion of the hindbrain and defines auditory, lateral line, and electroreceptive nuclei neurons. Upstream of *Atoh1* is *Lmx1a/b* and *Gfp7* that interact with several other bHLH genes (*Ascl1, Ptf1a, Neurog1/2, Olig3*). Without the expression of *Lmx1a/b*, many dorsal features, such as the roof plate, do not form. Beyond initial input from the periphery, sensory systems have at least one, and up to several, additional higher order projections into various brain regions which serve to relay and process environmental information. More globally, central integration of multimodal information is established by tertiary connections between the various systems. An example of such tertiary connectivity is observed in nearly every sensory system which centrally converges extensively with the vestibular system.

Central sensory neurons will depend on unique combination of neurons that may depend on a single gene (*Atoh1* defines auditory, lateral line and electroreception) that compares to an interplay of several other genes that interact (*Aslc1, Ptf1a, Neurog1/2* and *Olig3*, among others). Upstream of the roof plate that depends on *Lmx1a/b* and *Gfp7* and will lose in specific downstream genes, *Atoh1*. Central connections are with one or up to three connections to reach the thalamus or directly reach the olfactory bulb. At least three sensory systems (olfaction, vision, trigeminal) extend in lamprey and vertebrates that are connected of five additional second order projections. Tertiary connections are best analyzed that show a strong interaction between olfaction and taste, vision and auditory, and shows a broad sensory input to expand nearly all vestibular input.

Each of the vertebrate sensory systems are reviewed here. Individual chapters will provide an overview of the cellular networks of each sensory system with particular focus on our current understanding of gene regulatory influences in the development of these features. These chapters will explore the functional principles which drive the relay of information from sensory cells at the periphery to central circuits, including cortical areas. We will conclude with a final chapter that offers a comparison gene regulatory network in vertebrates to out-group species, such as lancelets and ascidians.

We hope that these chapters will give a better understanding of sensory system assembly and function as they reliably relay environmental information from the periphery to central processing centers and their context in the light of evolution.

We acknowledge the support by NIH/NIA for R01 AG060504 (BF) and R03 DC015333 (KE).

Bernd Fritzsch
Karen L. Elliott

1 The Senses
Perspectives from Brain, Sensory Ganglia, and Sensory Cell Development in Vertebrates

Bernd Fritzsch, Karen L. Elliott

CONTENTS

1.1 INTRODUCTION OF PRIMARY NEUROSENSORY ORGANIZATION

Responding to an environment is a demand placed on all organisms and necessitates the ability to sense external stimuli. For organisms that have complex interactions with their environment, such as vertebrates, this ability has been evolutionary optimized to ensure precise methods of discriminating sensory information. Such optimization has been driven by coordinated development of a centralized processing center that is connected to peripheral sensory organs by way of peripheral neurons. Many studies have investigated the origins of the nervous system (Layden, 2019). Current evidence suggests a common ancestor for cnidarians and bilaterian nervous systems (Galliot et al., 2009). In chordates, the central nervous system (CNS) develops from the dorsal ectoderm. However, the urbilaterian ancestor likely had a similar organization to protostomes with a ventrally positioned central nervous system (Gerhart, 2000). The evolution of deuterostomes saw an inversion of that body plan. Multiple ideas have been proposed to explain these differing dorso-ventral developmental inversion schemes and originated primarily from comparative data on neuron, heart, and mouth location (Arendt and Nübler-Jung, 1997; Fritzsch et al., 2017; Gee, 2007). Among protostomes, the central nervous system is located ventrally, whereas the heart is found dorsally in the body (Gerhart, 2000). In contrast, among deuterostomes, the central nervous system is located dorsally and the heart is ventral (Gerhart, 2000). The ancestral deuterostome mouth is hypothesized to have originally formed dorsally following the inversion of the body plan, suggesting the mouth had to migrate ventrally (Lacalli, 2008). Closer examination showed that this inversion of body plan may have happened in different steps, as shown in the lateral positioning of the mouth in developing lancelets, which migrates ventrally during development (Kaji et al., 2016; Lacalli, 2008). The tunicates, which are the sister group to vertebrates, form their mouth directly from the neuropore, the opening at the rostral end of the neural tube (Veeman and Reeves, 2015), confirming the independent connection of the neuropores with the opening of the gut (Veeman et al., 2010) to forming a new ventral opening of vertebrates (Figure 1.1). Movement of the deuterostome mouth ventrally allows for the entire gut to remain ventrally, whereas in protostomes, the gut passes from the ventral mouth through an opening in the brain to then run dorsally in the organism (Gerhart, 2000). Elimination of the gut passing through the brain in deuterostomes would have allowed for the evolution of larger brains in vertebrates and thus acquisition of more complex processing centers. Gaining these new and more complex sensory processing occurred following duplication and diversification of gene regulatory networks (Fritzsch and Elliott, 2017).

Development from a single cell to a complex adult individual foremost requires establishing the two body axes: antero-posterior and dorso-ventral (Meinhardt, 2015b). Several genes are involved in this two-step process (Figure 1.1): First, factors define the antero-posterior organization. Wnt signaling has been shown to be a key component in establishing the antero-posterior axis across various organisms, including Cnidarians (Hobmayer et al., 2000; Meinhardt, 2015b). Also, *Hox* genes become expressed in a time-dependent, segmental pattern along the antero-posterior axis to further specify regions along the length of this axis (Wacker et al., 2004). Second, additional factors establish the dorso-ventral organization. Bilaterally-symmetrical protostomes and deuterostomes express *Chordin* and *Noggin* dorsally, which inhibit the expression of *BMPs* (Meinhardt, 2015a). In addition, *BMPs* antagonize the expression of *Chordin* (Lele et al., 2001;

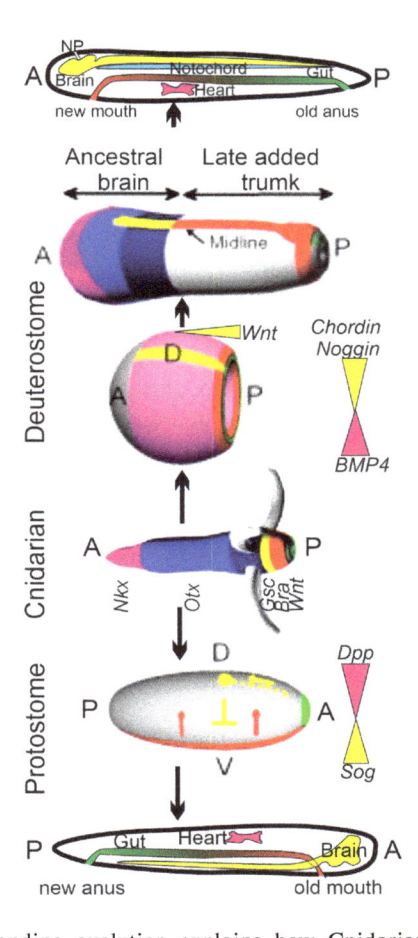

FIGURE 1.1 Understanding evolution explains how Cnidarians, with polarized *Wnt* expression (green) and the 'heart' (*Nkx*, lilac) is transformed independently to generate a ring-to-rod transformation by downregulation of *BMP2/4* in Deuterostomes. This contrasts with a ventral midline (red, *Sog*) that results a dorsal signaling (green, *Dpp*) in Protostomes. Cnidarians start with *Wnt* expression at the posterior part, followed by *Bra, Gsc, Otx,* and *Nkx* (most anterior). Deuterostome development follows the Spemann organizer that will induce a symmetry break by *Wnt* and suppression of *BMP2/4* to generate a strip of ectoderm that produces the brain and spinal cord, followed by convergent extension. Earliest expression of heart genes (*Nkx*) is anterior to the head in hagfish and migrates ventrally during vertebrate development; a master reorganization consistent with heart homology between Deuterostomes and Cnidarians. Note the unique opening of the neuropore (*NP*) in chordates. Nearly identical genes are present among Protostomes, but they have a different expression of *Dpp* (the *BMP* homolog): In contrast to being downregulated in Deuterostomes under dorsal expressing genes (*Chordin, Noggin,* lilac), they are upregulated from dorsal *Dpp* (lilac, Protostomes) and specify the dorsal expression of the heart. The A/P is different, flipping the old mouth in Protostomes into the old anus in Deuterostomes. (Modified from Layden, 2019; Meinhardt, 2015b.)

Xue et al., 2014). Thus, *Chordin* and *Noggin* define dorsal and *BMPs* define ventral. In protostomes, *Chordin* and *BMP* are referred to as *short gastrulation (Sog)* and *decapentaplegic (Dpp)*, respectively (Meinhardt, 2015a). Thus in *Drosophila* and other protostomes, ventral identity is defined by *Sog* and dorsal by *Dpp* (Heingård and Janssen, 2020). Like *Chordin*, *Sog* is an antagonist of *Dpp* expression (Akiyama-Oda and Oda, 2006; Heingård and Janssen, 2020; Irish et al., 1987) and in spiders, *Dpp* has been shown to repress *Sog* expression (Akiyama-Oda and Oda, 2006). Additional genes, such as *Nodal*, *Fgfs*, and *Otx*, play further roles in body patterning (Meinhardt, 2015b). These genes involved in body axes patterning will go on to play important roles in the induction of the nervous system and will be discussed below.

This chapter will explore the different genes that define the dorsal portion of the brain and spinal cord, the sensory neurons, and the different sensory cells. We will discuss neural induction of the brain and spinal cord and the formation of the neural crest and placodes that give rise to components of the various sensory systems.

1.2 NEURAL INDUCTION

Neural induction involves genes known to be important for earliest steps of general body patterning. Initially *BMP*, *Wnt*, *Activin*, and *Nodal* are expressed throughout the ectoderm (Lee et al., 2014; Moody et al., 2013; Rogers et al., 2009). The transition from ectodermal tissue to neural ectoderm requires the downregulation of *BMP* by *Noggin*, *Chordin*, and *Follistatin* and the downregulation of *Wnt* by *Dkk* and *Cerberus* (Piccolo et al., 1999; Rogers et al., 2009; Semënov et al., 2001). *Fgfs* also contribute to this process (Lee et al., 2014).

Following initial neural induction, at least three genes, *Geminin (Gmnn)*, *Zic*, and *Foxd4* become expressed that further define and stabilize the neural ectoderm by inhibiting *BMP* and *Wnt* signaling as well as upregulation of other pro-neural genes (Aruga and Hatayama, 2018; Lee et al., 2014; Moody et al., 2013; Sankar et al., 2016; Yan et al., 2009) (Figure 1.2). Overexpression of *Gmnn* leads to expression of *Zic1* and, while there is no direct interaction of these two genes, they were shown to have overlapping and cooperative roles in neuronal progenitor formation and maintenance (Sankar et al., 2016). Similarly, *Foxd4* promotes the proliferation of neuronal precursors (Moody et al., 2013) and regulates expression of the *Gmnn* and *Zic* genes (Lee et al., 2014; Yan et al., 2009).

1.2.1 Formation of the Neural Plate and Neural Tube

The neural plate is induced following the inhibition of the epidermally expressed *BMPs* and *Wnts* along with the upregulation of *Gmnn*, *Foxd4*, and *Zic* genes (Lee et al., 2014; Sadler, 2005; Sankar et al., 2016). The expression of *Gmnn*, *Foxd4*, and *Zic* genes in neural precursor cells leads to the upregulation of downstream of *Sox* genes (Lee et al., 2014; Sankar et al., 2016) (Figure 1.2). Expression of *Sox* genes results in the transformation of these precursor cells into a neural plate

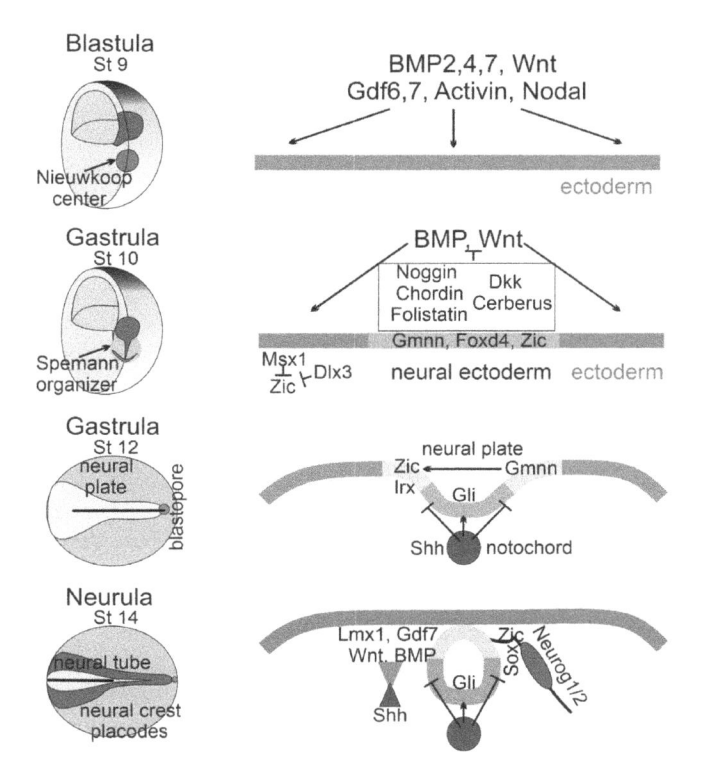

FIGURE 1.2 Comparison of the critical steps identified in frogs (left) and the distinct steps of neural induction and formation (right). Initially, *Nodal, Activin, BMP*, and *Wnt* are expressed throughout the ectodermal tissue. Spemann's organizer induces neural induction by expressing *Chordin/Noggin/Follistatin*, which act against *BMP2/4. Gmnn, Zic,* and *Foxd4* become expressed, upstream from *Irx* and upregulate *Sox* expression. An early expression of *Shh* drives *Gli* expression ventral and is counteracted by the roof plate/choroid plexus. *Lmx1a/b* drives the dorsal expression of *Wnts* and *BMPs*, including *Gdf7*. Additional downstream genes include the bHLH genes (*Atoh1, Neurog1/2, Ascl1, Olig3*). *Neurog1/2* drives the generation of sensory neurons from neural crest and placodes. (Modified from Aruga and Hatayama, 2018; Glover et al., 2018; Lee et al., 2000; Mishima et al., 2009; Sankar et al., 2016.)

stem cell (Lee et al., 2014). The *SoxB1* family (*Sox1, Sox2, Sox3*), together with *Gmnn, Foxd4*, and *Zic2*, are essential for the continued proliferation of undifferentiated neural stem cells (Lee et al., 2014). The explosive proliferation of these neural stem and the coordination with convergent-extension leads to the folding of the neural plate into the neural tube (Figure 1.1; Sadler, 2005; Schoenwolf and Alvarez, 1989). The following transition from a neural stem cell into a neuronal progenitor cell requires the expression of *Sox11* (Lee et al., 2014). For a neuronal progenitor cell to exit the cell cycle and proceed to differentiation, the initial upregulation of *Gmnn, Foxd4, Zic*, and *Sox* genes must be downregulated (Aruga and Hatayama, 2018; Imayoshi and Kageyama, 2014; Kageyama et al., 2019; Lee et al., 2014; Reiprich and Wegner, 2015; Sankar et al., 2016). Following the down-regulation of the various proliferative genes, bHLH genes involved in neuronal differentiation in the CNS become expressed in dorso-ventral expression patterns within the developing neural tube to regulate fates of neurons within various subdomains (Glover et al., 2018; van der Heijden and Zoghbi, 2020; Wilson and Maden, 2005).

1.2.2 PATTERNING OF THE NEURAL TUBE

After the neural plate fuses to form the neural tube, the newly formed dorsal roof plate begins to express *BMPs* and *Wnts*, while the ventral floorplate and underlying notochord expresses *Shh* (Ulloa and Martí, 2010; Wilson and Maden, 2005). Although many genes play a role in patterning the neural tube, *BMPs*, *Wnts*, and *Shh* play a primary role (Wilson and Maden, 2005). Expression of *Wnts* and *BMPs* in the dorsal neural tube is driven by the dorsal expression of *Lmx1a/b* (Chizhikov et al., 2019; 2021; Glover et al., 2018; Mishima et al., 2009). *Lmx1b* expression itself in the roof plate is regulated by *Zic1*, *Zic3*, and *Zic4* (Winata et al., 2013; Winata and Korzh, 2018). While the timing of *Wnt* and *BMP* expression during neural tube formation varies between mice and frogs (Aruga and Hatayama, 2018; Barratt and Arkell, 2018), these two genes, as well as *Fgfs* and retinoids, are critical in defining the dorsal part of the brainstem and spinal cord (Hernandez-Miranda et al., 2017). Loss of *Wnt* or *BMP* signaling negatively affects dorsal progenitors. Mice null for *Gdf7*, a *BMP* family member, do not form the more dorsal structures (Lee et al., 1998; 2000).

Determination of dorso-ventral identity within the spinal cord and brainstem is further governed by the expression of various homeodomain and bHLH transcription factors in restricted areas along the dorso-ventral axis (Hernandez-Miranda et al., 2017; Lai et al., 2016; Le Dréau and Martí, 2012; van der Heijden and Zoghbi, 2020). The dorsal brainstem and spinal cord are subdivided into eight or six domains of neuronal progenitor populations, respectively (Hernandez-Miranda et al., 2017). The bHLH gene, *Olig3*, is expressed in the three to four dorsal-most domains of progenitors along the length of the hindbrain and spinal cord and is dependent upon Wnt signaling (Hernandez-Miranda et al., 2017; Lai et al., 2016). Expression of additional genes further subdivides these domains. For instance, the bHLH gene, *Atoh1*, is expressed in the most dorsal domain (dA1) throughout the brainstem and spinal cord (Bermingham et al., 2001; Hernandez-Miranda et al., 2017). *Atoh1*-positive cells generate the rhombic lip of the hindbrain, superficial migratory neuron streams, and cerebellar granule cells that contribute to auditory, vestibular, and proprioceptive networks (Wang et al., 2005). As expected from the expression pattern, loss of *Atoh1* eliminates most cochlear nuclei neurons (Fritzsch et al., 2006). Additional genes expressed in the *Olig3* domain ventral to *Atoh1* include *Neurog1* (dA2), *Ascl1* (dA3-dB1), and *Ptf1a* (dA4-dB1) (Storm et al., 2009; van der Heijden and Zoghbi, 2020). Independently, homeodomain transcription factors are also expressed in a dorso-ventral pattern, overlapping with some of the bHLH transcription factors

(Hernandez-Miranda et al., 2017). For example, the dorsal-most subdomain (dA1), that also expresses *Atoh1*, expresses *Pax3*, the next subdomain (dA2) expresses *Pax3* and *Pax7*, and the third subdomain (dA3) expresses *Pax3*, *Pax6*, and *Pax7* (Hernandez-Miranda et al., 2017). Additional genes are uniquely expressed in progenitor subdomains, such as *Barhl1* in dA1 progenitors, *Lhx* in dA2, *Tbx* in dA3, *Foxd3/Foxp2* in dA2 and dA4, and *Phox2b* in dB2 (Hernandez-Miranda et al., 2017). Meanwhile, genes such as *Pou4f1* are expressed in multiple subdomains (dA1-dA3, dB3) (Hernandez-Miranda et al., 2017). Interestingly, *Atoh1*, which defines the dorsal most progenitor population (dA1), is also expressed in a more ventral subpopulation (dB2) during their maturation (Hernandez-Miranda et al., 2017; van der Heijden and Zoghbi, 2020).

Compared with the hindbrain (dA1-4; dB1-4), the spinal cord (dL1-3; dL 4-6) has fewer domains (Hernandez-Miranda et al., 2017). While some expression domains extend between the hindbrain and spinal cord, such as *Olig3* and *Atoh1*, not all do (Hernandez-Miranda et al., 2017). For instance, *Phox2b* is expressed in rhombomere 2-6 of the hindbrain (dB2) but not in rhombomere 7 or the spinal cord (Hernandez-Miranda et al., 2017). This unique hindbrain domain (dB2) also later expresses *Atoh1*, in an expression pattern that is unequal to that of the spinal cord (Hernandez-Miranda et al., 2017). A second unique domain in the hindbrain (dA4) expresses *Ptf1a*, *Foxd3*, and *FoxP2* among a few other genes (Hernandez-Miranda et al., 2017; Yamada et al., 2007). Loss of *Ptf1a* results in the loss of dA4 and dB1 neurons in combination with the expansion of dA3 and dB3 neurons, leading to the eventual misspecification of somatosensory and viscerosensory nuclei neurons in the hindbrain (Iskusnykh et al., 2016).

The boundaries and identities of these various hindbrain and spinal cord subdomains are further specified by cross regulation and feedback loops of the various genes expressed. For instance, there is a reciprocal cross regulation of *Atoh1-Neurog1* to sharpen the boundaries of *Atoh1* and *Neurog1/2* in the spinal cord (Lai et al., 2016). In *Atoh1* null mice, the dA1 neurons are lost and *Neurog1/2* expression expands and generates additional dA2 neurons (Gowan et al., 2001). Moreover, *Neurog1/2* and *Ascl1* cross regulate each other and *Ascl1* and *Ptf1a* also cross regulate to define additional subdomains (Lai et al., 2016).

Transition from neuronal progenitor to neuronal differentiation requires the interaction of proneural bHLH genes and additional bHLH genes, the Class I *Hes/Hey* genes (Imayoshi and Kageyama, 2014; Kageyama et al., 2019) and the Class V *ID* genes (Fritzsch et al., 2015; Jahan et al., 2015; Sakamoto et al., 2020). Proliferation of neurosensory precursors is driven by *Hes*, *ID*, *Sox2*, and *Myc* genes and the transition to differentiated cells is controlled by the balance of *Notch* signaling molecules and the proneural bHLH genes (Lai et al., 2016). Oscillation of a proneural bHLH gene and a repressive one, such as *Ascl1* and *Hes1*, through cross regulation of each other controls the timing of neurogenesis (Kageyama et al., 2019; Lai et al., 2016). This oscillation maintains proliferation of neuronal progenitors, whereas the eventual loss of *Hes1* expression and subsequent sustained expression of a proneural bHLH gene leads to neuronal differentiation (Kageyama et al., 2019).

The expression of downstream bHLH genes, such as *Neurod1*, adds to these complex interactions. For instance, *Atoh1* expression extends along the roof plate to the

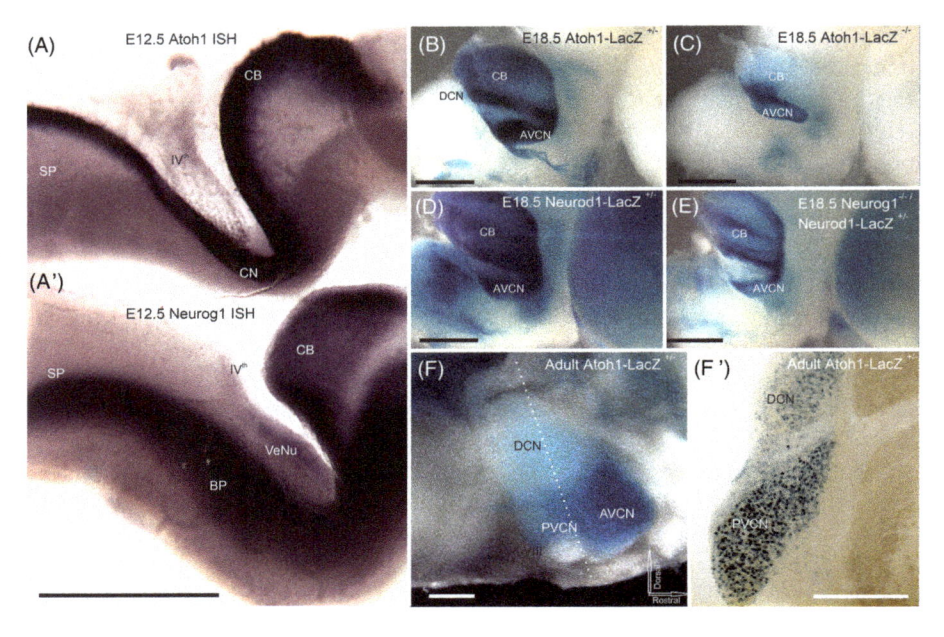

FIGURE 1.3 Gene expression and supporting interactions among brainstem nuclei. The dorsal-most expression in the brainstem is *Atoh1* (A) and ventral to it is a limited expression of *Neurog1* in E12.5 mice (A'). The effect of *Atoh1* loss is demonstrated in *Atoh1* lacZ null compared to a control expressing one allele (B), (C). Loss of one allele of *Neurod1* (D) alone shows no effect. The loss of one allele of *Neurod1* combined with *Neurog1* null mice shows a distinct interaction of *Neurog1* compared to Atoh1 (C), (E). In the adult system, the level of *Atoh1* is changing and is downregulated in the dorsal cochlear nucleus (DCN) compared to the ventral cochlear nuclei (AVCN, PVCN). (F), (F'), BP, basal plate; CB, cerebellum; SP. spinal cord. (Modified from Fritzsch et al., 2006; Pan et al., 2009.)

cerebellum, parallel to the slightly more ventral expression of *Neurog1/2*, and the loss of dorsal neurons in *Atoh1* null mice (Fritzsch et al., 2006; Wang et al., 2005) results in a reduced cerebellum and spinal cord (Figure 1.3). *Neurod1* negatively regulates *Atoh1* expression during cerebellum and gut proliferation, and manipulating *Neurod1* expression may help to counteract Medulloblastoma (Cheng et al., 2020; Kersigo et al., 2021; Li et al., 2019; Pan et al., 2009). Furthermore, there is a near identical expression of *Atoh1* and *Neurod1* in the cerebellum and in auditory nuclei (Pan et al., 2009). *Atoh1* shows a much higher level of expression in the auditory nuclei and counteracts with *Neurod1*, indicating a differential regulation of expression in auditory nuclei (Elliott et al., 2021).

Uniquely among vertebrates is the formation of neural crest and placodes. These are the origin of the neurosensory cells and neurons of olfactory, vision, trigeminal, taste, vestibular, auditory, lateral line, and electroreceptive systems, which provide input into the brain. These will be highlighted in the following sections.

1.3 PLACODE AND NEURAL CREST DEVELOPMENT INTO SENSORY CELLS AND NEURONS

Once the neural plate is fused and generates the roof plate, neural crest and placodes develop from immediately adjacent ectodermal tissue. Neural crest and placodal cells give rise to the formation of all or part of the various sensory systems. The work of Northcutt and Gans provided a novel perspective of the organization of the neural crest and placodes (Northcutt and Gans, 1983), supporting the idea that neural crest and neurogenic placodes evolved from the epidermal nerve plexus of ancestral deuterostomes (Fritzsch and Northcutt, 1993; Green et al., 2015; Holland, 2020; Moody and LaMantia, 2015; Northcutt, 2005; O'Neill et al., 2012; Plouhinec et al., 2017; Schlosser, 2015, 2021; Thiery et al., 2020). Development of neural crest and placodes is highly conserved across vertebrate species (Moody and LaMantia, 2015; Schlosser, 2021).

Neural crest induction occurs at the lateral edge of the neural plate (Hong and Saint-Jeannet, 2005). Following closure of the neural tube, neural crest cells migrate throughout the embryo, contributing to a wide range of tissues including neurons, craniofacial skeleton, smooth muscles, and melanocytes (Knecht and Bronner-Fraser, 2002; Rogers et al., 2012). *BMP, Wnt, Fgf*, and *Notch* signaling are all essential for the formation of neural crest cells (Hong and Saint-Jeannet, 2005; Knecht and Bronner-Fraser, 2002; Rogers et al., 2012). Upon induction of the neural crest, many crest-specific genes are expressed, including *LSox5, Sox8, Sox9*, and *Sox10* (Hong and Saint-Jeannet, 2005). While there are specific species differences in the onset and sequence of expression, these *Sox* genes are important for the specification, migration, and differentiation of neural crest cells (Hong and Saint-Jeannet, 2005). Additional genes that are upregulated include *Snail, Slug, Pax3/7, Hairy2, Msx1/2, Dlx5*, and *Gbx2* (Knecht and Bronner-Fraser, 2002; Rogers et al., 2012).

Cranial placodes develop from a panplacodal region at the neural border zone and give rise to the anterior pituitary gland, the olfactory epithelium, the lens of the eye, the large neurons in the trigeminal, facial, glossopharyngeal, and vagus nerves, the inner ear containing the vestibular and auditory epithelia and sensory neurons, and finally in many aquatic animals, the lateral line and electroreceptive organs and sensory neurons (Moody and LaMantia, 2015; Schlosser, 2021; Schlosser et al., 2014; Xu et al., 1997). The panplacodal region broadly expresses *Eya1, Six1*, and *Six4* (Schlosser, 2010; Xu et al., 2021). During development, the panplacodal region begins to subdivide. Differential expression of transcription factors along the antero-posterior axis gives rise to anterior and posterior compartments, which are further subdivided into individual placodes (Schlosser and Ahrens, 2004). The anterior compartment, which gives rise to the adenohypophyseal, olfactory, and lens placodes, expresses *Six3, Pitx, Pax6, Anf, FoxE*, and *Dmrt* (Lleras-Forero et al., 2013; Schlosser, 2015). In contrast, the posterior compartment, which gives rise to the otic, lateral line, and epibranchial placodes, expresses *Pax2, Pax8, Sox2*, and *Sox3* (Schlosser, 2015).

1.3.1 OLFACTORY RECEPTORS ARE FOUND ON OLFACTORY SENSORY NEURONS WHICH PROJECT DIRECTLY TO THE FOREBRAIN

The olfactory system is present in all vertebrates and its development is highly conserved (Schlosser et al., 2014). It is derived from an anterior bilateral placode that invaginates, forming the olfactory epithelium (Figure 1.4). The olfactory epithelium contains olfactory sensory neurons (OSNs) expressing odorant receptors (Buck and Axel, 1991). These OSNs remain as a part of the sensory epithelium and their axons project centrally to the olfactory bulb in the forebrain. Several genes are involved in the development of the olfactory placode and in its differentiation into OSNs. *Sox2* and *Pax6*, which are expressed in many proliferating cells, are expressed in the olfactory epithelium during development and persisting into adulthood (Guo et al., 2010; Panaliappan et al., 2018). In addition, *Foxg1* is important for olfactory neurogenesis (Kawauchi et al., 2009). The olfactory epithelium has direct contact with the environment and thus retaining the ability to regenerate damaged neurons involves the persistence of olfactory stem cells and the need for continued *Sox2* and *Pax6* expression (Guo et al., 2010). Early loss of *Sox2* completely ablates the formation of the olfactory epithelium (Dvorakova et al., 2020). Downstream of *Sox2* in the olfactory epithelium is *Hes5*, which is needed for neural progenitor induction, as well as *Neurog1* and *Neurod1*, which are expressed in intermediate neuronal precursors and committed neuronal cells, respectively (Panaliappan et al., 2018). In addition, *Sox2*, together with *Meis1*, regulates *Ascl1*, and *Ascl1* regulates *Neurog1* expression (Cau et al., 2002; Schwob et al., 2017; Tucker et al., 2010). Furthermore, *Sox2* restricts expression of *BMP4* (Panaliappan et al., 2018). MicroRNAs are also important in olfactory development as the loss of *Dicer1* resulted in cell death and reduces the olfactory epithelium (Kersigo et al., 2011).

The more than 1000 different odorant receptors are arranged in dorsoventral zones in the olfactory epithelium (Tan and Xie, 2018). Each of these unique odorant receptors, located on an OSN, projects specific odorant information to a given

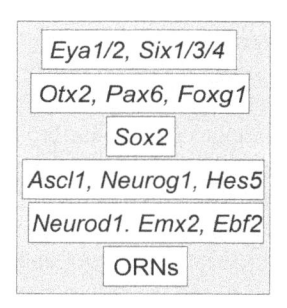

FIGURE 1.4 Olfactory receptor development depends first on *Eya1/Six1*, followed by *Otx2, Pax6,* and *Foxg1*. Evidence is provided for the importance of *Sox2* by using *Foxg1*-cre to eliminate *Sox2* and block olfactory invagination. Downstream of *Sox2* is the expression of *Ascl1, Neurog1,* and *Hes5* to proliferate neuronal cells. Further downstream of *Ascl1/Neurog1/Hes5* is expression of *Neurod1* that interacts with another set of downstream genes to generate all olfactory sensory neurons.

glomerulus (Fritzsch et al., 2019). The particular odorant receptor and the level of gene expression in an OSN assist the navigation of the OSN growth cone to a particular glomerulus. For example, a set of genes, *Kirrel*, *Eph*, and *Prp2*, are responsible for establishing these proper connections of odorant receptors to glomeruli (Fritzsch et al., 2019; Imai, 2020; Mombaerts et al., 1996; Nishizumi and Sakano, 2015). An additional and unique set of odorant receptors develops from the otic placode and generates the vomeronasal system. These neurons express gonadotropin releasing hormone (GnRH) and migrate along the developing olfactory nerve to the hypothalamus and nervus terminalis (Maier et al., 2014; Whitlock, 2004). These neurons have been described in various animals (de Caprona and Fritzsch, 1983; Münz et al., 1982; Von Bartheld, 2004) and may play a role in olfactory modulation (Maier et al., 2014).

1.3.2 The Retina and Lens have Different Embryonic Origins

The development of the visual system focuses on two areas: (1) evolution of rhodopsin to generate rods and cones and (2) the evagination of the brain in vertebrates that interacts with the ectoderm to induce a lens placode (Figure 1.5). Lancelets and ascidians have opsin positive cells and lamella cells (Holland, 2020), but these cells do not interact with the surrounding ectoderm to evaginate an optic retina and associated ganglion neurons (Lamb, 2013; Pantzartzi et al., 2017). The evagination of the retina from a signal frontal position near the neuropore to expand to a bilateral retina (Lamb et al., 2007) is incompletely understood.

Evagination of a ventral portion of the diencephalon forms the optic vesicle, from which the neural and retinal layers develop (Graw, 2010; Lamb, 2013). This optic vesicle initiates induction of the lens placode (Graw, 2010; Lamb, 2013). Both retina and lens development require *Pax6* expression (Moody and LaMantia, 2015; Pichaud and Desplan, 2002). The necessity of *Pax6* for eye development is highly conserved from insects to vertebrates (Pichaud and Desplan, 2002). *Pax6* regulates expression of *Eya1/Six1*, which are additional crucial genes for eye development (Xu et al., 1997). Another *Pax* gene, *Pax2* (Figure 1.5), is important for the

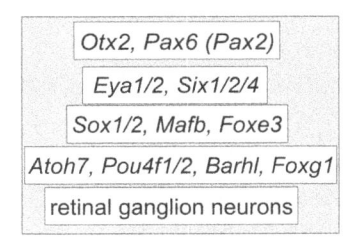

FIGURE 1.5 Eye development depends on *Pax6*, which is upstream of *Eya/Six* to define the lens to upregulate downstream genes. Note that *Pax6* is independent of *Pax2* that defines coloboma defects. Downstream of *Sox1/2*, *Mafb*, and *Foxe3* is *Atoh7* which defines the retinal ganglion neurons and upregulates several downstream genes, such as *Pou4f1/2*, *Barhl1*. An interesting feature is the expression in half of the retina of *Foxg1*.

development of retinal ganglion cells (Favor et al., 1996; Kozmik et al., 2003; Schwarz et al., 2000). Retinal ganglion cell development also depends upon *Atoh7* expression, which is downstream of *Pax6* and *Neurog2* (Ghiasvand et al., 2011; Mao et al., 2013; Wu et al., 2021). Rods and cones depend on *Neurod1* (Brzezinski and Reh, 2015; Dennis et al., 2019). Additional genes are important for various aspects of retinal development, such as *Sox2*, *Foxg1*, *Pou4f1/2*, *Hes* genes, and in addition, microRNAs (Dvorakova et al., 2020; Hatini et al., 1994; Huang et al., 2014; Kageyama et al., 2019; Kersigo et al., 2011; Lassiter et al., 2014; Tian et al., 2008). Additional genes, such as *Eph/Ephrin*, *Wnt3*, and *Ryk*, are involved in guiding retinal projections to the lateral geniculate body and the superior colliculus (Fritzsch et al., 2019; Zhang et al., 2017).

1.3.3 CRANIAL GANGLION NEURONS DEVELOP FROM BOTH PLACODES AND NEURAL CREST

Trigeminal (V), facial (VII), glossopharyngeal (IX), and vagal (X) neurons develop from placodal and neural crest cells (Lassiter et al., 2014; Northcutt, 2005; O'Neill et al., 2012). These neurons project centrally to specific brainstem nuclei (trigeminal, V; solitary tract, VII, IX, X) and peripherally to sensory cells. The trigeminal nerve develops more anteriorly and comprises placodal and neural crest cells, whereas the facial, glossopharyngeal, vagal neurons develop more posteriorly from the epibranchial placodes and neural crest, with fewer cells in mammalian facial neurons (Ayer-Le Lievre and Le Douarin, 1982; D'Amico-Martel and Noden, 1983; Fritzsch et al., 1997). For the trigeminal somatosensory system (V), there is a difference in size between the ophthalmic neurons (Figure 1.6) that develop from the trigeminal placode as compared to the larger maxillary and mandibular branches that develop from neural crest (Erzurumlu et al., 2010).

Induction of the trigeminal and epibranchial placodes depends critically on the expression of *Eya1/2/Six1/4*. These genes are upstream of *Pax3* and *Pax2*,

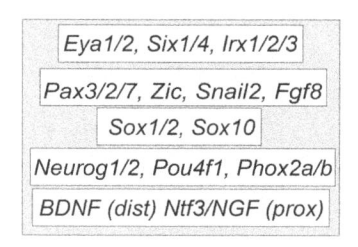

Eya1/2, Six1/4, Irx1/2/3

Pax3/2/7, Zic, Snail2, Fgf8

Sox1/2, Sox10

Neurog1/2, Pou4f1, Phox2a/b

BDNF (dist) Ntf3/NGF (prox)

FIGURE 1.6 Cranial ganglion neurons depend on *Eya/Six*, with Irx also needed for normal formation. *Pax3* (V), *Pax2* (embryonic placode; VII, IX, X), and *Pax7* (neural crest) interact with *Zic, Snail2,* and *Fgf8*. Downstream of *Sox1/2* (Placode) and *Sox10* (neural crest) are needed to drive *Notch* signaling to upregulate *Neurog1* (proximal V) and *Neurog2* (distal VII, IX, X) expression in neurons. Further interactions are known for *Pou4f1* for neuronal development and *Phox2a/b* for embryonic placode development. These neurons also require neurotrophins that have a distinct distribution of *BDNF* for distal and *Ntf3/NGF*, with few *BDNF*, for proximal dependency.

which help define the rostral (*Pax3*, V) and caudal (*Pax2/8*, VII, IX, X) placodes (Erzurumlu et al., 2010; Schlosser, 2010). Meanwhile, neural crest cells depend on *Pax7* (Basch et al., 2006). Additional genes are important in the development of the placodes contributing to the different cranial ganglia. For instance, *Neurog1* is necessary for trigeminal sensory ganglia (Ma et al., 1998), whereas *Neurog2* is necessary for sensory ganglia that develop from the epibranchial placodes (VII, IX, X) (Fode et al., 1998). Downstream of *Neurog1/2* are *Neurod1*, *Isl1*, and *Pou4f1* genes, which promote the differentiation of placodally derived sensory neurons (Fode et al., 1998; Huang et al., 2001; Kim et al., 2001; Ma et al., 1998; Thiery et al., 2020). In addition, the epibranchial placodally derived neurons (VII, IX, X) depend upon *Phox2a/b* expression, which is downstream of *Eya1/Six1* (Zou et al., 2004). In trigeminal ganglion neurons, additional genes such as *Notch*, *Hes*, *Rbpj*, *Fgf8*, *PDGF*, and *Wnts* are important (Fekete et al., 1998; Lassiter et al., 2014; Riddiford and Schlosser, 2016). There are also differential gene expressions within the trigeminal ganglia. Ophthalmic and part of the maxillary trigeminal neurons express *Tbx3*, maxillary and mandibular trigeminal neurons express *Oc2*, and mandibular trigeminal neurons express *Oc1/Hmx1* (Erzurumlu et al., 2010). Moreover, differential expression of neurotrophins (*BDNF*, *Ntf3*) enables discrete innervation of different regions of trigeminal innervation (Erzurumlu et al., 2010). Additional downstream genes expressed in trigeminal neurons, such as *Hox2*, *Lmx1b*, *Slit1/2/Robo2*, and *Neuropilin*, further define distinct projections (Erzurumlu et al., 2010).

While some sensory neurons provide sensory input without contacting sensory cells, others rely on input from receptor cells at the periphery (Abraira and Ginty, 2013; Harold et al., 2019). Sensory Merkel cells depend on *Atoh1*, *Pou4f3*, and *Gfi1* (Maricich et al., 2009; Ostrowski et al., 2015). Other trigeminal sensory cells depend on *Piezo2* expression (Bagriantsev et al., 2014). Taste buds depend on *Eya1*, *Sox2*, and *Shh* (Germanà et al., 2009; Ohmoto et al., 2017; Okubo et al., 2006) to generate at least two distinct sensory receptors (Kinnamon and Finger, 2019; Roper and Chaudhari, 2017), suggesting an independent development of sensory cells.

1.3.4 Otic Placode Development uses Common and Unique Gene Regulatory Networks

While the vestibular system is highly conserved across vertebrates, the auditory system is much more variable (Fritzsch et al., 2013). Both systems are housed within the inner ear. The inner ear develops from the otic placode (Schlosser, 2010). The otic placode uniquely depends on *Foxi3* (Birol et al., 2016) and *Fgf3/10* (Urness et al., 2011), but shares the requirement of *Eya1/2/Six1/2* with many other sensory placodes (Li et al., 2020; Zou et al., 2004). Additionally, *Pax2/8* (Bouchard et al., 2010), *Fgf8* and *Fgfr2b* (Ladher et al., 2010; Pauley et al., 2003; Pirvola et al., 2000; Urness et al., 2011; Wright and Mansour, 2003), *Gbx2* (Steventon et al., 2012), *Gata3* (Duncan and Fritzsch, 2013; Karis et al., 2001), and *Shh* (Riccomagno et al., 2002) are required for inner ear development (Figure 1.7). For instance, the ear never develops beyond the otocyst stage with the

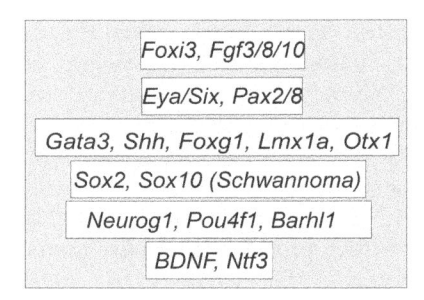

FIGURE 1.7 *Foxi3* expression is unique for the otic placode. Additional important genes are *Eya/Six*, *Pax2/8*, and *Fgf3/8/10*. Downstream of these genes are *Gata3, Shh, Foxg1, Lmx1a*, and *Otx1/2*. *Notch* signaling, downstream of *Sox2*, allows for the expression of *Neurog1*, *Pou4f1*, *Barhl1*, and *Neurod1* in developing otic neurons. *Sox10* is needed for Schwann cells. Neurotrophins, *BDNF* and *Ntf3*, are also important for neurosensory development in the ear.

loss of both *Pax2* and *Pax8* (Bouchard et al., 2010). During development, the inner ear is patterned antero-posteriorly, dorso-ventrally, and proximal-distally (Fekete and Wu, 2002; Groves and Fekete, 2012). These differential patterning gives rise to the different features of the inner ear, for example, sensory vs non-sensory tissue, or vestibular vs auditory structures (Wu and Kelley, 2012). *Shh* specifies the ventral portion of the ear and in its absence, ventral structures, such as the cochlea, are absent (Bok et al., 2005; Riccomagno et al., 2002; Wu and Kelley, 2012). In contrast, *BMP4* specifies dorsal vestibular structures (Chang et al., 2008; Ohyama et al., 2010). Other genes, such as *Foxg1* (Hwang et al., 2009; Pauley et al., 2006), *Lmx1a* (Huang et al., 2018; Mann et al., 2017; Nichols et al., 2008; 2020), *Otx1/2* (Fritzsch et al., 2001; Morsli et al., 1999; Steventon et al., 2012), and *n-Myc* (Domínguez-Frutos et al., 2011; Kopecky et al., 2011; Vendrell et al., 2015), affect specific sensory epithelia and/or structures of the inner ear. Notch signaling and retinoids are also involved in inner ear patterning (Gnedeva et al., 2020; Kageyama et al., 2019; Lassiter et al., 2014; Ono et al., 2020).

Inner ear neurogenesis requires the induction of *Neurog1* by *Sox2*, which in turn downregulates *Sox2* expression (Dvorakova et al., 2020; Evsen et al., 2013; Kiernan et al., 2005; Steevens et al., 2019). *Neurog1* (Ma et al., 2000; Matei et al., 2005) upregulates *Neurod1* (Kim et al., 2001; Macova et al., 2019), and several other bHLH genes (Jahan et al., 2010; Krüger et al., 2006). *Pou4f1* is also involved in neuronal development and in addition, for proper pathfinding of inner ear afferents (Huang et al., 2001). Additional genes, such as *Npr1*, *Prickle1*, *Fzd3/6*, and *Vangl2*, have been implicated in inner ear afferent central and/or peripheral pathfinding (Appler and Goodrich, 2011; Coate and Kelley, 2013; Duncan et al., 2019; Ghimire et al., 2018; Ghimire and Deans, 2019; Gu et al., 2003; Salehi et al., 2017; Yang et al., 2011). The peripheral target of inner ear neurons is the mechanosensory hair cell and will be discussed below.

Neurotrophins are important for survival of neurosensory components of the inner ear. *BDNF* and *Ntf3* are expressed in the inner ear but are differentially distributed. *BDNF* is primarily expressed in the vestibular part and the cochlear apex, whereas *Ntf3* is expressed in all of the cochlea with a gradient as well as in the utricle and saccule (Fariñas et al., 2001; Fritzsch et al., 2004; 2016; Tessarollo et al., 2004). Loss of both neurotrophins results in a delayed loss of all cochlear hair cells (Kersigo and Fritzsch, 2015).

1.3.5 HAIR CELLS OF THE INNER EAR, LATERAL LINE, AND ELECTRORECEPTION HAVE A SHARED DEVELOPMENTAL PROGRAM

Hair cell development is best understood for the inner ear, though many of the mechanisms and gene regulatory networks involved in inner ear hair cell development are conserved for hair cell development in the lateral line and electroreception systems. Prosensory specification depends upon *Sox2* expression (Dvorakova et al., 2020; Evsen et al., 2013; Kiernan et al., 2005; Steevens et al., 2019) and *Notch/Hes* genes specify hair cells and supporting cells within these prosensory regions (Kageyama et al., 2019). All mechanosensory hair cells require the bHLH gene, *Atoh1*, for their differentiation (Bermingham et al., 1999; Fritzsch et al., 2005; Pan et al., 2012). Downstream of *Atoh1* are *Pou4f3* (Xiang et al., 2003), *Gfi1* (Hertzano et al., 2004), and *Barhl1* (Chellappa et al., 2008); however, some types of hair cells do not need *Pou4f3* or *Gfi1* expression for their differentiation (Hertzano et al., 2004; Matern et al., 2020). Additional genes affect only specific hair cells when deleted. For example, loss of *Srrm4*, which regulates *REST*, results in the loss of all vestibular and auditory inner hair cells, but not the auditory outer hair cells (Nakano et al., 2012; 2020). Another gene, *Emx2*, has differential effect on the various hair cells, leading to a partial loss of outer hair cells and vestibular hair cells in its absence (Holley et al., 2010; Jiang et al., 2017). In addition, *Emx2* affects the orientation of the stereocilia at the apical surface of the hair cells (Holley et al., 2010; Jiang et al., 2017). *Emx2* plays a similar role for hair cell orientation of the lateral line and in addition, *Emx2* was shown to correlate with selective innervation by lateral line neurons (Lozano-Ortega et al., 2018). This is consistent with segregated vestibular innervation to the cerebellum (Balmer and Trussell, 2019; Maklad and Fritzsch, 2003). Finally, only outer hair cells express *Insm1* and is the earliest known gene to differentiate the two populations of auditory hair cells; however, its loss did not appear to affect outer hair cell differentiation (Lorenzen et al., 2015).

An independent development of the lateral line and electroreception neuronal projections (Baker, 2019; Elliott and Fritzsch, 2021; Schlosser, 2015; Wullimann and Grothe, 2013) is comparable to the projections of inner ear vestibular and, in amniotes, auditory fibers. These fibers form branches that reach the lateral line and electroreception targets. Lateral line and electroreceptive neurons depend on *Neurog1*, as with those of the inner ear (Fritzsch and Elliott, 2017), and innervate lateral line mechanosensory hair cells (*Atoh1*) and electroreceptive 'hair cells' (*Atoh1*; Baker, 2019).

1.4 SUMMARY AND CONCLUSION

In summary, the vertebrate brain and spinal cord develop from a neural plate that invaginates to form a neural tube. This neural tube forms the future CNS that receives information from all eight peripheral sensory systems. Across the various vertebrate species, a conserved set of genes is involved in the development of the CNS, the sensory neurons, and the peripheral sensory cell receptors. *Sox2* is an early regulator of proliferation within the developing brain and across many of the sensory systems. Upregulation of bHLH genes, such as *Neurog1*, *Neurod1*, *Atoh1*, or *Atoh7*, is necessary for the commitment of cells into neurons or sensory cells. The following chapters will detail the eight sensory systems, highlighting their peripheral sensory neurons and cells, as well as central projections of these sensory neurons and additional innervation to higher order processing centers.

ACKNOWLEDGEMENTS

We wish to thank all of our past and present lab members: Jeremy Duncan, Clayton Gordy, Israt Jahan, Jennifer Kersigo, Benjamin Kopecky, Ning Pan, and Tian Yang.

REFERENCES

Abraira, V.E., Ginty, D.D., 2013. The sensory neurons of touch. Neuron 79, 618–639.

Akiyama-Oda, Y., Oda, H.J.D., 2006. Axis specification in the spider embryo: dpp is required for radial-to-axial symmetry transformation and sog for ventral patterning. Development 133, 2347–2357.

Appler, J.M., Goodrich, L.V., 2011. Connecting the ear to the brain: Molecular mechanisms of auditory circuit assembly. Progress in Neurobiology 93, 488–508.

Arendt, D., Nübler-Jung, K., 1997. Dorsal or ventral: similarities in fate maps and gastrulation patterns in annelids, arthropods and chrodates. Mechanisms of Development 61, 7–21.

Aruga, J., Hatayama, M., 2018. Comparative genomics of the Zic family genes. Zic Family. Springer, pp. 3–26.

Ayer-Le Lievre, C.S., Le Douarin, N.M., 1982. The early development of cranial sensory ganglia and the potentialities of their component cells studied in quail-chick chimeras. Developmental Biology 94, 291–310.

Bagriantsev, S.N., Gracheva, E.O., Gallagher, P.G., 2014. Piezo proteins: regulators of mechanosensation and other cellular processes. Journal of Biological Chemistry 289, 31673–31681.

Baker, C.V., 2019. The Development and Evolution of Lateral Line Electroreceptors: Insights from Comparative Molecular Approaches, Electroreception: Fundamental Insights from Comparative Approaches. Springer, pp. 25–62.

Balmer, T.S., Trussell, L.O., 2019. Selective targeting of unipolar brush cell subtypes by cerebellar mossy fibers. Elife 8, e44964.

Barratt, K.S., Arkell, R.M., 2018. ZIC2 in Holoprosencephaly. Zic Family. Springer, pp. 269–299.

Basch, M.L., Bronner-Fraser, M., García-Castro, M.I., 2006. Specification of the neural crest occurs during gastrulation and requires Pax7. Nature 441, 218–222.

Bermingham, N.A., Hassan, B.A., Price, S.D., Vollrath, M.A., Ben-Arie, N., Eatock, R.A., Bellen, H.J., Lysakowski, A., Zoghbi, H.Y., 1999. Math1: an essential gene for the generation of inner ear hair cells. Science 284, 1837–1841.

Bermingham, N.A., Hassan, B.A., Wang, V.Y., Fernandez, M., Banfi, S., Bellen, H.J., Fritzsch, B., Zoghbi, H.Y., 2001. Proprioceptor pathway development is dependent on Math1. Neuron 30, 411–422.

Birol, O., Ohyama, T., Edlund, R.K., Drakou, K., Georgiades, P., Groves, A.K., 2016. The mouse Foxi3 transcription factor is necessary for the development of posterior placodes. Developmental Biology 409, 139–151.

Bok, J., Bronner-Fraser, M., Wu, D.K., 2005. Role of the hindbrain in dorsoventral but not anteroposterior axial specification of the inner ear. Development 132, 2115–2124.

Bouchard, M., de Caprona, D., Busslinger, M., Xu, P., Fritzsch, B., 2010. Pax2 and Pax8 cooperate in mouse inner ear morphogenesis and innervation. BMC Developmental Biology 10, 89.

Brzezinski, J.A., Reh, T.A., 2015. Photoreceptor cell fate specification in vertebrates. Development 142, 3263–3273.

Buck, L., Axel, R., 1991. A novel multigene family may encode odorant receptors: a molecular basis for odor recognition. Cell 65, 175–187.

Cau, E., Casarosa, S., Guillemot, F., 2002. Mash1 and Ngn1 control distinct steps of determination and differentiation in the olfactory sensory neuron lineage. Development 129, 1871–1880.

Chang, W., Lin, Z., Kulessa, H., Hebert, J., Hogan, B.L., Wu, D.K., 2008. Bmp4 is essential for the formation of the vestibular apparatus that detects angular head movements. PLoS Genet 4, e1000050.

Chellappa, R., Li, S., Pauley, S., Jahan, I., Jin, K., Xiang, M., 2008. Barhl1 regulatory sequences required for cell-specific gene expression and autoregulation in the inner ear and central nervous system. Molecular and Cellular Biology 28, 1905–1914.

Cheng, Y., Liao, S., Xu, G., Hu, J., Guo, D., Du, F., Contreras, A., Cai, K.Q., Peri, S., Wang, Y., 2020. NeuroD1 dictates tumor cell differentiation in medulloblastoma. Cell Reports 31, 107782.

Chizhikov, V.V., Iskusnykh, I.Y., Fattakhov, N., Fritzsch, B., 2021. Lmx1a and Lmx1b are redundantly required for the development of multiple components of the mammalian auditory system. Neuroscience 452, 247–264.

Chizhikov, V.V., Iskusnykh, I.Y., Steshina, E.Y., Fattakhov, N., Lindgren, A.G., Shetty, A.S., Roy, A., Tole, S., Millen, K.J., 2019. Early dorsomedial tissue interactions regulate gyrification of distal neocortex. Nature Communications 10.

Coate, T.M., Kelley, M.W., 2013. Making connections in the inner ear: recent insights into the development of spiral ganglion neurons and their connectivity with sensory hair cells. Seminars in Cell and Developmental Biology 24, 460–469.

D'Amico-Martel, A., Noden, D.M., 1983. Contributions of placodal and neural crest cells to avian cranial peripheral ganglia. The American Journal of Anatomy 4, 445–468.

de Caprona, M.-D.C., Fritzsch, B., 1983. The development of the retinopetal nucleus olfacto-retinalis of two cichlid fish as revealed by horseradish peroxidase. Developmental Brain Research 11, 281–301.

Dennis, D.J., Han, S., Schuurmans, C., 2019. bHLH transcription factors in neural development, disease, and reprogramming. Brain Research 1705, 48–65.

Domínguez-Frutos, E., López-Hernández, I., Vendrell, V., Neves, J., Gallozzi, M., Gutsche, K., Quintana, L., Sharpe, J., Knoepfler, P.S., Eisenman, R.N., 2011. N-myc controls proliferation, morphogenesis, and patterning of the inner ear. Journal of Neuroscience 31, 7178–7189.

Duncan, J.S., Fritzsch, B., 2013. Continued expression of GATA3 is necessary for cochlear neurosensory development. PloS ONE 8, e62046.

Duncan, J.S., Fritzsch, B., Houston, D.W., Ketchum, E.M., Kersigo, J., Deans, M.R., Elliott, K.L., 2019. Topologically correct central projections of tetrapod inner ear afferents require Fzd3. Scientific Reports 9, 10298.

Dvorakova, M., Macova, I., Bohuslavova, R., Anderova, M., Fritzsch, B., Pavlinkova, G., 2020. Early ear neuronal development, but not olfactory or lens development, can proceed without SOX2. Developmental Biology 457, 43–56.

Elliott, K.L., Fritzsch, B., 2020. Evolution and development of lateral line and electroreception: an integrated perception of neurons, hair cells and brainstem nuclei, in: Fritzsch, B. (Ed.), The Senses. Elsevier, pp. 95–115.

Elliott, K.L., Pavlinkova, G., Chizhikov, V.V., Yamoah, E.N., Fritzsch, B., 2021. Neurog1, Neurod1, and Atoh1 are essential for spiral ganglia, cochlear nuclei, and cochlear hair cell development. Faculty Review 10, 47.

Erzurumlu, R.S., Murakami, Y., Rijli, F.M., 2010. Mapping the face in the somatosensory brainstem. Nature Reviews Neuroscience 11, 252–263.

Evsen, L., Sugahara, S., Uchikawa, M., Kondoh, H., Wu, D.K., 2013. Progression of neurogenesis in the inner ear requires inhibition of Sox2 transcription by neurogenin1 and neurod1. Journal of Neuroscience 33, 3879–3890.

Fariñas, I., Jones, K.R., Tessarollo, L., Vigers, A.J., Huang, E., Kirstein, M., de Caprona, D.C., Coppola, V., Backus, C., Reichardt, L.F., Fritzsch, B., 2001. Spatial shaping of cochlear innervation by temporally regulated neurotrophin expression. Journal of Neuroscience 21, 6170–6180.

Favor, J., Sandulache, R., Neuhäuser-Klaus, A., Pretsch, W., Chatterjee, B., Senft, E., Wurst, W., Blanquet, V., Grimes, P., Spörle, R., 1996. The mouse Pax21Neu mutation is identical to a human PAX2 mutation in a family with renal-coloboma syndrome and results in developmental defects of the brain, ear, eye, and kidney. Proceedings of the National Academy of Sciences 93, 13870–13875.

Fekete, D.M., Muthukumar, S., Karagogeos, D., 1998. Hair cells and supporting cells share a common progenitor in the avian inner ear. Journal of Neuroscience 18, 7811–7821.

Fekete, D.M., Wu, D.K., 2002. Revisiting cell fate specification in the inner ear. Current Opinion in Neurobiology Neurobiol 12, 35–42.

Fode, C., Gradwohl, G., Morin, X., Dierich, A., LeMeur, M., Goridis, C., Guillemot, F., 1998. The bHLH protein NEUROGENIN 2 is a determination factor for epibranchial placode–derived sensory neurons. Neuron 20, 483–494.

Fritzsch, B., Elliott, K.L., 2017. Gene, cell, and organ multiplication drives inner ear evolution. Developmental Biology 431, 3–15.

Fritzsch, B., Elliott, K.L., Glover, J.C., 2017. Gaskell revisited: new insights into spinal autonomics necessitate a revised motor neuron nomenclature. Cell and Tissue Research 370, 195–209.

Fritzsch, B., Elliott, K.L., Pavlinkova, G., 2019. Primary sensory map formations reflect unique needs and molecular cues specific to each sensory system. F1000Research 8, 345.

Fritzsch, B., Jahan, I., Pan, N., Elliott, K.L., 2015. Evolving gene regulatory networks into cellular networks guiding adaptive behavior: an outline how single cells could have evolved into a centralized neurosensory system. Cell and Tissue Research 359, 295–313.

Fritzsch, B., Kersigo, J., Yang, T., Jahan, I., Pan, N., 2016. Neurotrophic factor function during ear development: expression changes define critical phases for neuronal viability, in: Dabdoub, A., Fritzsch, B., Popper, A., Fay, R. (Eds.), The Primary Auditory Neurons of the Mammaliancochlea., Springer, pp. 49–84.

Fritzsch, B., Matei, V., Nichols, D., Bermingham, N., Jones, K., Beisel, K., Wang, V., 2005. Atoh1 null mice show directed afferent fiber growth to undifferentiated ear sensory epithelia followed by incomplete fiber retention. Developmental dynamics: An Official Publication of the American Association of Anatomists 233, 570–583.

Fritzsch, B., Northcutt, G., 1993. Cranial and spinal nerve organization in amphioxus and lampreys: evidence for an ancestral craniate pattern. Cells Tissues Organs 148, 96–109.

Fritzsch, B., Pan, N., Jahan, I., Duncan, J.S., Kopecky, B.J., Elliott, K.L., Kersigo, J., Yang, T., 2013. Evolution and development of the tetrapod auditory system: An organ of Corticentric perspective. Evolution & Development 15, 63–79.

Fritzsch, B., Pauley, S., Feng, F., Matei, V., Nichols, D., 2006. The molecular and developmental basis of the evolution of the vertebrate auditory system. International Journal of Comparative Psychology 19, 1–25.

Fritzsch, B., Sarai, P., Barbacid, M., Silos-Santiago, I., 1997. Mice with a targeted disruption of the neurotrophin receptor trkB lose their gustatory ganglion cells early but do develop taste buds. International Journal of Developmental Neuroscience 15, 563–576.

Fritzsch, B., Signore, M., Simeone, A., 2001. Otx1 null mutant mice show partial segregation of sensory epithelia comparable to lamprey ears. Development Genes and Evolution 211, 388–396.

Fritzsch, B., Tessarollo, L., Coppola, E., Reichardt, L.F., 2004. Neurotrophins in the ear: their roles in sensory neuron survival and fiber guidance. Progress in Brain Research 146, 265–278.

Galliot, B., Quiquand, M., Ghila, L., de Rosa, R., Miljkovic-Licina, M., Chera, S., 2009. Origins of neurogenesis, a cnidarian view. Developmental Biology 332, 2–24.

Gee, H., 2007. Before the Backbone: Views on the Origin of the Vertebrates. Springer Science & Business Media.

Gerhart, J., 2000. Inversion of the chordate body axis: Are there alternatives? Proceedings of the National Academy of Sciences 97, 4445–4448.

Germanà, A., Montalbano, G., Guerrera, M.C., Laura, R., Levanti, M., Abbate, F., de Carlos, F., Vega, J.A., Ciriaco, E., 2009. Sox-2 in taste bud and lateral line system of zebrafish during development. Neuroscience Letters 467, 36–39.

Ghiasvand, N.M., Rudolph, D.D., Mashayekhi, M., Brzezinski, J.A., Goldman, D., Glaser, T., 2011. Deletion of a remote enhancer near ATOH7 disrupts retinal neurogenesis, causing NCRNA disease. Nature Neuroscience 14, 578–586.

Ghimire, S.R., Deans, M.R., 2019. Frizzled3 and Frizzled6 Cooperate with Vangl2 to Direct Cochlear Innervation by type II Spiral Ganglion Neurons. Journal of Neuroscience 39, 8013–8023.

Ghimire, S.R., Ratzan, E.M., Deans, M.R., 2018. A non-autonomous function of the core PCP protein VANGL2 directs peripheral axon turning in the developing cochlea. Development 145, dev159012.

Glover, J.C., Elliott, K.L., Erives, A., Chizhikov, V.V., Fritzsch, B., 2018. Wilhelm His' lasting insights into hindbrain and cranial ganglia development and evolution. Developmental Biology 444, S14–S24.

Gnedeva, K., Wang, X., McGovern, M.M., Barton, M., Tao, L., Trecek, T., Monroe, T.O., Llamas, J., Makmura, W., Martin, J.F., Groves, A.K., Warchol, M., Segil, N., 2020. Organ of Corti size is governed by Yap/Tead-mediated progenitor self-renewal. Proceedings of the National Academy of Sciences of the United States of America 117, 13552–13561.

Gowan, K., Helms, A.W., Hunsaker, T.L., Collisson, T., Ebert, P.J., Odom, R., Johnson, J.E., 2001. Crossinhibitory activities of Ngn1 and Math1 allow specification of distinct dorsal interneurons. Neuron 31, 219–232.

Graw, J., 2010. Eye development. Current Topics in Developmental Biology 90, 343–386.

Green, S.A., Simoes-Costa, M., Bronner, M.E., 2015. Evolution of vertebrates as viewed from the crest. Nature 520, 474–482.

Groves, A.K., Fekete, D.M., 2012. Shaping sound in space: the regulation of inner ear patterning. Development 139, 245–257.

Gu, C., Rodriguez, E.R., Reimert, D.V., Shu, T., Fritzsch, B., Richards, L.J., Kolodkin, A.L., Ginty, D.D., 2003. Neuropilin-1 conveys semaphorin and VEGF signaling during neural and cardiovascular development. Developmental Cell 5, 45–57.

Guo, Z., Packard, A., Krolewski, R.C., Harris, M.T., Manglapus, G.L., Schwob, J.E., 2010. Expression of pax6 and sox2 in adult olfactory epithelium. Journal of Comparative Neurology 518, 4395–4418.

Harold, A., Amako, Y., Hachisuka, J., Bai, Y., Li, M.Y., Kubat, L., Gravemeyer, J., Franks, J., Gibbs, J.R., Park, H.J., Ezhkova, E., Becker, J.C., Shuda, M., 2019. Conversion of Sox2-dependent Merkel cell carcinoma to a differentiated neuron-like phenotype by T antigen inhibition. Proceedings of the National Academy of Sciences of the United States of America 116, 20104–20114.

Hatini, V., Tao, W., Lai, E., 1994. Expression of winged helix genes, BF-1 and BF-2, define adjacent domains within the developing forebrain and retina. Journal of Neurobiology 25, 1293–1309.

Heingård, M., Janssen, R., 2020. The forkhead box containing transcription factor FoxB is a potential component of dorsal-ventral body axis formation in the spider Parasteatoda tepidariorum. Development Genes and Evolution 230, 65–73.

Hernandez-Miranda, L.R., Müller, T., Birchmeier, C., 2017. The dorsal spinal cord and hindbrain: From developmental mechanisms to functional circuits. Developmental Biology 432, 34–42.

Hertzano, R., Montcouquiol, M., Rashi-Elkeles, S., Elkon, R., Yücel, R., Frankel, W.N., Rechavi, G., Möröy, T., Friedman, T.B., Kelley, M.W., 2004. Transcription profiling of inner ears from Pou4f3 ddl/ddl identifies Gfi1 as a target of the Pou4f3 deafness gene. Human Molecular Genetics 13, 2143–2153.

Hobmayer, B., Rentzsch, F., Kuhn, K., Happel, C.M., von Laue, C.C., Snyder, P., Rothbächer, U., Holstein, T.W.J.N., 2000. WNT signalling molecules act in axis formation in the diploblastic metazoan Hydra. Nature 407, 186–189.

Holland, L., 2020. Invertebrate origins of vertebrate nervous systems. Evolutionary Neuroscience. Elsevier, pp. 51–73.

Holley, M., Rhodes, C., Kneebone, A., Herde, M.K., Fleming, M., Steel, K.P., 2010. Emx2 and early hair cell development in the mouse inner ear. Developmental Biology340, 547–556.

Hong, C.-S., Saint-Jeannet, J.-P., 2005. Sox proteins and neural crest development. Seminars in Cell and Developmental Biology. Elsevier, pp. 694–703.

Huang, Y., Hill, J., Yatteau, A., Wong, L., Jiang, T., Petrovic, J., Gan, L., Dong, L., Wu, D.K., 2018. Reciprocal negative regulation between Lmx1a and Lmo4 is required for inner ear formation. Journal of Neuroscience 38, 5429–5440.

Huang, L., Hu, F., Xie, X., Harder, J., Fernandes, K., Zeng, X.-y., Libby, R., Gan, L., 2014. Pou4f1 and pou4f2 are dispensable for the long-term survival of adult retinal ganglion cells in mice. PLoS ONE 9, e94173.

Huang, E.J., Liu, W., Fritzsch, B., Bianchi, L.M., Reichardt, L.F., Xiang, M., 2001. Brn3a is a transcriptional regulator of soma size, target field innervation and axon pathfinding of inner ear sensory neurons. Development 128, 2421–2432.

Hwang, C.H., Simeone, A., Lai, E., Wu, D.K., 2009. Foxg1 is required for proper separation and formation of sensory cristae during inner ear development. Developmental Dynamics 238, 2725–2734.

Imai, T., 2020. Odor coding in the olfactory bulb, in: Fritzsch, B. (Ed.), The Senses. Elsevier, pp. 640–649.

Imayoshi, I., Kageyama, R., 2014. bHLH factors in self-renewal, multipotency, and fate choice of neural progenitor cells. Neuron 82, 9–23.

Irish, V.F., Gelbart, W.M.J.G., 1987. The decapentaplegic gene is required for dorsal-ventral patterning of the Drosophila embryo. Genes and Development1, 868–879.

Iskusnykh, I.Y., Steshina, E.Y., Chizhikov, V.V., 2016. Loss of Ptf1a leads to a widespread cell-fate misspecification in the brainstem, affecting the development of somatosensory and viscerosensory nuclei. Journal of Neuroscience 36, 2691–2710.

Jahan, I., Pan, N., Elliott, K.L., Fritzsch, B., 2015. The quest for restoring hearing: understanding ear development more completely. BioEssays 37, 1016–1027.

Jahan, I., Pan, N., Kersigo, J., Fritzsch, B., 2010. Neurod1 suppresses hair cell differentiation in ear ganglia and regulates hair cell subtype development in the cochlea. PloS ONE 5, e11661.

Jiang, T., Kindt, K., Wu, D.K., 2017. Transcription factor Emx2 controls stereociliary bundle orientation of sensory hair cells. Elife 6, e23661.

Kageyama, R., Shimojo, H., Ohtsuka, T., 2019. Dynamic control of neural stem cells by bHLH factors. Neuroscience Research 138, 12–18.

Kaji, T., Reimer, J.D., Morov, A.R., Kuratani, S., Yasui, K., 2016. Amphioxus mouth after dorso-ventral inversion. Zoological Letters 2, 1–14.

Karis, A., Pata, I., van Doorninck, J.H., Grosveld, F., de Zeeuw, C.I., de Caprona, D., Fritzsch, B., 2001. Transcription factor GATA-3 alters pathway selection of olivocochlear neurons and affects morphogenesis of the ear. Journal of Comparative Neurology 429, 615–630.

Kawauchi, S., Kim, J., Santos, R., Wu, H.-H., Lander, A.D., Calof, A.L., 2009. Foxg1 promotes olfactory neurogenesis by antagonizing Gdf11. Development 136, 1453–1464.

Kersigo, J., D'Angelo, A., Gray, B.D., Soukup, G.A., Fritzsch, B., 2011. The role of sensory organs and the forebrain for the development of the craniofacial shape as revealed by Foxg1-cre-mediated microRNA loss. Genesis 49, 326–341.

Kersigo, J., Fritzsch, B., 2015. Inner ear hair cells deteriorate in mice engineered to have no or diminished innervation. Front Aging Neuroscience 7, 33.

Kersigo, J., Gu, L., Xu, L., Pan, N., Vijayakuma, S., Jones, T., Shibata, S.B., Fritzsch, B., Hansen, M.R., 2021. Effects of Neurod1 Expression on Mouse and Human Schwannoma Cells. The Laryngoscope 131, E259–E270.

Kiernan, A.E., Pelling, A.L., Leung, K.K., Tang, A.S., Bell, D.M., Tease, C., Lovell-Badge, R., Steel, K.P., Cheah, K.S., 2005. Sox2 is required for sensory organ development in the mammalian inner ear. Nature 434, 1031–1035.

Kim, W.-Y., Fritzsch, B., Serls, A., Bakel, L.A., Huang, E.J., Reichardt, L.F., Barth, D.S., Lee, J.E., 2001. NeuroD-null mice are deaf due to a severe loss of the inner ear sensory neurons during development. Development 128, 417–426.

Kinnamon, S.C., Finger, T.E., 2019. Recent advances in taste transduction and signaling. F1000Research 8, 2117.

Knecht, A.K., Bronner-Fraser, M., 2002. Induction of the neural crest: a multigene process. Nature Reviews Genetics 3, 453–461.

Kopecky, B., Santi, P., Johnson, S., Schmitz, H., Fritzsch, B., 2011. Conditional deletion of N-Myc disrupts neurosensory and non-sensory development of the ear. Developmental Dynamics 240, 1373–1390.

Kozmik, Z., Daube, M., Frei, E., Norman, B., Kos, L., Dishaw, L.J., Noll, M., Piatigorsky, J., 2003. Role of Pax genes in eye evolution: A cnidarian PaxB gene uniting Pax2 and Pax6 functions. Developmental Cell 5, 773–785.

Krüger, M., Schmid, T., Krüger, S., Bober, E., Braun, T., 2006. Functional redundancy of NSCL-1 and NeuroD during development of the petrosal and vestibulocochlear ganglia. European Journal of Neuroscience 24, 1581–1590.

Lacalli, T.C., 2008. Basic features of the ancestral chordate brain: a protochordate perspective. Brain Research Bulletin 75, 319–323.

Ladher, R.K., O'Neill, P., Begbie, J., 2010. From shared lineage to distinct functions: the development of the inner ear and epibranchial placodes. Development 137, 1777–1785.

Lai, H.C., Seal, R.P., Johnson, J.E., 2016. Making sense out of spinal cord somatosensory development. Development 143, 3434–3448.

Lamb, T.D., 2013. Evolution of phototransduction, vertebrate photoreceptors and retina. Progress in Retinal and Eye Research 36, 52–119.

Lamb, T.D., Collin, S.P., Pugh, E.N., 2007. Evolution of the vertebrate eye: opsins, photoreceptors, retina and eye cup. Nature Reviews Neuroscience 8, 960–976.

Lassiter, R.N., Stark, M.R., Zhao, T., Zhou, C.J., 2014. Signaling mechanisms controlling cranial placode neurogenesis and delamination. Developmental Biology 389, 39–49.

Layden, M.J., 2019. Origin and Evolution of Nervous Systems, Old Questions and Young Approaches to Animal Evolution. Springer, pp. 151–171.

Le Dréau, G., Martí, E., 2012. Dorsal–ventral patterning of the neural tube: a tale of three signals. Developmental Neurobiology 72, 1471–1481.

Lee, K.J., Dietrich, P., Jessell, T.M., 2000. Genetic ablation reveals that the roof plate is essential for dorsal interneuron specification. Nature 403, 734–740.

Lee, H.-K., Lee, H.-S., Moody, S.A., 2014. Neural transcription factors: from embryos to neural stem cells. Molecules and Cells 37, 705.

Lee, K.J., Mendelsohn, M., Jessell, T.M., 1998. Neuronal patterning by BMPs: a requirement for GDF7 in the generation of a discrete class of commissural interneurons in the mouse spinal cord. Genes & development 12, 3394–3407.

Lele, Z., Nowak, M., Hammerschmidt, M., 2001. Zebrafish admp is required to restrict the size of the organizer and to promote posterior and ventral development. Developmental Dynamics 222, 681–687.

Li, H.J., Ray, S.K., Pan, N., Haigh, J., Fritzsch, B., Leiter, A.B., 2019. Intestinal Neurod1 expression impairs paneth cell differentiation and promotes enteroendocrine lineage specification. Scientific Reports 9, 1–11.

Li, J., Zhang, T., Ramakrishnan, A., Fritzsch, B., Xu, J., Wong, E.Y., Loh, Y.-H.E., Ding, J., Shen, L., Xu, P.-X., 2020. Dynamic changes in cis-regulatory occupancy by Six1 and its cooperative interactions with distinct cofactors drive lineage-specific gene expression programs during progressive differentiation of the auditory sensory epithelium. Nucleic Acids Research 48, 2880–2896.

Lleras-Forero, L., Tambalo, M., Christophorou, N., Chambers, D., Houart, C., Streit, A.J.D., 2013. Neuropeptides: developmental signals in placode progenitor formation. 26, 195–203.

Lorenzen, S.M., Duggan, A., Osipovich, A.B., Magnuson, M.A., García-Añoveros, J., 2015. Insm1 promotes neurogenic proliferation in delaminated otic progenitors. Mechanisms of Development 138 Pt 3, 233–245.

Lozano-Ortega, M., Valera, G., Xiao, Y., Faucherre, A., López-Schier, H., 2018. Hair cell identity establishes labeled lines of directional mechanosensation. PLoS Biol 16, e2004404.

Ma, Q., Anderson, D.J., Fritzsch, B., 2000. Neurogenin 1 null mutant ears develop fewer, morphologically normal hair cells in smaller sensory epithelia devoid of innervation. Journal of the Association for Research in Otolaryngology 1, 129–143.

Ma, Q., Chen, Z., del Barco Barrantes, I., de la Pompa, J.L., Anderson, D.J., 1998. Neurogenin1 is essential for the determination of neuronal precursors for proximal cranial sensory ganglia. Neuron 20, 469–482.

Macova, I., Pysanenko, K., Chumak, T., Dvorakova, M., Bohuslavova, R., Syka, J., Fritzsch, B., Pavlinkova, G., 2019. Neurod1 is essential for the primary tonotopic organization and related auditory information processing in the midbrain. Journal of Neuroscience 39, 984–1004.

Maier, E.C., Saxena, A., Alsina, B., Bronner, M.E., Whitfield, T.T., 2014. Sensational placodes: neurogenesis in the otic and olfactory systems. Developmental Biology 389, 50–67.

Maklad, A., Fritzsch, B., 2003. Partial segregation of posterior crista and saccular fibers to the nodulus and uvula of the cerebellum in mice, and its development. Brain Research Developmental Brain Research 140, 223–236.

Mann, Z.F., Galvez, H., Pedreno, D., Chen, Z., Chrysostomou, E., Żak, M., Kang, M., Canden, E., Daudet, N., 2017. Shaping of inner ear sensory organs through antagonistic interactions between Notch signalling and Lmx1a. Elife 6, e33323.

Mao, C.-A., Cho, J.-H., Wang, J., Gao, Z., Pan, P., Tsai, W.-W., Frishman, L.J., Klein, W.H., 2013. Reprogramming amacrine and photoreceptor progenitors into retinal ganglion cells by replacing Neurod1 with Atoh7. Development 140, 541–551.

Maricich, S.M., Wellnitz, S.A., Nelson, A.M., Lesniak, D.R., Gerling, G.J., Lumpkin, E.A., Zoghbi, H.Y., 2009. Merkel cells are essential for light-touch responses. Science 324, 1580–1582.

Matei, V., Pauley, S., Kaing, S., Rowitch, D., Beisel, K.W., Morris, K., Feng, F., Jones, K., Lee, J., Fritzsch, B., 2005. Smaller inner ear sensory epithelia in Neurog 1 null mice are related to earlier hair cell cycle exit. Developmental Dynamics 234, 633–650.

Matern, M.S., Milon, B., Lipford, E.L., McMurray, M., Ogawa, Y., Tkaczuk, A., Song, Y., Elkon, R., Hertzano, R.J.D., 2020. GFI1 functions to repress neuronal gene expression in the developing inner ear hair cells. Development 147, dev186015.

Meinhardt, H., 2015a. Dorsoventral patterning by the Chordin-BMP pathway: a unified model from a pattern-formation perspective for drosophila, vertebrates, sea urchins and nematostella. Developmental Biology 405, 137–148.

Meinhardt, H., 2015b. Models for patterning primary embryonic body axes: the role of space and time, Seminars in Cell & Developmental Biology. Elsevier, pp. 103–117.

Mishima, Y., Lindgren, A.G., Chizhikov, V.V., Johnson, R.L., Millen, K.J., 2009. Overlapping function of Lmx1a and Lmx1b in anterior hindbrain roof plate formation and cerebellar growth. Journal of Neuroscience 29, 11377–11384.

Mombaerts, P., Wang, F., Dulac, C., Chao, S.K., Nemes, A., Mendelsohn, M., Edmondson, J., Axel, R., 1996. Visualizing an olfactory sensory map. Cell 87, 675–686.

Moody, S.A., Klein, S.L., Karpinski, B.A., Maynard, T.M., LaMantia, A.-S., 2013. On becoming neural: what the embryo can tell us about differentiating neural stem cells. American Journal of Stem Cells 2, 74.

Moody, S.A., LaMantia, A.-S., 2015. Transcriptional regulation of cranial sensory placode development, Current Topics in Developmental Biology. Elsevier, pp. 301–350.

Morsli, H., Tuorto, F., Choo, D., Postiglione, M.P., Simeone, A., Wu, D.K., 1999. Otx1 and Otx2 activities are required for the normal development of the mouse inner ear. Development 126, 2335–2343.

Münz, H., Claas, B., Stumpf, W., Jennes, L., 1982. Centrifugal innervation of the retina by luteinizing hormone releasing hormone (LHRH)-immunoreactive telencephalic neurons in teleostean fishes. Cell and Tissue Research 222, 313–323.

Nakano, Y., Jahan, I., Bonde, G., Sun, X., Hildebrand, M.S., Engelhardt, J.F., Smith, R.J., Cornell, R.A., Fritzsch, B., Bánfi, B., 2012. A mutation in the Srrm4 gene causes alternative splicing defects and deafness in the Bronx waltzer mouse. PLoS Genet 8, e1002966.

Nakano, Y., Wiechert, S., Fritzsch, B., Bánfi, B., 2020. Inhibition of a transcriptional repressor rescues hearing in a splicing factor–deficient mouse. Life Science Alliance 3, e202000841.

Nichols, D.H., Bouma, J.E., Kopecky, B.J., Jahan, I., Beisel, K.W., He, D.Z., Liu, H., Fritzsch, B., 2020. Interaction with ectopic cochlear crista sensory epithelium disrupts basal cochlear sensory epithelium development in Lmx1a mutant mice. Cell and Tissue Research, 1–14.

Nichols, D.H., Pauley, S., Jahan, I., Beisel, K.W., Millen, K.J., Fritzsch, B., 2008. Lmx1a is required for segregation of sensory epithelia and normal ear histogenesis and morphogenesis. Cell and Tissue Research 334, 339–358.

Nishizumi, H., Sakano, H., 2015. Decoding and deorphanizing an olfactory map. Nature Neuroscience 18, 1432–1433.

Northcutt, R., 2005. The new head hypothesis revisited. Journal of Experimental Zoology Part B: Molecular and Developmental Evolution 304, 274–297.

Northcutt, R.G., Gans, C., 1983. The genesis of neural crest and epidermal placodes: a reinterpretation of vertebrate origins. The Quarterly Review of Biology 58, 1–28.

O'Neill, P., Mak, S.-S., Fritzsch, B., Ladher, R.K., Baker, C.V., 2012. The amniote paratympanic organ develops from a previously undiscovered sensory placode. Nature Communications 3, 1–11.

Ohmoto, M., Ren, W., Nishiguchi, Y., Hirota, J., Jiang, P., Matsumoto, I., 2017. Genetic Lineage Tracing in Taste Tissues Using Sox2-CreERT2 Strain. Chem Senses 42, 547–552.

Ohyama, T., Basch, M.L., Mishina, Y., Lyons, K.M., Segil, N., Groves, A.K., 2010. BMP signaling is necessary for patterning the sensory and nonsensory regions of the developing mammalian cochlea. The Journal of Neuroscience: The Official Journal of the Society for Neuroscience 30, 15044–15051.

Okubo, T., Pevny, L.H., Hogan, B.L., 2006. Sox2 is required for development of taste bud sensory cells. Genes & Development 20, 2654–2659.

Ono, K., Sandell, L.L., Trainor, P.A., Wu, D.K., 2020. Retinoic acid synthesis and autoregulation mediate zonal patterning of vestibular organs and inner ear morphogenesis. Development 147, dev192070.

Ostrowski, S.M., Wright, M.C., Bolock, A.M., Geng, X., Maricich, S.M., 2015. Ectopic Atoh1 expression drives Merkel cell production in embryonic, postnatal and adult mouse epidermis. Development 142, 2533–2544.

Panaliappan, T.K., Wittmann, W., Jidigam, V.K., Mercurio, S., Bertolini, J.A., Sghari, S., Bose, R., Patthey, C., Nicolis, S.K., Gunhaga, L., 2018. Sox2 is required for olfactory pit formation and olfactory neurogenesis through BMP restriction and Hes5 upregulation. Development 145.

Pan, N., Jahan, I., Kersigo, J., Duncan, J.S., Kopecky, B., Fritzsch, B., 2012. A novel Atoh1 "self-terminating" mouse model reveals the necessity of proper Atoh1 level and duration for hair cell differentiation and viability. PLoS ONE 7, e30358.

Pan, N., Jahan, I., Lee, J.E., Fritzsch, B., 2009. Defects in the cerebella of conditional Neurod1 null mice correlate with effective Tg (Atoh1-cre) recombination and granule cell requirements for Neurod1 for differentiation. Cell and Tissue Research 337, 407–428.

Pantzartzi, C.N., Pergner, J., Kozmikova, I., Kozmik, Z., 2017. The opsin repertoire of the European lancelet: a window into light detection in a basal chordate. International Journal of Developmental Biology 61, 763–772.

Pauley, S., Lai, E., Fritzsch, B., 2006. Foxg1 is required for morphogenesis and histogenesis of the mammalian inner ear. Developmental Dynamics 235, 2470–2482.

Pauley, S., Wright, T.J., Pirvola, U., Ornitz, D., Beisel, K., Fritzsch, B., 2003. Expression and function of FGF10 in mammalian inner ear development. Developmental dynamics 227, 203–215.

Piccolo, S., Agius, E., Leyns, L., Bhattacharyya, S., Grunz, H., Bouwmeester, T., De Robertis, E., 1999. The head inducer Cerberus is a multifunctional antagonist of Nodal, BMP and Wnt signals. Nature 397, 707–710.

Pichaud, F., Desplan, C., 2002. Pax genes and eye organogenesis. Current Opinion in Genetics & Development 12, 430–434.

Pirvola, U., Spencer-Dene, B., Xing-Qun, L., Kettunen, P., Thesleff, I., Fritzsch, B., Dickson, C., Ylikoski, J., 2000. FGF/FGFR-2 (IIIb) signaling is essential for inner ear morphogenesis. Journal of Neuroscience 20, 6125–6134.

Plouhinec, J.-L., Medina-Ruiz, S., Borday, C., Bernard, E., Vert, J.-P., Eisen, M.B., Harland, R.M., Monsoro-Burq, A.H., 2017. A molecular atlas of the developing ectoderm defines neural, neural crest, placode, and nonneural progenitor identity in vertebrates. PLoS Biology 15, e2004045.

Reiprich, S., Wegner, M., 2015. From CNS stem cells to neurons and glia: Sox for everyone. Cell and tissue research 359, 111–124.

Riccomagno, M.M., Martinu, L., Mulheisen, M., Wu, D.K., Epstein, D.J., 2002. Specification of the mammalian cochlea is dependent on Sonic hedgehog. Genes & Development 16, 2365–2378.

Riddiford, N., Schlosser, G., 2016. Dissecting the pre-placodal transcriptome to reveal presumptive direct targets of Six1 and Eya1 in cranial placodes. Elife 5, e17666.

Rogers, C., Jayasena, C., Nie, S., Bronner.,. 2012. Neural crest specification: tissues, signals, and transcription factors. Wiley Interdisciplinary Reviews: Developmental Biology 1, 52–68.

Rogers, C.D., Moody, S.A., Casey, E.S., 2009. Neural induction and factors that stabilize a neural fate. Birth Defects Research Part C: Embryo Today: Reviews 87, 249–262.

Roper, S.D., Chaudhari, N., 2017. Taste buds: cells, signals and synapses. Nature Reviews Neuroscience 18, 485–497.

Sadler, T., 2005. Embryology of neural tube development, American Journal of Medical Genetics Part C: Seminars in Medical Genetics. Wiley Online Library, pp. 2–8.

Sakamoto, S., Tateya, T., Omori, K., Kageyama, R., 2020. Id genes are required for morphogenesis and cellular patterning in the developing mammalian cochlea. Developmental Biology 460, 164–175.

Salehi, P., Ge, M.X., Gundimeda, U., Michelle Baum, L., Lael Cantu, H., Lavinsky, J., Tao, L., Myint, A., Cruz, C., Wang, J., Nikolakopoulou, A.M., Abdala, C., Kelley, M.W., Ohyama, T., Coate, T.M., Friedman, R.A., 2017. Role of Neuropilin-1/Semaphorin-3A signaling in the functional and morphological integrity of the cochlea. PLoS Genet 13, e1007048.

Sankar, S., Yellajoshyula, D., Zhang, B., Teets, B., Rockweiler, N., Kroll, K.L., 2016. Gene regulatory networks in neural cell fate acquisition from genome-wide chromatin association of Geminin and Zic1. Scientific Reports 6, 37412.

Schlosser, G., 2010. Chapter Four - Making Senses: Development of Vertebrate Cranial Placodes, in: Kwang, J. (Ed.), International Review of Cell and Molecular Biology. Academic Press, pp. 129–234.

Schlosser, G., 2015. Vertebrate cranial placodes as evolutionary innovations—the ancestor's tale, Current Topics in Developmental Biology. Elsevier, pp. 235–300.

Schlosser, G., 2021. Development of Sensory and Neurosecretory Cell Types: Vertebrate cranial placodes, Vol. 1.

Schlosser, G., Ahrens, K., 2004. Molecular anatomy of placode development in Xenopus laevis. Developmental Biology 271, 439–466.

Schlosser, G., Patthey, C., Shimeld, S.M., 2014. The evolutionary history of vertebrate cranial placodes II. Evolution of ectodermal patterning. Developmental Biology 389, 98–119.

Schoenwolf, G.C., Alvarez, I.S., 1989. Roles of neuroepithelial cell rearrangement and division in shaping of the avian neural plate. Development 106, 427–439.

Schwarz, M., Cecconi, F., Bernier, G., Andrejewski, N., Kammandel, B., Wagner, M., Gruss, P., 2000. Spatial specification of mammalian eye territories by reciprocal transcriptional repression of Pax2 and Pax6. Development 127, 4325–4334.

Schwob, J.E., Jang, W., Holbrook, E.H., Lin, B., Herrick, D.B., Peterson, J.N., Hewitt Coleman, J., 2017. Stem and progenitor cells of the mammalian olfactory epithelium: Taking poietic license. Journal of Comparative Neurology 525, 1034–1054.

Semënov, M.V., Tamai, K., Brott, B.K., Kühl, M., Sokol, S., He, X., 2001. Head inducer Dickkopf-1 is a ligand for Wnt coreceptor LRP6. Current Biology 11, 951–961.

Steevens, A.R., Glatzer, J.C., Kellogg, C.C., Low, W.C., Santi, P.A., Kiernan, A.E., 2019. SOX2 is required for inner ear growth and cochlear nonsensory formation before sensory development. Development 146.

Steventon, B., Mayor, R., Streit, A., 2012. Mutual repression between Gbx2 and Otx2 in sensory placodes reveals a general mechanism for ectodermal patterning. Developmental Biology 367, 55–65.

Storm, R., Cholewa-Waclaw, J., Reuter, K., Bröhl, D., Sieber, M., Treier, M., Müller, T., Birchmeier, C., 2009. The bHLH transcription factor Olig3 marks the dorsal neuroepithelium of the hindbrain and is essential for the development of brainstem nuclei. Development 136, 295–305.

Tan, L., Xie, X.S., 2018. A near-complete spatial map of olfactory receptors in the mouse main olfactory epithelium. Chemical Senses 43, 427–432.

Tessarollo, L., Coppola, V., Fritzsch, B., 2004. NT-3 replacement with brain-derived neurotrophic factor redirects vestibular nerve fibers to the cochlea. Journal of Neuroscience 24, 2575–2584.

Thiery, A., Buzzi, A.L., Streit, A., 2020. Cell fate decisions during the development of the peripheral nervous system in the vertebrate head. Current Topics in Developmental Biology 139, 127–167.

Tian, N.M., Pratt, T., Price, D.J., 2008. Foxg1 regulates retinal axon pathfinding by repressing an ipsilateral program in nasal retina and by causing optic chiasm cells to exert a net axonal growth-promoting activity. Development 135, 4081–4089.

Tucker, E.S., Lehtinen, M.K., Maynard, T., Zirlinger, M., Dulac, C., Rawson, N., Pevny, L., LaMantia, A.-S., 2010. Proliferative and transcriptional identity of distinct classes of neural precursors in the mammalian olfactory epithelium. Development 137, 2471–2481.

Ulloa, F., Martí, E., 2010. Wnt won the war: Antagonistic role of Wnt over Shh controls dorsoventral patterning of the vertebrate neural tube. Developmental Dynamics: An Official Publication of the American Association of Anatomists 239, 69–76.

Urness, L.D., Bleyl, S.B., Wright, T.J., Moon, A.M., Mansour, S.L., 2011. Redundant and dosage sensitive requirements for Fgf3 and Fgf10 in cardiovascular development. Developmental Biology 356, 383–397.

van der Heijden, M.E., Zoghbi, H.Y., 2020. Development of the brainstem respiratory circuit. Wiley Interdisciplinary Reviews: Developmental Biology 9, e366.

Veeman, M.T., Newman-Smith, E., El-Nachef, D., Smith, W.C., 2010. The ascidian mouth opening is derived from the anterior neuropore: reassessing the mouth/neural tube relationship in chordate evolution. Developmental Biology 344, 138–149.

Veeman, M., Reeves, W., 2015. Quantitative and in toto imaging in ascidians: Working toward an image-centric systems biology of chordate morphogenesis. Genesis 53, 143–159.

Vendrell, V., López-Hernández, I., Alonso, M.B.D., Feijoo-Redondo, A., Abello, G., Gálvez, H., Giráldez, F., Lamonerie, T., Schimmang, T., 2015. Otx2 is a target of N-myc and acts as a suppressor of sensory development in the mammalian cochlea. Development 142, 2792–2800.

Von Bartheld, C.S., 2004. The terminal nerve and its relation with extrabulbar "olfactory" projections: lessons from lampreys and lungfishes. Microscopy Research and Technique 65, 13–24.

Wacker, S.A., Jansen, H.J., McNulty, C.L., Houtzager, E., Durston, A.J., 2004. Timed interactions between the Hox expressing non-organiser mesoderm and the Spemann organiser generate positional information during vertebrate gastrulation. Developmental Biology 268, 207–219.

Wang, V.Y., Rose, M.F., Zoghbi, H.Y., 2005. Math1 expression redefines the rhombic lip derivatives and reveals novel lineages within the brainstem and cerebellum. Neuron 48, 31–43.

Whitlock, K.E., 2004. Development of the nervus terminalis: origin and migration. Microscopy Research and Technique 65, 2–12.

Wilson, L., Maden, M., 2005. The mechanisms of dorsoventral patterning in the vertebrate neural tube. Developmental Biology 282, 1–13.

Winata, C.L., Kondrychyn, I., Kumar, V., Srinivasan, K.G., Orlov, Y., Ravishankar, A., Prabhakar, S., Stanton, L.W., Korzh, V., Mathavan, S., 2013. Genome wide analysis reveals Zic3 interaction with distal regulatory elements of stage specific developmental genes in zebrafish. PLoS Genet 9, e1003852.

Winata, C.L., Korzh, V., 2018. Zebrafish Zic genes mediate developmental signaling, Zic Family. Springer, pp. 157–177.

Wright, T.J., Mansour, S.L., 2003. Fgf3 and Fgf10 are required for mouse otic placode induction. Development 130, 3379–3390.

Wu, F., Bard, J.E., Kann, J., Yergeau, D., Sapkota, D., Ge, Y., Hu, Z., Wang, J., Liu, T., Mu, X., 2021. Single cell transcriptomics reveals lineage trajectory of retinal ganglion cells in wild-type and Atoh7-null retinas. Nature Communications 12, 1–20.

Wu, D.K., Kelley, M.W., 2012. Molecular mechanisms of inner ear development. Cold Spring Harbor Perspectives in Biology 4, a008409.

Wullimann, M.F., Grothe, B., 2013. The central nervous organization of the lateral line system, The Lateral Line System. Springer, pp. 195–251.

Xiang, M., Maklad, A., Pirvola, U., Fritzsch, B., 2003. Brn3c null mutant mice show long-term, incomplete retention of some afferent inner ear innervation. BMC Neuroscience 4, 2.

Xue, Y., Zheng, X., Huang, L., Xu, P., Ma, Y., Min, Z., Tao, Q., Tao, Y., Meng, A.J.N.c., 2014. Organizer-derived Bmp2 is required for the formation of a correct Bmp activity gradient during embryonic development. Nature Communications 5, 1–14.

Xu, J., Li, J., Zhang, T., Jiang, H., Ramakrishnan, A., Fritzsch, B., Shen, L., Xu, P.X., 2021. Chromatin remodelers and lineage-specific factors interact to target enhancers to establish proneurosensory fate within otic ectoderm. Proceedings of the National Academy of Sciences of the United States of America 118, 1–12.

Xu, P.-X., Woo, I., Her, H., Beier, D.R., Maas, R.L., 1997. Mouse Eya homologues of the Drosophila eyes absent gene require Pax6 for expression in lens and nasal placode. Development 124, 219–231.

Yamada, M., Terao, M., Terashima, T., Fujiyama, T., Kawaguchi, Y., Nabeshima, Y.-i., Hoshino, M., 2007. Origin of climbing fiber neurons and their developmental dependence on Ptf1a. Journal of Neuroscience 27, 10924–10934.

Yan, B., Neilson, K.M., Moody, S.A., 2009. foxD5 plays a critical upstream role in regulating neural ectodermal fate and the onset of neural differentiation. Developmental Biology 329, 80–95.

Yang, T., Kersigo, J., Jahan, I., Pan, N., Fritzsch, B., 2011. The molecular basis of making spiral ganglion neurons and connecting them to hair cells of the organ of Corti. Hearing Research 278, 21–33.

Zhang, C., Kolodkin, A.L., Wong, R.O., James, R.E., 2017. Establishing Wiring Specificity in Visual System Circuits: From the Retina to the Brain. Annual Review of Neuroscience 40, 395–424.

Zou, D., Silvius, D., Fritzsch, B., Xu, P.-X., 2004. Eya1 and Six1 are essential for early steps of sensory neurogenesis in mammalian cranial placodes. Development 131, 5561–5572.

2 Development of the Olfactory System
From Sensory Neurons to Cortical Projections

CONTENTS

2.1 INTRODUCTION: ORGANIZATION OF THE MAMMALIAN OLFACTORY SYSTEM

In mammals, the olfactory system detects airborne chemicals as odorants. Environmental odorants have to be actively delivered to the olfactory epithelium (OE) in the nasal cavity by rhythmic inhalation or sniffing. Odorants generated by food intake can be delivered by the exhalation process (retronasal smell) in humans (Shepherd, 2006). Odorants are then dissolved into the olfactory mucus, which is secreted from the Bowman's gland in the OE. The dissolved odorants are then detected by the odorant receptors (ORs) expressed by olfactory sensory neurons (OSNs). OSNs extend multiple olfactory cilia into the olfactory mucus from their dendritic knobs. ORs are localized at the olfactory cilia and subsequent signal transduction occurs within this compartmentalized structure. ORs in the main

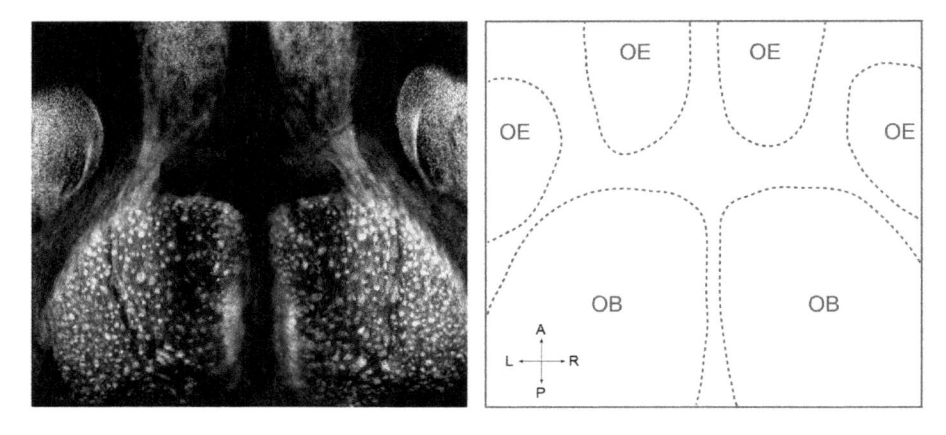

FIGURE 2.1   The mouse olfactory system. An OMP-GFP knock-in mouse, in which all mature OSNs are labeled with EGFP. OE and OB were cleared with BABB and imaged with a confocal microscope. OSN somata are scattered in the OE. OSN axons converge onto ~1,000 sets of glomeruli in the medial and lateral surface of the OBs. Dorsal view of the OE and OB. A, anterior; P, posterior; L, left; R, right. (Image modified from Imai, 2011.)

olfactory system are G-protein coupled receptors (GPCRs) with seven transmembrane domains. The crystal structures of various GPCRs have been reported to date; however, the crystal structure of vertebrate ORs has yet to be determined and remains as the major challenge in the field. All the ORs are considered to activate a heterotrimeric G-protein, G_{olf}, which then stimulates adenylyl cyclase type III. The cAMP then gates cyclic nucleotide-gated (CNG) channels, which contains CNGA2 as an essential subunit. Calcium influx through the CNG channels gate chloride channels to mediate chloride efflux, and together depolarizes membrane potentials (reviewed in Firestein, 2001). Action potentials generated in OSNs are propagated along axons to the glomeruli of the olfactory bulb (OB), in which glutamatergic neurotransmission occurs between OSNs and mitral and tufted (M/T) cells (Figure 2.1). M/T cells project axons to various areas of the brain, including the anterior olfactory nucleus, piriform cortex, lateral entorhinal cortex, olfactory tubercle, cortical amygdala, tenia tecta, etc., which are collectively called the olfactory cortex (reviewed in Imai, 2014; Mori and Sakano, 2011).

Odorants emitted from the environment or conspecifics are functionally and chemically diverse. Moreover, unlike visual or auditory stimuli, chemical information is discrete in nature. ORs have evolved to cover a huge variety of chemicals, resulting in the largest gene family of up to 2,000 genes among vertebrates. There are ~1,000 types of functional ORs in mice and ~390 in humans. In addition, there are hundreds of OR pseudogenes, reflecting the dynamic expansion and shrinkage of the OR gene repertoire during evolution (Niimura and Nei, 2007; Zhang and Firestein, 2002). After the discovery of ORs in 1991 (Buck and Axel, 1991), molecular, cellular, and circuit logics of the olfactory system have been extensively studied in mice. Importantly, odor information detected by ~1,000 sets of ORs is processed in discrete parallel circuits in the olfactory system. To ensure

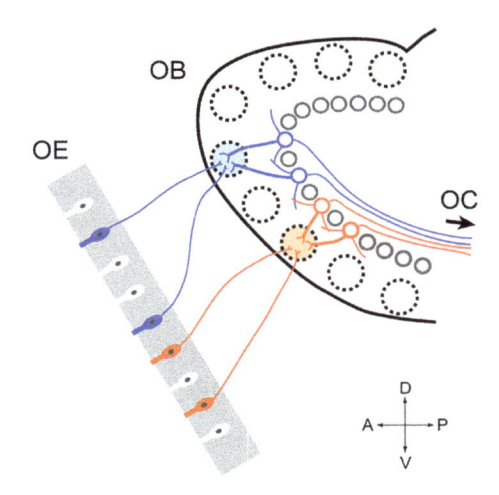

FIGURE 2.2 Organization of the olfactory system. Each OSN in the OE expresses just one type of OR in a mono-allelic manner (*one neuron – one receptor rule*). OSNs expressing the same type of OR converge their axons to a common set of glomeruli in the OB (*one glomerulus – one receptor rule*). In the OB, each M/T cell connects their primary dendrite to just one glomerulus to receive excitatory sensory inputs (*one M/T cell – one glomerulus rule*). M/T cells connecting to the same glomerulus are called sister M/T cells. M/T cell axons project to the olfactory cortex. Sagittal image. A, anterior; P, posterior; D, dorsal; V, ventral.

parallel processing, there are three important principles in the organization of the olfactory system (Figure 2.2).

Firstly, each OSN expresses just one type of OR out of ~1,000 repertoires, which is known as the *'one neuron – one receptor rule'* (Malnic et al., 1999; Serizawa et al., 2004). In other words, there are 1,000 distinct types of OSNs, each of which expresses just one type of OR. This is a stark contrast to the taste system, in which all the tastants are detected and categorized by just five types of taste receptor cells (for sweet, umami, bitter, sour, and salty) (Ache and Young, 2020; Roper and Chaudhari, 2017).

Secondly, OSNs expressing the same type of ORs converge their axons onto a common glomerulus in the olfactory bulb, known as the *'one glomerulus – one receptor rule'* (Mombaerts et al., 1996). This enables segregated sensory inputs to the OB. In mice, there are mirror-symmetric glomerular maps in the medial and lateral surface of the olfactory bulb, each of which is composed of ~1,000 sets of glomeruli. The location of glomeruli for a given OR is largely stereotyped, while there are local permutations among individuals.

Thirdly, each M/T cells receives direct excitatory inputs from just one glomerulus through its primary dendrite, which could be designated here as *'one M/T cell – one glomerulus rule'*. As a result, distinct chemical information detected by an OR will be processed in a segregated pathway from OSNs to a specific set of M/T cells. Each glomerulus is innervated by 20–50 M/T cells, which are called 'sister' M/T cells (Ke et al., 2013; Kikuta et al., 2013; Sosulski et al., 2011).

In this chapter, I will describe the odor coding mechanisms at different stages of the olfactory system based on the above three principles (Section 2.2). The evolutionary aspects will be discussed based on the olfactory receptor genes (Section 2.3 and 2.4). I then describe developmental mechanisms that establish the three principles in the mammalian olfactory system (Sections 2.5–2.10). I will mainly describe our knowledge in the most extensively studied mammalian model organism, mouse, but some aspects will be compared with other vertebrate species with an evolutionary viewpoint.

2.2 PHYSIOLOGY AND CODING LOGICS AT DIFFERENT STAGES OF THE OLFACTORY CIRCUITS

In mice, nearly 5% of genes in the genome are dedicated to ORs. However, this number is still much smaller than the total number of possible odorants animals would encounter in their life, which is estimated to be at least hundreds of thousands. A key to understanding this discrepancy is the ligand specificity of ORs. Each OR typically interacts with multiple types of odorants. Similarly, each OR is recognized by multiple types of ORs (Malnic et al., 1999). Some ORs are narrowly tuned but others are more broadly tuned to a variety of odorants in a heterologous assay system (Saito et al., 2009). While it has been previously considered that odorants 'activate' ORs and OSNs, recent studies have demonstrated that ORs have variable levels of basal activities and an odorant can act as an inverse agonist for some ORs (Inagaki et al., 2020). Thus, each odorant is represented as the combinatorial activation and inhibition patterns of ORs at the level of OSNs (combinatorial receptor code).

In nature, an odor is often composed of multiple odorants. When multiple odorants are presented, some odorants can suppress OSN responses to other odorants, known as antagonism (Inagaki et al., 2020; Oka et al., 2004; Pfister et al., 2020; Xu et al., 2020; Zak et al., 2020). Moreover, some odorants can enhance the responses to other odorants, known as synergy (Inagaki et al., 2020; Xu et al., 2020). As a result, OSN responses to a mixture of odorants can be smaller or larger than the linear sum of responses to its components. Thus, the perception of natural odors is already extensively modulated at the most peripheral level (Kurian et al., 2021).

In physiological conditions, OSNs respond not only to odorants but also to mechanical stimuli produced by the nasal airflow, i.e., sniffing (Grosmaitre et al., 2007). In mammals, one sniff is a unit for odor information processing in the brain (Kepecs et al., 2006), and the mechanosensory signals serve as an important pacemaker (Iwata et al., 2017).

OSN responses in the OE are then converted to the odor map in the glomerular layer of the OB (Johnson and Leon, 2000; Mombaerts et al., 1996; Mori et al., 2006; Wachowiak and Cohen, 2001). Odor information is represented by spatial patterns of activity, as well as both activation and inhibition, in glomeruli (Inagaki et al., 2020). Similar odorants tend to activate glomeruli in similar areas of the OB (Rubin and Katz, 1999; Uchida et al., 2000). Therefore, there is a chemotopic representation in the OB, even if chemotopy is not necessarily evident at a finer scale (Chae et al., 2019; Ma et al., 2012). Different parts of the OB mediate distinct innate behaviors. For example, the dorsal domain of the OB mediates innate fear responses (Kobayakawa et al., 2007), while other parts of the OB can mediate learned fear responses. Furthermore, some pheromone signals are mediated by the ventral OB (Lin et al., 2005).

Inputs from OSN axons are then relayed to the second-order neurons in the OB. However, the odor map in the OB is more than just a spatial map. Due to the sniff-coupled mechanosensation of OSNs (Grosmaitre et al., 2007), M/T cells show rhythmic neuronal activity without odors (Iwata et al., 2017). Odor stimuli change not only the firing rate but also the timing of activity within a sniff cycle in M/T cells (Dhawale et al., 2010; Shusterman et al., 2011; Spors and Grinvald, 2002). In particular, the temporal patterns of activity in M/T cells are important for the concentration-invariant representation of an odor 'identity' (Iwata et al., 2017). Recent studies using optogenetic activation of glomeruli indicated that the temporal sequence of glomerular activation within a sniff cycle is critical for odor identity coding (Chong et al., 2020; Smear et al., 2013). Indeed, the temporal patterns of activity, but not the response amplitude, are concentration-invariant in M/T cells (Imai, 2020; Iwata et al., 2017). It is also suggested that glomeruli which are activated earlier within the sniff cycle are invariant and have more impacts on odor identity coding, known as primacy code hypothesis (Chong et al., 2020; Hopfield, 1995; Wilson et al., 2017).

In the OB, ~99% of neurons are interneurons. Interneurons play important roles to reformat odor inputs in the OB (Imai, 2014; Wilson and Mainen, 2006). For example, gain control is mediated by juxtaglomerular interneurons and parvalbumin-expressing interneurons. Periglomerular short axon cells mediate lateral inhibition among glomeruli and granule cells mediate lateral inhibition among individual M/T cells (Economo et al., 2016; Yokoi et al., 1995). Periglomerular neurons and granule cells regulate theta and gamma oscillations, respectively (Fukunaga et al., 2014). Many of them are also regulated by the excitatory top-down inputs from the olfactory cortices (Boyd et al., 2012; Markopoulos et al., 2012).

Mitral and Tufted cells are anatomically and functionally distinct. Single-cell RNA sequencing indicated that there may be more subtypes within mitral and tufted cells (Zeppilli et al., 2020). Tufted cells receive direct glutamatergic inputs from OSNs and have relatively low-threshold, short-latency odor responses, fire strongly in phase with sniff cycles, and are less influenced by OB interneurons. On the other hand, Mitral cells receive more indirect sensory inputs via tufted cells and show higher-threshold and longer-latency responses, possibly due to more inhibitory modulations (Fukunaga et al., 2012; Gire et al., 2012; Igarashi et al., 2012). Moreover, responses of M/T cells are extensively modulated by various types of interneurons (i.e., juxtaglomerular interneurons, short axon cells, and granule cells), centrifugal feedback from the olfactory cortex, and neuromodulations (Imai, 2014).

The activity of M/T cells is transmitted to a variety of brain regions (e.g., anterior olfactory nucleus, olfactory tubercle, tenia tecta, piriform cortex, lateral entorhinal cortex, and cortical amygdala), which are together called the olfactory cortex (Figure 2.3). Each of these areas seems to have distinct functions. For example, the anterior olfactory nucleus mediates the coordination between two hemispheres (Kikuta et al., 2010; Yan et al., 2008); the cortical amygdala mediates

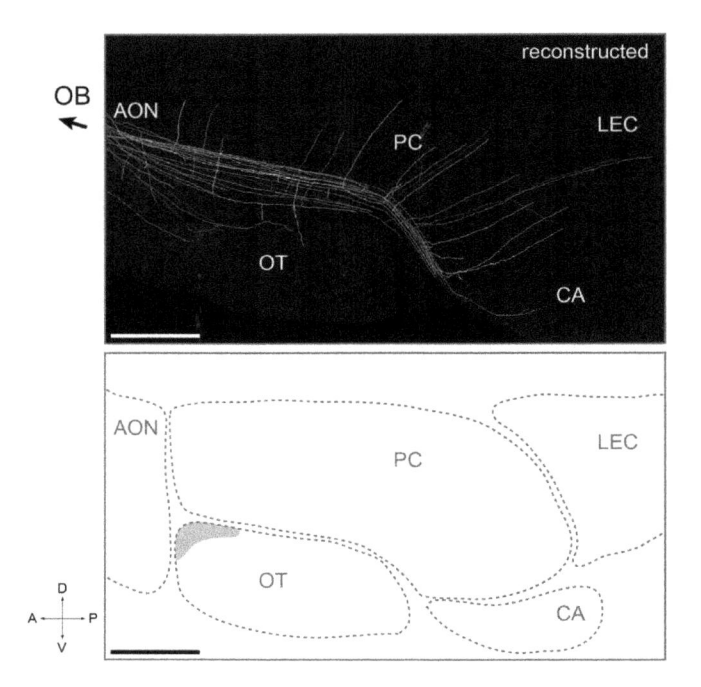

FIGURE 2.3 Axonal projections of M/T cells. Multicolor Tetbow AAV vectors were locally injected into the OB. Fourteen M/T cell axons were traced and reconstructed in the olfactory cortex. Tufted cells project axons to the AONpE and the cap region (shaded) of the olfactory tubercle. Topographic projection is seen in AONpE. Mitral cells project axons to all the other regions of the olfactory cortex. Mitral cells show differential patterns of axon collaterals, with no obvious topography. Image from the ventrolateral surface of the brain. AON, anterior olfactory nucleus; CA, cortical amygdala; LEC, lateral entorhinal cortex; LOT, lateral olfactory tract; OT, olfactory tubercle; PC, piriform cortex. A scale bar, 1 mm. (Modified from Sakaguchi et al., 2018.)

innate olfactory behaviors (Root et al., 2014); the amygdalo-piriform transition area mediates stress hormone responses to predator odors (Kondoh et al., 2016); and the piriform cortex mediates olfactory discrimination and learning (Giessel and Datta, 2014; Wilson and Sullivan, 2011).

The axonal projection profiles of mitral and tufted cells are segregated: Tufted cells project to the anterior olfactory nucleus (AONpE) and the cap region of the olfactory tubercle, while mitral cells project to all the other regions (Hirata et al., 2019; Igarashi et al., 2012). The axonal projection profiles of M/T cells are different in different cortical areas. Only in the AONpE, Tufted cells are arranged into a topographic projection, which may be important for the glomerulus-specific precise coordination and unity between left and right OBs (Grobman et al., 2018; Yan et al., 2008). In the cortical amygdala, inputs from different glomerulus are segregated, possibly representing distinct valence and stereotyped innate behaviors produced by these regions (Sosulski et al., 2011). In the piriform cortex, axons from each M/T cell are highly scattered with no obvious topography

(Igarashi et al., 2012; Sosulski et al., 2011). Inputs from multiple glomeruli converge onto individual pyramidal neurons (Apicella et al., 2010; Miyamichi et al., 2011). Moreover, responses of pyramidal neurons in the piriform cortex are sensitive to the temporal sequence of glomerular activation (Chong et al., 2020; Haddad et al., 2013). Thus, these neurons read out the specific combinations of the spatiotemporal patterns of glomerular activity. In the piriform cortex, the inputs from the recurrent network is another important source of activity in pyramidal neurons (Blazing and Franks, 2020; Wilson and Sullivan, 2011). The recurrent circuit implements some important features in odor perception, such as concentration-invariance (Bolding and Franks, 2018), pattern completion (Bolding et al., 2020), and odor categorization (Pashkovski et al., 2020). The auto-associative network may also be useful for olfactory working memory (Zhang et al., 2019).

Odor representation remains unchanged after learning in the piriform cortex; however, odor value affects the odor representation in the downstream regions, orbitofrontal cortex (OFC), and medial prefrontal cortex (mPFC) (Wang et al., 2020). Olfactory information is also conveyed from the lateral entorhinal cortex to the hippocampus for associative memory (Igarashi et al., 2014; Li et al., 2017). The olfactory tubercle is located in the ventral striatum and represents an odor-induced motivation for approach vs. avoidance behavior based on associative learning (Millman and Murthy, 2020; Murata et al., 2015). Odor-induced innate and learned fear signals are integrated and processed in a part of central amygdala (Isosaka et al., 2015). In humans, the integration of olfactory and taste information occurs in the dorsal and anterior part of insular (Fadool and Kolling, 2020; Shepherd, 2006).

2.3 RECEPTOR GENES AND EVOLUTION OF THE OLFACTORY SYSTEM

OR genes are present in all vertebrates including fish, amphibians, reptiles, birds, and mammals, and the origin of OR genes can be traced back to the latest common ancestor of chordates, including amphioxus (Niimura, 2009). ORs are absent in ascidians, which is correlated well to the unique organization of the mouth (Kaji et al., 2016; Veeman et al., 2010) and the lack of clear olfactory organs (Holland, 2020). Insects have distinct types of olfactory receptors, but they are ionotropic receptors (Benton et al., 2009; Clyne et al., 1999; Sato et al., 2008; Vosshall et al., 1999), rather than GPCRs. Thus, a lot of similarities seen between vertebrate and insect olfactory systems are the results of convergent evolution (Imai et al., 2010).

Different aspects of olfactory function are linked to specific receptor types with different evolutionary traits. ORs can be divided into class I and class II based on the sequence similarity (Niimura and Nei, 2007; Zhang and Firestein, 2002). Class I genes were first identified in fish and frog that have persisted throughout the evolution of most vertebrate taxa. In contrast, class II genes are specific to terrestrial animals and account for ~90% of the mammalian OR repertoires. It is, therefore, suggested that class I and class II ORs are utilized for water-soluble and more volatile odorants, respectively (Bear et al., 2016; Nei et al., 2008).

While OR genes are preserved in all vertebrates, the number of receptors ranges from ~10 to ~2,000. Primates have 300–400 functional OR genes. In humans, 840 OR genes were found but among them only ~390 genes have intact coding sequences. Even within humans, various polymorphisms are found for ORs, many of which may be evolved under different selective pressure in different geographic areas (Mainland et al., 2014). OR genes are dispersed in the genome, except for chromosomes 20 and Y. (Nei et al., 2008) and 40% of human OR are found in chromosome 11. African and Asian elephants have the largest number of coding genes (~2,000) and pseudogenes (~2,200) that is twice as many when compared to dogs and five times as many when compared to humans (Niimura et al., 2014). On the other hand, marine mammals typically have <100 functional ORs as a result of the evolutionary loss. Dolphins do not smell and no longer maintain OR genes: They only have 12 intact OR genes and ~100 pseudogenes (Kishida et al., 2015; Niimura et al., 2014). Rodents have 1000–1500 ORs (Nei et al., 2008; Zhang and Firestein, 2002).

In addition to the OR family, there is another type of receptor family, TAARs, in the main olfactory system (Liberles and Buck, 2006). TAARs are specialized for the detection of volatile amines (Dewan et al., 2013; Li et al., 2013; Pacifico et al., 2012). ORs and TAARs are expressed in a restricted area within the OE, thus forming zones in the OE (Miyamichi et al., 2005; Ressler et al., 1993; Vassar et al., 1993). Class I ORs and TAARs are only expressed in the dorsal zone of the OE, while class II ORs are expressed in both dorsal and ventral zones. OSNs in the dorsal and ventral zones of the OE project axons to the dorsal (D) and ventral (V) domains of the OB, respectively. Within the dorsal domain of the OB, OSNs expressing different types of receptors have distinct glomerular territories: DI for class I ORs, DII for class II ORs, and DIII for TAARs (Kobayakawa et al., 2007; Pacifico et al., 2012). Different domains of the OB may be linked to distinct innate olfactory behaviors (Inokuchi et al., 2017; Kobayakawa et al., 2007).

The overall structure of olfactory circuits is conserved across vertebrate species. In zebrafish, different classes of OSNs, namely ciliated OSNs expressing OMP and microvillar OSNs expressing Trpc2, project axons to distinct domains of the OB (Sato et al., 2005). Majority of OSNs express just one OR, and OSNs expressing the same type of OR converge their axons to common glomeruli (Sato et al., 2007). As a result, the odor map is formed in the glomerular layer of the OB (Friedrich and Korsching, 1997). Taking advantage of its small size, the connectome of mitral cells is beginning to be elucidated (Miyasaka et al., 2014; Wanner and Friedrich, 2020). Like mice, zebrafish mitral cells receive inputs from single glomerulus and send stereotyped projection to some, but more divergent projection to other cortical regions. In the lamprey, mitral-like and tufted-like cells show segregated projections to distinct regions or the pallium, suggesting that parallel odor information processing is a conserved feature of the vertebrate olfactory system (Suryanarayana et al., 2021).

In summary, ORs (class I and class II) and TAARs mediate odor detection in the main olfactory system in mice. There is a zonal organization in the OE and OB, based on the receptor class and the expression zones of the receptors. This may be linked to distinct innate behaviors mediated by these receptors.

2.4 VOMERONASAL RECEPTORS AND OLFACTORY SUBSYSTEMS

The rodent olfactory system has additional chemosensory receptor families and olfactory subsystems (Bear et al., 2016; Munger et al., 2009). The vomeronasal organ (VNO) and its projection target, accessory olfactory bulb (AOB) constitute the accessory olfactory system. Vomeronasal sensory neurons (VSNs) in the VNO express three different types of chemosensory receptors, V1Rs, V2Rs, and formyl peptide receptors (FPRs)(Dulac and Axel, 1995; Herrada and Dulac, 1997; Matsunami and Buck, 1997; Riviere et al., 2009). The VNO also shows a zonal organization: VSNs in the apical layer of the VNO express V1Rs or FPRs, G_{i2}, and project axons to the rostral half of the AOB, whereas VNSs in the basal layer express V2Rs, G_o, and project to the caudal AOB. Unlike the main olfactory system, VSNs expressing a given receptor project axons to multiple glomeruli in the AOB (Belluscio et al., 1999; Del Punta et al., 2002). As many of the glomeruli are composed of heterogeneous VSNs, inputs from different receptors may be partially converged at the level of mitral cells. Some of V1Rs are known to detect volatile and non-volatile small molecular-weight compounds, whereas some V2Rs detect peptide and protein ligands, such as MUPs, ESPs, and MHC peptides (Chamero et al., 2007; Kimoto et al., 2005; Leinders-Zufall et al., 2004). FPRs detect formyl peptides and may mediate avoidance of sick conspecifics (Bufe et al., 2019). The sensitivity and selectivity of vomeronasal receptors seems to be higher than those for ORs (Leinders-Zufall et al., 2000).

Many V2R-expressing VSNs express members of another multigene family, H2-Mv, which is known as a non-classical class I MHC (Ishii et al., 2003; Loconto et al., 2003). Together with its co-receptor, β2-microglobin, H2Mv facilitates the functional expression of V2Rs (Leinders-Zufall et al., 2009).

Some atypical sensory neurons in the OE express non-GPCR receptors, such as GC-D and MS4As, and project axons to the 'necklace glomeruli' that are located at the posterior end of the main OB (Greer et al., 2016; Hu et al., 2007). GC-D and MS4As are exceptions to the *one neuron – one receptor rule*. They are expressed in the same neurons and respond to carbon dioxide, carbon disulfide, fatty acids, and volatile pheromones (Greer et al., 2016; Hu et al., 2007; Munger et al., 2010).

The receptor repertoire and olfactory subsystems demonstrate extensive species-specific diversification under various environmental challenges (Nei et al., 2008). V1Rs and V2Rs are entirely lost in some species. The accessory olfactory system seems to be non-functional in primates, as most of the V1R/V2Rs and a key transduction channel, Trpc2, are missing. In bony fish and amphibians, ORs, V2Rs, and TAARs comprise the major components of the olfactory system (Korsching, 2020; Nei et al., 2008).

So far, only a small subset of receptors has been deorphanized for olfactory receptors (ORs, TAARs, V1Rs, V2Rs, and FPRs). New deorphanization strategies *in vitro* and *in vivo* are being developed, and a comprehensive description of receptor-ligand interactions should facilitate our understanding of olfactory physiology, behavior, and evolution (Dey and Matsunami, 2011; Jiang et al., 2015; Lee et al., 2019; Saito et al., 2009; von der Weid et al., 2015).

In summary, GPCR and non-GPCR chemosensory receptors mediate a variety of species-specific chemosensory functions in both the vomeronasal and olfactory subsystems. The identification of their ligands should facilitate our understanding of animal behavior and hormonal regulation based on chemical communication.

2.5   GENERATION OF OSNs AND SINGULAR OR GENE CHOICE

A remarkable feature of the mammalian olfactory system is that receptor-specific neuronal circuits are constructed from the periphery to the central brain, despite the dynamic changes of receptor repertoire that occurred during evolution. From this section, I will describe developmental mechanisms to achieve this goal, which has been unveiled in mice during the past 30 years.

During development, OE and OB develop from different parts of the embryo. The OE is derived from the olfactory placode, whereas the OB is a part of the central nervous system. The olfactory placode is one of the cranial sensory placodes that gives rise to several specialized sensory organs (anterior pituitary gland, OE, lens, auditory and vestibular organs) and sensory ganglia of the trigeminal, facial, glossopharyngeal, and vagus cranial nerves. A set of transcription factors, including *Eya1/Six1, Otx2, Pax6, Emx2,* and *Ebf2*, regulate the induction of olfactory placode. Additionally, retinoic acid (*RA*), *Fgf8, Shh,* and *BMP4* secreted from adjacent mesenchymal cells define the axis of the OE and induce nasal cavity formation (Moody and LaMantia, 2015). These factors together allow for the upregulation of specific genes required for the generation of OSNs (e.g., *Sox2, Ascl1, Neurog1, Neurod1,* and *Foxg1*) (Cau et al., 2002; Dvorakova et al., 2020; Kawauchi et al., 2009; Panaliappan et al., 2018; Tucker et al., 2010). In addition, microRNA plays several critical roles in neuronal induction (Kersigo et al., 2011). The OE and VNO develop from the olfactory placode that also gives rise to a set of gonadotropin-releasing hormone (GnRH)-positive neurons that migrate into the hypothalamus (Wierman et al., 2011; Wray, 2010). Hypothalamic GnRH neurons play crucial roles in reproduction and are also modified by olfactory inputs (Boehm et al., 2005; Yoon et al., 2005). In some fish species, GnRH neurons in the olfactory system project to the retina (Crapon de Caprona and Fritzsch, 1983).

The OE is composed of multiple cell types. In addition to mature and immature OSNs, there are two types of basal stem cells, the horizontal basal cells (HBCs) and globose basal cells (GBCs), sustentacular (supporting) cells in the apical surface of the OE, and cells comprising the bowman's gland. All of these cells are generated from HBCs, a common multipotent stem cell type (Schwob et al., 2017). Wnt signaling plays a critical role to make a neuronal fate choice from HBCs to GBCs (Fletcher et al., 2017). OSN lineage develops from transit-amplifying cells (GBC$_{TA-OSN}$; Ascl1+) through a second transit-amplifying progenitor and the intermediate precursor (GBC$_{INP}$; *Neurog1*$^+$ and/or *Neurod1*$^+$). Daughter cells from GBC$_{INP}$ differentiate into immature OSNs (GAP43$^+$) and then mature OSNs (OMP$^+$). OSNs are regenerated and replaced throughout the life of animals. The renewal of OSNs is enhanced by OSN injury, following the same differentiation trajectory (Gadye et al., 2017).

The first important event after the terminal cell division is the OR gene choice (Hanchate et al., 2015). In the OE, expression of each OR gene is restricted within a zone, which are continuous and overlapping from dorsomedial to ventrolateral

(Miyamichi et al., 2005; Ressler et al., 1993; Vassar et al., 1993). However, an OR is expressed in a punctate pattern within a zone. A mature OSN expresses a single functional OR gene in a mono-allelic manner, forming the basis of odor coding (*one neuron – one receptor rule*) (Chess et al., 1994; Malnic et al., 1999). This is also an important basis for the OR-instructed axonal projection in the OB. Singular OR gene expression was also observed among OR transgenes having the same regulatory sequences (Serizawa et al., 2000; Vassalli et al., 2002), suggesting that OR gene expression is a result of stochastic choices from 2,000 possible alleles in the genome. This is useful to accommodate new and/or polymorphic OR genes generated during evolution. The molecular mechanisms of the singular OR gene choice have been extensively studied during the last two decades (Monahan and Lomvardas, 2015).

OR genes are distributed throughout most chromosomes, but many of them are located close to each other, forming OR gene clusters. Cis-regulatory enhancer elements were found in many of the OR gene clusters (Serizawa et al., 2003). These are required for the expression of multiple OR genes in the cluster in cis. A typical OR gene promoter has two conserved sequences: O/E motifs for the Olf/EBF family of transcription factors (Olf1-4) and homeodomain sites for *Lhx2* and *Emx2*. These motifs are essential for the OR gene expression (Vassalli et al., 2002).

To ensure the singular OR gene choice, it is important to silence all the other OR genes. Prior to the OR gene choice, heterochromatin compacts and silence the entire OR genes in the genome. OR gene loci are decorated with histone H3 lysine 9 trimethylation (H3K9m39) and H4K20me3, which are characteristic of constitutive heterochromatin (Magklara et al., 2011). However, after OR gene choice, the heterochromatin mark is absent only at the chosen OR gene allele. The chosen OR gene locus instead has H3K4me3, a hallmark of the transcriptionally active euchromatin.

The singular OR gene choice is ensured by the combination of stochastic activation and the subsequent feedback regulation (Serizawa et al., 2003) (Figure 2.4). To stochastically activate an OR gene, a histone demethylase, LSD1, plays a critical role. LSD1 mediates the demethylation of H3K9 in the OR gene locus. Knockout experiments indicate that LSD1 is required for the activation but not the maintenance of the OR gene expression (Lyons et al., 2013). Once a functional OR gene is activated, feedback regulation prevents further activation process to ensure the singular expression. The feedback signal does not require G-protein signals from ORs (Imai et al., 2006). Due to the poor folding of the OR proteins, the translated OR proteins triggers an unfolded protein response (UPR), activating a kinase, Perk, which in turn phosphorylates the translation-initiation factor eIF2a (Dalton et al., 2013). The phosphorylated eIF2a halts translation of most transcripts, but facilitates the translation of ATF5, which then stabilizes the expression of the chosen OR, prevents further activation of other ORs, and facilitates OSN maturation. Due to dynamic evolutional changes, significant fractions of OR genes in the genome are pseudogenes. If the first choice was a pseudogene, the UPR does not occur, and gene choice continues until a functional OR gene is activated (Serizawa et al., 2003).

It remains elusive how chromatin demethylation occurs exclusively at just one OR gene allele at a time. As a cis-regulatory enhancer can interact with one OR gene

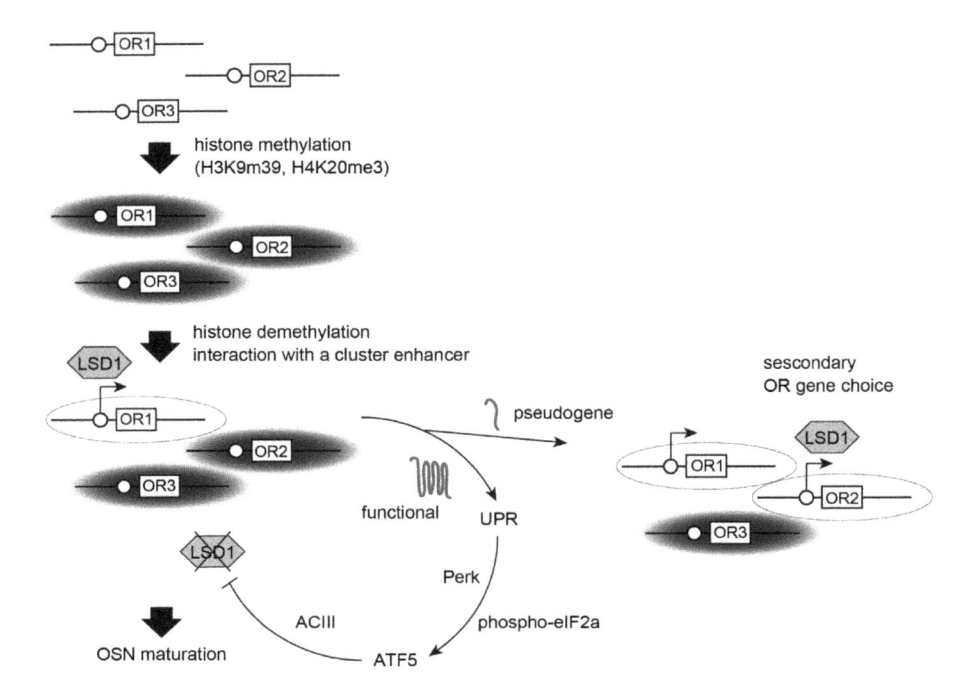

FIGURE 2.4 Singular OR gene choice. Prior to OR gene choice, OR gene loci in the genome are silenced by heterochromatin harboring H3K9m39 and H4K20me3 (black). A histone demethylase, LSD1 mediates the stochastic activation of an OR gene locus. Physical interaction between a cluster enhancer and an OR gene promoter may also play an important role in the stochastic OR gene choice. Once a functional OR gene is activated (indicated by an arrow), the translated OR proteins trigger an unfolded protein response (UPR), which mediates the feedback regulation. When an OR pseudogene is chosen, however, feedback regulation does not occur, and the gene choice continues until a functional OR gene is activated.

within a cluster at a time, this may limit the chance of co-activation (Lomvardas et al., 2006; Serizawa et al., 2003). It is also known that enhancers from different OR gene clusters in different chromosomes can physically interact with a chosen OR gene locus in trans (Markenscoff-Papadimitriou et al., 2014), suggesting a possible role for trans-chromosomal interactions in the singular OR gene choice.

In summary, the *one neuron – one receptor rule* in the olfactory system is ensured by stochastic activation of OR genes via histone demethylation and feedback regulation by the translated OR protein via UPR.

2.6   COARSE TARGETING OF OSN AXONS

Odor information detected by ~1,000 types of OSNs is then sorted into ~1,000 sets of glomeruli in the OB (*one glomerulus – one receptor rule*). Thus, a major challenge in the OSN axonal projection is how to sort OSN axons based on the expressed OR type. In the retinotopic visual map formation in the tectum or

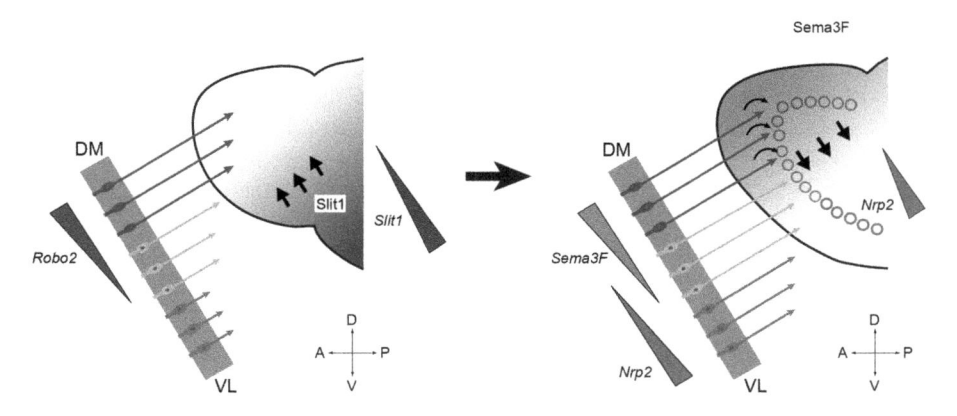

FIGURE 2.5 Dorsal-ventral (D-V) projection of OSN axons. During development, Robo2-high OSNs in the dorsomedial OE project axons, which are repelled by Slit1 in the ventral OB. These OSNs secrete Sema3F at their axon terminals. Nrp2-high OSNs in the ventrolateral OE project axons later and are repelled by Sema3F. This sequential regulation of OSN projections establish the D-V pattern of glomeruli in the OB. Sema3F secreted from dorsomedial OSN axons are also important to localize Nrp2-high M/T cells to the ventral domain of the OB. Saggital view of the OB. DM, dorsomedial; VL, ventrolateral. (Modified from Imai, 2011.)

superior colliculus, axon guidance molecules expressed in a graded manner (e.g., ephrin-A and B) regulate the coarse targeting of axons. This is then further refined based on spontaneous neuronal activity generated in the retina (Feldheim and O'Leary, 2010; Huberman et al., 2008). Similarly, the formation of the discrete olfactory map is also controlled by the combination of coarse axon targeting and local axon sorting.

One parameter that defines the coarse targeting of axons is the cell type and positional information of the OSNs in the OE. OSNs in the dorsomedial and ventrolateral zones project axons to the D and V domains of the OB, respectively. The D domain of the OB is comprised of three subdomains, DI, DII, and DIII, for class I ORs, class II ORs, and TAARs, respectively (Kobayakawa et al., 2007; Pacifico et al., 2012; Tsuboi et al., 2006). Within the V domain of the OB, dorsal-ventral locations of the glomeruli are correlated with the dorsomedial-ventrolateral expression zones of the ORs in the OE (Miyamichi et al., 2005).

Neuropilin-2 and Robo2 are expressed in a graded manner in the OE and play key roles in the OSN projection along the D-V axis of the OB (Figure 2.5). Nrp2 is expressed in a ventrolateral-high and dorsomedial-low gradient, whereas Robo-2 is expressed in a counter gradient (Cho et al., 2007; Nguyen-Ba-Charvet et al., 2008; Norlin et al., 2001). During development, OSNs in the dorsomedial zone project axons earlier than the ventrolateral ones (Takeuchi et al., 2010). At this stage, Slit1, a repulsive ligand for Robo2, is expressed in the septum and ventral OB. As a result, the early-arriving dorsomedial OSNs axons are confined to the D domain of the OB. At a later stage, Nrp2-high OSNs in the ventrolateral zone project axons to the OB. Sema3F coding for a repulsive ligand for Nrp2 is expressed in the dorsomedial

OSNs, but not in the OB. A conditional knockout experiment indicates that Sema3F secreted from early-arriving OSN axons is important to guide Nrp2-positive late-arriving OSN axons to the ventral OBs. As a result, the temporal sequence of the OSN projection is converted to the dorsal-ventral gradient in the OB with the aid of OB-derived Slit1 and OSN-derived Sema3F (Takeuchi et al., 2010).

Contrary to D-V patterning, the anterior-posterior (A-P) positioning of glomeruli is independent of expression zones of the ORs in the OE. The first mechanistic insight came from OR swapping experiments: When an OR coding sequence was replaced with that of another OR gene, OSN projection sites shifted, often along the A-P axis (Feinstein et al., 2004; Mombaerts et al., 1996; Wang et al., 1998). Thus, ORs have an instructive role in OSN projection. However, the OR-instructed OSN projection occurs independently of odor-evoked neuronal activity (Belluscio et al., 1998; Lin et al., 2000; Zheng et al., 2000). An OR mutant without G-protein coupling failed to form glomeruli in the OB, and this was rescued by a constitutively active G_s protein (Imai et al., 2006). Genetic manipulations to decrease and increase cAMP levels led to anterior and posterior shifts of glomeruli, suggesting that cAMP levels are a determinant of A-P projection position (Imai et al., 2006). Mice deficient for adenylyl cyclase type III show distorted topography of the glomerular map (Chesler et al., 2007; Dal Col et al., 2007; Zou et al., 2007). Basal activity, rather than ligand-dependent GPCR activity was correlated with the A-P positioning of glomeruli (Nakashima et al., 2013).

The OR-derived cAMP signals regulate transcriptional levels of Nrp1 positively and its repulsive ligand Sema3A negatively, forming complementary expression. Axon-axon interactions mediated by Sema3A and Nrp1 facilitate pre-target axon sorting, which can occur without OB. Together with the OB-derived Sema3A, this mechanism establishes the anterior-posterior positioning of axons (Imai et al., 2009) (Figure 2.6). Nrp1-high OSNs project axons to the posterior OB and this pattern is perturbed in Sema3A knockout (Schwarting et al., 2000). Plexin-A1 is expressed in a complementary manner to Nrp1, suggesting its role in axonal projection to anterior OB (Nakashima et al., 2013).

In summary, coarse targeting of OSN axons depends on graded guidance cues and their receptors. The D-V axis is determined by the positional information of OSNs in the OE and the A-P axis is determined by OR-derived cAMP signals. Axon guidance is controlled not only by the axon-target interactions, but also by repulsive axon-axon interactions (Imai and Sakano, 2011).

2.7 LOCAL SORTING OF OSN AXONS

After forming a coarse map in the olfactory bulb, OSN axons need to be further segregated to form discrete glomerular structures. Each OSN sends an unbranched axon into a glomerulus. A hallmark of glomerular organization is the coalescence of homotypic OSN axons expressing the same OR. However, this process occurs in the absence of the postsynaptic neurons (Bulfone et al., 1998).

When neuronal activity in OSNs was silenced, local axon sorting was perturbed forming multiple ectopic glomeruli (Imamura and Rodriguez Gil, 2020; Schwob et al., 2020; Yu et al., 2004; Zheng et al., 2000). As the silencing OSNs with Kir2.1

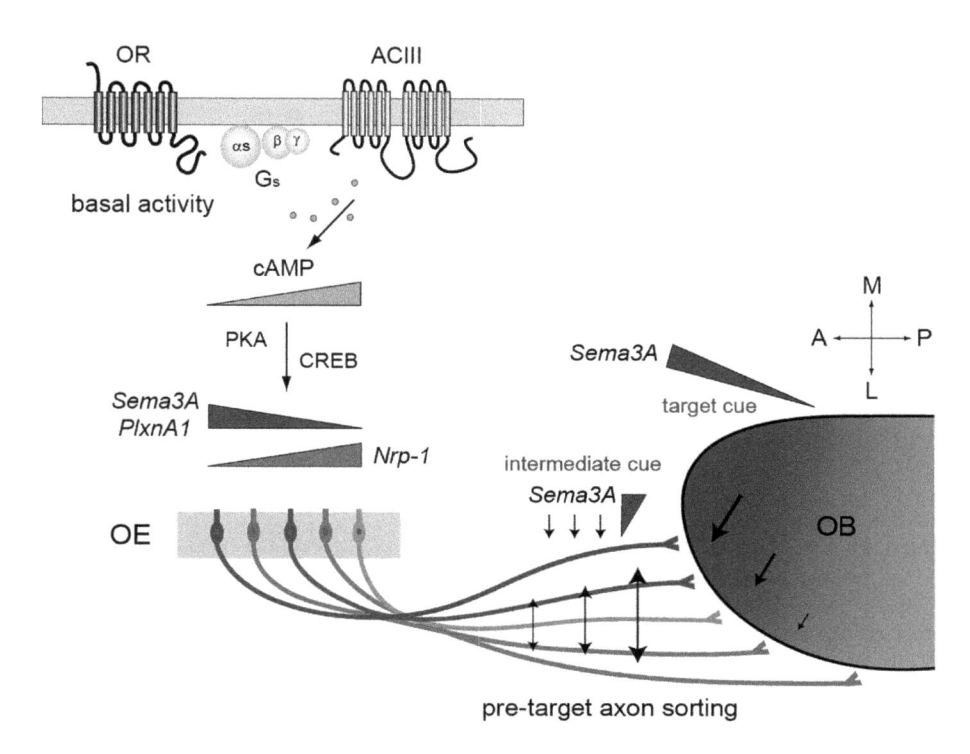

FIGURE 2.6   Anterior-posterior (A-P) projection of OSN axons. Each OR has a unique level of basal activity without odors. The basal activity positively regulates the transcriptional levels of Nrp1 via the cAMP pathway. Sema3A and PlxnA1 are negatively regulated by the basal activity. Repulsive axon-axon interactions mediated by Sema3A and Nrp1, together with the intermediate and target cues, establish the A-P positioning of glomeruli. Horizontal cartoon of the OB. A, anterior; P, posterior; M, medial; L, lateral. (Modified from Imai, 2011.)

shows a more severe phenotype than CNGA2 knockout, spontaneous activity may play a more important role. In retinotopic map formation, the spontaneous neuronal activity generated in the retina mediates the refinement of the map based on the Hebbian mechanisms, in which neurons that fire together wire together (Huberman et al., 2008; Shatz, 1992). However, the olfactory system utilizes a distinct mechanism to control local axon sorting (Figure 2.7).

Neuronal activity in OSNs regulates the expression of a set of adhesion molecules, Kirrel2 and Kirrel3, positively and negatively, respectively. Different levels of OR-dependent neuronal activity define OR-specific levels of Kirrel2 and 3. Kirrel2 and 3 are homophilic but not heterophilic. Therefore, the complementary expression of Kirrel2/3 leads to the fasciculation of like axons (Serizawa et al., 2006). The neuronal activity also regulates the expression of EphA5 and ephrin-A5, positively and negatively, respectively. Axonal ephrin-A5 and EphA5 are assumed to mediate repulsive interactions. Therefore, EphA5/ephrin-A5 facilitates the segregation of heterotypic axons (Serizawa et al., 2006). There are additional types of cell surface molecules that show glomerulus-specific expression patterns, e.g., BIG2, Pcdh10,

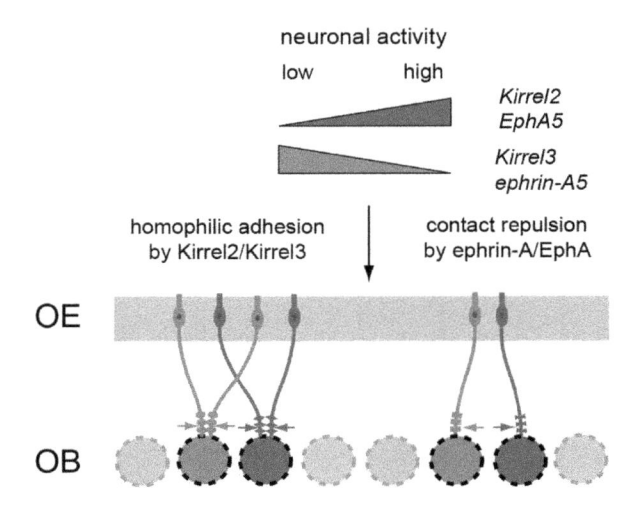

FIGURE 2.7 Local axon sorting to form a discrete glomerular map. At a later stage of OSN projection, neuronal activity regulates the expression of Kirrel2 and EphA5 positively, and Kirrel3 and ephrin-A5 negatively. Homophilic adhesion by Kirrel2 and Kirrel3 is thought to facilitate fasciculation of like axons. In contrast, repulsive interaction by ephrin-A5 and EphA5 is thought to facilitate the segregation of heterotypic axons. Note that this scheme may be too simplified: Different patterns of neuronal activity regulate different sets of cell surface molecules. (Modified from Imai, 2011.)

and Sema7A (Kaneko-Goto et al., 2008; Nakashima et al., 2019). It is not just the firing rate of OSNs that determines the expression level of these molecules. Phasic and tonic firing regulate different sets of cell surface molecules. This may indicate that multiple signaling pathways tuned to different firing modes are engaged for different sets of cell surface molecules (Nakashima et al., 2019). As a result, each OSN type expresses a unique pattern of cell surface molecules. It remains to be determined how the OR defines unique patterns of spontaneous activity in OSNs.

While sensory-evoked activity plays a limited role in the initial map formation, it plays an important role at a later stage. OSNs often mistarget axons and form ectopic glomeruli during the first two weeks after birth. However, the ectopic glomeruli are eliminated at later stages. Sensory-evoked signals mediate the elimination (Zou et al., 2004), possibly by facilitating activity-dependent competition among OSNs (Yu et al., 2004; Zhao and Reed, 2001).

OSNs are continuously regenerated throughout life. Regenerated OSNs have to project axons to the pre-existing glomeruli. This process also involves various types of axon-axon interactions described above. Genetic perturbation of early-born OSNs impairs the projection of later-born OSNs (Ma et al., 2014; Wu et al., 2018), suggesting that the projection of later-born OSNs depends more on axon-axon interactions. This may be a reason for the poor recovery of the olfactory map after severe OSN injury (Costanzo, 2000; Murai et al., 2016; St John and Key, 2003).

In summary, the local sorting of OSN axons to form glomerular structures is also a result of axon-axon interactions, namely fasciculation and segregation mediated by axon sorting molecules. OR-specific patterns of axon sorting molecules are regulated by neuronal activity.

2.8 NEUROGENESIS, MIGRATION, AND AXONAL PROJECTION OF M/T CELLS

Unlike the OE that develops from the olfactory placode, the OB is a rostral part of the central nervous system. The most rostral part of the telencephalon starts to invaginate at E12.5 to form the OB. M/T cells develop in three different steps: neurogenesis, migration, and dendritic remodeling (Treloar et al., 2010). Similar to pyramidal neurons in the cerebral cortex, M/T cells are generated from stem cells, known as radial glial cells that are located in the surface of the ventricle (ventricular zone). M/T cells are generated during E10-17. Like cortical pyramidal neurons, the birthdate is a determinant of the M/T cell types. The earliest-generated population becomes M/T cells in the AOB. Mitral cells in the OB are generated between E10-13, and then tufted and external tufted cells are generated between E13-17 (Hirata et al., 2019). Within the mitral cell population, earlier and later-born neurons tend to locate in the D and V domains of the OB, respectively (Imamura et al., 2011). After neurogenesis, M/T cells start to migrate radially and then tangentially toward the surface of the OB. Radial migration of M/T cells is in part regulated by OSN-derived factors in the OB. In normal conditions, Nrp2-positive M/T cells are located in the posteroventral OB and mediate the attractive olfactory behavior via the medial amygdala. However, Nrp2-positive M/T cells mis-migrate to the more anterodorsal part when Sema3F is specifically knocked out in OSNs (Inokuchi et al., 2017), suggesting that Sema3F secreted from OSN axons regulate the correct migration of M/T cells and the formation of innate circuits.

M/T cells also start to extend axons soon after the neurogenesis. M/T cell axons fasciculate to form the lateral olfactory tract (LOT). There are guidepost cells called LOT cells that help guidance of M/T cell axons (Sato et al., 1998). Early-born and later-born M/T cells are segregated within LOT and project to different parts of the olfactory cortex (Hirata et al., 2019; Inaki et al., 2004). Several axon guidance molecules are involved in the formation of the LOT and the axonal projections of M/T cells (e.g., Sema3F-Nrp2, Slit-Robo, Netrin-Dcc) (de Castro, 2009; Fouquet et al., 2007; Inokuchi et al., 2017; Kawasaki et al., 2006). At a later stage, M/T cells extend multiple dendrites toward the surface of the OB. Dendritic remodeling occurs at an early postnatal stage as will be discussed in more detail (Malun and Brunjes, 1996). In the cerebral cortex, neurons generated from the same progenitor tend to share the cortical columns; however, cell lineage is not a determinant for the dendrite wiring specificity of M/T cells (Sanchez-Guardado and Lois, 2019).

In the OB, ~99% of neurons are interneurons, among which 95% are granule cells. The generation of OB interneurons starts during embryonic stages but persists throughout life. Embryonically generated OB interneurons are derived from the lateral ganglionic eminence (LGE) and dorsal telencephalon, whereas postnatally they are derived from the subventricular zone (SVZ) of the lateral ventricle (Lledo et al., 2008). Different types of interneurons are generated from progenitors located in distinct microdomains and are defined with specific sets of transcription factors (Emx1, Gsh2, Nkx2.1, Nkx2.6, Gli1, and Zic). In the embryo, different subtypes of interneurons are generated at different timing (Batista-Brito et al., 2008). In parallel, a subpopulation of embryonically generated progenitors gives rise to neural stem cells for postnatal neurogenesis, while retaining their subtype specificity (Fuentealba et al., 2015). Newly generated OB interneurons migrate toward the OB through the rostral migratory stream (RMS) and then radially toward the appropriate destination within the OB. In the adult, a significant fraction of OB interneurons have postnatal origins (Imayoshi et al., 2008). Adult-born OB interneurons are required for flexible olfactory behavior based on cortical feedback (Sakamoto et al., 2014; Wu et al., 2020a).

In summary, M/T cells are generated from the ventricular zone of the rostral part of telencephalon, whereas OB interneurons are supplied from LGE and dorsal telencephalon during development and from the SVZ in the adult. OSN-derived factors (e.g., Sema3F) play important role for correct migration of some OB neurons.

2.9 DENDRITE REMODELING OF M/T CELLS

As mentioned in the previous section, newly generated M/T cells initially extend multiple primary dendrites toward the protoglomerular structure. During the first postnatal week, however, each M/T cell strengthen one winner and weaken all the other primary dendrite to form singular connectivity to a glomerulus (*one M/T cell – one glomerulus rule*) (Malun and Brunjes, 1996) (Figure 2.8). In the *Drosophila* olfactory system, matching between axons of sensory neurons and dendrites of projection neurons are defined by the cell surface molecules, e.g., Semaphorins and Teneurins (Hong and Luo, 2014). In mice, however, M/T cell connections can be newly allocated for OSNs expressing an artificially-introduced receptor (e.g., rat OR and β2-adrenergic receptor), suggesting a non-deterministic mechanism for OSN-M/T pairing (Belluscio et al., 2002; Feinstein et al., 2004). Thus, the postnatal dendrite remodeling process is critical to ensure the *one M/T cell – one glomerulus rule*. As somata of sister M/T cells are scattered in the mitral cell layer (Ke et al., 2013), the dendrite wiring is not just the connection to a nearest glomeruli. Even when multiple glomeruli are formed for an OR or OSN map is perturbed, each M/T cell still connects to just one glomerulus, excluding precise molecular matching between axons and dendrites as a possibility (Ma et al., 2014; Nishizumi et al., 2019).

Dendrite remodeling of M/T cells occurs without sensory-evoked nor spontaneous neuronal activity transmitted from OSNs (Fujimoto et al., 2019; Lin et al., 2000). Instead, the remodeling is controlled by the spontaneous activity generated within the OB. The dendro-dendritic glutamatergic neurotransmission among M/T cells is the origin of the spontaneous neuronal activity in the OB (Fujimoto et al., 2019). The remodeling is controlled by stabilization signals to strengthen the winner and destabilization signal to eliminate the loser dendrites. Stabilization is mediated by the concomitant inputs of BMP signaling and NMDAR-dependent Rac1 activity (Aihara et al., 2020). On the other hand, destabilization is mediated by activity-dependent

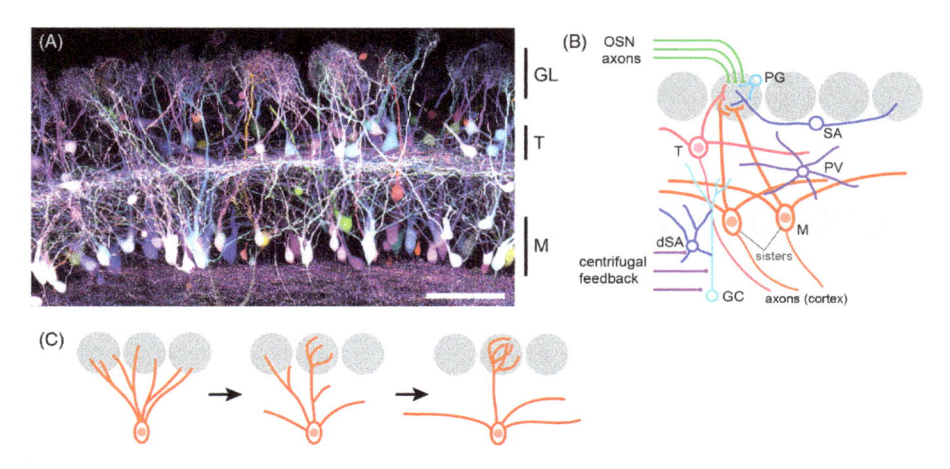

FIGURE 2.8 Development of the OB circuits. (A) Multicolor labeling of M/T cells in the OB. Tetbow plasmids were introduced to M/T cells using in utero electroporation. Note that each M/T cell connects to just one glomerulus. GL, glomerular layer; T, tufted cells; M, mitral cells. (B) A diagram of the OB circuits. In addition to M/T cells, there are various types of GABAergic interneurons, including periglomerular neurons (PG), short axon cells (SA), parvalbumin-positive interneurons (PV), deep short axon cells (dSA), and granule cells (GC). (C) Dendrite remodeling of M/T cells during the first postnatal week. An M/T cell initially extends multiple primary dendrites to the protoglomeruli. Later on, however, an M/T cell strengthens one winner dendrite and eliminates all the other loser dendrites to establish a single primary dendrite. ([A] is modified from Sakaguchi et al., 2018. [B] and [C] are modified from Imai, 2014.)

RhoA signaling (Fujimoto et al., 2019). During the remodeling process, activity-dependent competition between winner and loser dendrites establishes just one winner. Glutamatergic inputs via NMDARs lead to neuron-wide increase of RhoA, causing the lateral inhibition of weaker dendrites (Fujimoto et al., 2019). Synapse formation between OSN axons and OB neurons occur in the absence of neuronal activity (Fujimoto et al., 2019; Lin et al., 2000; Ma et al., 2014), but its maturation requires neuronal activity (Aihara et al., 2020).

In summary, spontaneous neuronal activity is generated in the developing OB, and activity-dependent competition within a neuron establishes the *one M/T cell – one glomerulus rule*.

2.10 MODULATION AND PLASTICITY OF THE OLFACTORY SYSTEM

While the overall architecture of the olfactory system is established by the early postnatal stages, various levels of plasticity persist throughout the life of animals to adapt to the ever-changing environment. Sensory stimuli affect the expression level of ORs and survival of OSNs (von der Weid et al., 2015; Zhao et al., 2013). In the vomeronasal organ, VSN responses to some male pheromones are suppressed by a female hormone, progesterone (Dey et al., 2015). Thus, hormonal regulation of olfaction might be an interesting topic for future study. In the OB, sensory-evoked activity modulates the excitability of M/T cells and short axon cells via axon initial segment (AIS) plasticity (Chand et al., 2015; George et al., 2021). In short axon cells, dopamine synthesis is modulated by the sensory-evoked activity (Baker et al., 1993). In the OB, the glomerulus-specific intrabulbar projection of tufted cell axons is controlled by the sensory-evoked activity (Marks et al., 2006). It will be important to investigate in the future how the sensory inputs modulate the cortical projection of M/T cells as well as odor-evoked innate and learned behaviors.

In the OB, learning-related plasticity has been extensively studied using chronic calcium imaging of awake behaving mice (Wu et al., 2020b). Chronic two-photon imaging during the learning process revealed that M/T cell responses are continuously updated to convey behaviorally relevant odor information to the olfactory cortex. For example, odor responses gradually decline over days when odors are passively presented to animals (Kato et al., 2012). In contrast, the representation of threatening stimuli is enhanced by experience (Kass et al., 2013). When mice perform fine or coarse odor discrimination tasks, odor representation changes in opposite ways to optimally separate the test odors (Chu et al., 2016; Doucette and Restrepo, 2008; Yamada et al., 2017). For this purpose, top-down inputs from the olfactory cortex and the plasticity of OB interneurons play key roles (Yamada et al., 2017). In the future, it will be important to investigate how the experience affects the structure and function of the cortical circuits.

2.11   SUMMARY AND CONCLUSION

This chapter, I first described how odor information is processed at different stages of the olfactory system. In the OE, the combinatorial receptor codes for both excitatory and inhibitory responses are the basis for odor discrimination. Moreover, odor mixture responses can be tuned by antagonism and synergy. In the OB, odor information is spatiotemporally represented in the glomeruli. Particularly, temporal patterns may be a key to understand the concentration-invariant perception of odor identity. In the piriform cortex, convergent inputs from multiple glomeruli are used to learn and discriminate odors, whereas OR-specific stereotyped inputs to cortical amygdala drive innate behaviors. Odor information processing in higher brain regions is beginning to be elucidated. Secondly, I described various types of vertebrate olfactory receptors (ORs, TAARs, V1Rs, V2Rs, FPRs, etc.) that have dynamically evolved in a species-specific manner. Lastly, I described how the OR-specific neuronal circuits are constructed from the peripheral to the more central part of the brain. The *'one neuron – one receptor rule'* is established by the stochastic activation of an OR gene locus followed by the feedback regulation to prevent further activation. The *'one glomerulus – one receptor rule'* is the result of the coarse axon targeting (D-V and A-P axes) and subsequent local axon sorting. Remarkably, axon-axon interactions play important roles in every aspect of OSN projections. The *'one M/T cell – one glomerulus rule'* is established during the dendrite remodeling process: Intracellular competition between winner and loser dendrites establishes the single winner. These mechanisms enable the flexible acquisition and loss of OR-specific circuits in the brain upon OR gene gain/loss during evolution. The OR-specific parallel discrete circuits enable the transformation of the molecular world to the neuronal ensembles to mediate stereotyped and flexible behaviors.

ACKNOWLEDGEMENT

This work was supported by Grants in Scientific Research on Innovation Areas (JP16H06456, JP21H00205, and 21H05696) from MEXT, Japan, CREST program (JPMJCR2021) from Japan Science and Technology Agency, JSPS KAKENHI (JP17H06261, JP16K14568, JP15H05572, and JP15K14336), the Mochida Memorial Foundation, and the Uehara Memorial Foundation. TI thanks Bernd Fritzsch and Marcus N. Leiwe for critical reading of this manuscript.

REFERENCES

Ache, B.W., and Young, J.M. (2020). Phylogeny of chemical sensitivity. In The Senses, B. Fritzsch, ed. (Elsevier, Amsterdam), pp. 4–23.

Aihara, S., Fujimoto, S., Sakaguchi, R., and Imai, T. (2020). BMPR-2 gates activity-dependent stabilization of primary dendrites during mitral cell remodeling. Cell Reports, *35*, 109276.

Apicella, A., Yuan, Q., Scanziani, M., and Isaacson, J.S. (2010). Pyramidal cells in piriform cortex receive convergent input from distinct olfactory bulb glomeruli. Journal of Neuroscience *30*, 14255–14260.

Baker, H., Morel, K., Stone, D.M., and Maruniak, J.A. (1993). Adult naris closure profoundly reduces tyrosine-hydroxylase expression in mouse olfactory-bulb. Brain Research *614*, 109–116.

Batista-Brito, R., Close, J., Machold, R., and Fishell, G. (2008). The distinct temporal origins of olfactory bulb interneuron subtypes. The Journal of Neuroscience: The Official Journal of the Society for Neuroscience *28*, 3966–3975.

Bear, D.M., Lassance, J.M., Hoekstra, H.E., and Datta, S.R. (2016). The evolving neural and genetic architecture of vertebrate olfaction. Current Biology *26*, R1039–R1049.

Belluscio, L., Gold, G.H., Nemes, A., and Axel, R. (1998). Mice deficient in G(olf) are anosmic. Neuron *20*, 69–81.

Belluscio, L., Koentges, G., Axel, R., and Dulac, C. (1999). A map of pheromone receptor activation in the mammalian brain. Cell *97*, 209–220.

Belluscio, L., Lodovichi, C., Feinstein, P., Mombaerts, P., and Katz, L.C. (2002). Odorant receptors instruct functional circuitry in the mouse olfactory bulb. Nature *419*, 296–300.

Benton, R., Vannice, K.S., Gomez-Diaz, C., and Vosshall, L.B. (2009). Variant ionotropic glutamate receptors as chemosensory receptors in drosophila. Cell *136*, 149–162.

Blazing, R.M., and Franks, K.M. (2020). Odor coding in piriform cortex: mechanistic insights into distributed coding. Current Opinion in Neurobiology *64*, 96–102.

Boehm, U., Zou, Z.H., and Buck, L.B. (2005). Feedback loops link odor and pheromone signaling with reproduction. Cell *123*, 683–695.

Bolding, K.A., and Franks, K.M. (2018). Recurrent cortical circuits implement concentration-invariant odor coding. Science *361*, 1088-+.

Bolding, K.A., Nagappan, S., Han, B.X., Wang, F., and Franks, K.M. (2020). Recurrent circuitry is required to stabilize piriform cortex odor representations across brain states. Elife *9*, e53125.

Boyd, A.M., Sturgill, J.F., Poo, C., and Isaacson, J.S. (2012). Cortical feedback control of olfactory bulb circuits. Neuron 76, 1161–1174.

Buck, L., and Axel, R. (1991). A novel multigene family may encode odorant receptors: a molecular basis for odor recognition. Cell 65, 175–187.

Bufe, B., Teuchert, Y., Schmid, A., Pyrski, M., Perez-Gomez, A., Eisenbeis, J., Timm, T., Ishii, T., Lochnit, G., Bischoff, M., et al. (2019). Bacterial MgrB peptide activates chemoreceptor Fpr3 in mouse accessory olfactory system and drives avoidance behaviour. Nature Communications 10, 4889.

Bulfone, A., Wang, F., Hevner, R., Anderson, S., Cutforth, T., Chen, S., Meneses, J., Pedersen, R., Axel, R., and Rubenstein, J.L.R. (1998). An olfactory sensory map develops in the absence of normal projection neurons or GABAergic interneurons. Neuron 21, 1273–1282.

Cau, E., Casarosa, S., and Guillemot, F. (2002). Mash1 and Ngn1 control distinct steps of determination and differentiation in the olfactory sensory neuron lineage. Development 129, 1871–1880.

Chae, H., Kepple, D.R., Bast, W.G., Murthy, V.N., Koulakov, A.A., and Albeanu, D.F. (2019). Mosaic representations of odors in the input and output layers of the mouse olfactory bulb. Nature Neuroscience 22, 1306-+.

Chamero, P., Marton, T.F., Logan, D.W., Flanagan, K., Cruz, J.R., Saghatelian, A., Cravatt, B.F., and Stowers, L. (2007). Identification of protein pheromones that promote aggressive behaviour. Nature 450, 899–U823.

Chand, A.N., Galliano, E., Chesters, R.A., and Grubb, M.S. (2015). A Distinct Subtype of Dopaminergic Interneuron Displays Inverted Structural Plasticity at the Axon Initial Segment. Journal of Neuroscience 35, 1573–1590.

Chesler, A.T., Zou, D.J., Le Pichon, C.E., Peterlin, Z.A., Matthews, G.A., Pei, X., Miller, M.C., and Firestein, S. (2007). A G protein/cAMP signal cascade is required for axonal convergence into olfactory glomeruli. Proceedings of the National Academy of Sciences of the United States of America 104, 1039–1044.

Chess, A., Simon, I., Cedar, H., and Axel, R. (1994). Allelic inactivation regulates olfactory receptor gene expression. Cell 78, 823–834.

Cho, J.H., Lepine, M., Andrews, W., Parnavelas, J., and Cloutier, J.F. (2007). Requirement for slit-1 and robo-2 in zonal segregation of olfactory sensory neuron axons in the main olfactory bulb. Journal of Neuroscience 27, 9094–9104.

Chong, E., Moroni, M., Wilson, C., Shoham, S., Panzeri, S., and Rinberg, D. (2020). Manipulating synthetic optogenetic odors reveals the coding logic of olfactory perception. Science 368, 1329-+.

Chu, M.W., Li, W.L., and Komiyama, T. (2016). Balancing the robustness and efficiency of odor representations during learning. Neuron 92, 174–186.

Clyne, P.J., Warr, C.G., Freeman, M.R., Lessing, D., Kim, J.H., and Carlson, J.R. (1999). A novel family of divergent seven-transmembrane proteins: Candidate odorant receptors in Drosophila. Neuron 22, 327–338.

Costanzo, R.M. (2000). Rewiring the olfactory bulb: changes in odor maps following recovery from nerve transection. Chemical Senses 25, 199–205.

Crapon de Caprona, M.D., and Fritzsch, B. (1983). The development of the retinopetal nucleus olfacto-retinalis of two cichlid fish as revealed by horseradish peroxidase. Brain Research 313, 281–301.

Dal Col, J.A., Matsuo, T., Storm, D.R., and Rodriguez, I. (2007). Adenylyl cyclase-dependent axonal targeting in the olfactory system. Development 134, 2481–2489.

Dalton, R.P., Lyons, D.B., and Lomvardas, S. (2013). Co-opting the unfolded protein response to elicit olfactory receptor feedback. Cell 155, 321–332.

de Castro, F. (2009). Wiring olfaction: the cellular and molecular mechanisms that guide the development of synaptic connections from the nose to the cortex. Frontiers in Neuroscience 3.

Del Punta, K., Puche, A., Adams, N.C., Rodriguez, I., and Mombaerts, P. (2002). A divergent pattern of sensory axonal projections is rendered convergent by second-order neurons in the accessory olfactory bulb. Neuron *35*, 1057–1066.

Dewan, A., Pacifico, R., Zhan, R., Rinberg, D., and Bozza, T. (2013). Non-redundant coding of aversive odours in the main olfactory pathway. Nature *497*, 486–489.

Dey, S., and Matsunami, H. (2011). Calreticulin chaperones regulate functional expression of vomeronasal type 2 pheromone receptors. Proceedings of the National Academy of Sciences of the United States of America *108*, 16651–16656.

Dey, S., Chamero, P., Pru, J.K., Chien, M.S., Ibarra-Soria, X., Spencer, K.R., Logan, D.W., Matsunami, H., Peluso, J.J., and Stowers, L. (2015). Cyclic regulation of sensory perception by a female hormone alters behavior. Cell *161*, 1334–1344.

Dhawale, A.K., Hagiwara, A., Bhalla, U.S., Murthy, V.N., and Albeanu, D.F. (2010). Non-redundant odor coding by sister mitral cells revealed by light addressable glomeruli in the mouse. Nature Neuroscience *13*, 1404–U1183.

Doucette, W., and Restrepo, D. (2008). Profound context-dependent plasticity of mitral cell responses in olfactory bulb. Plos Biology *6*, 2266–2285.

Dulac, C., and Axel, R. (1995). A novel family of genes encoding putative pheromone receptors in mammals. Cell *83*, 195–206.

Dvorakova, M., Macova, I., Bohuslavova, R., Anderova, M., Fritzsch, B., and Pavlinkova, G. (2020). Early ear neuronal development, but not olfactory or lens development, can proceed without SOX2. Developmental Biology *457*, 43–56.

Economo, M.N., Hansen, K.R., and Wachowiak, M. (2016). Control of Mitral/Tufted Cell Output by Selective Inhibition among Olfactory Bulb Glomeruli. Neuron *91*, 397–411.

Fadool, D.A., and Kolling, L.J. (2020). Role of olfaction for eating behavior. In The Senses, B. Fritzsch, ed. (Elsevier, Amsterdam), pp. 675–716.

Feinstein, P., Bozza, T., Rodriguez, I., Vassalli, A., and Mombaerts, P. (2004). Axon guidance of mouse olfactory sensory neurons by odorant receptors and the beta 2 adrenergic receptor. Cell *117*, 833–846.

Feldheim, D.A., and O'Leary, D.D.M. (2010). Visual map development: bidirectional signaling, bifunctional guidance molecules, and competition. Cold Spring Harbor Perspectives in Biology *2*, a001768.

Firestein, S. (2001). How the olfactory system makes sense of scents. Nature *413*, 211–218.

Fletcher, R.B., Das, D., Gadye, L., Street, K.N., Baudhuin, A., Wagner, A., Cole, M.B., Flores, Q., Choi, Y.G., Yosef, N., *et al.* (2017). Deconstructing olfactory stem cell trajectories at single-cell resolution. Cell Stem Cell *20*, 817–830.e8.

Fouquet, C., Di Meglio, T., Ma, L., Kawasaki, T., Long, H., Hirata, T., Tessier-Lavigne, M., Chedotal, A., and Nguyen-Ba-Charvet, K.T. (2007). Robo1 and Robo2 control the development of the lateral olfactory tract. Journal of Neuroscience *27*, 3037–3045.

Friedrich, R.W., and Korsching, S.I. (1997). Combinatorial and chemotopic odorant coding in the zebrafish olfactory bulb visualized by optical imaging. Neuron *18*, 737–752.

Fuentealba, L.C., Rompani, S.B., Parraguez, J.I., Obernier, K., Romero, R., Cepko, C.L., and Alvarez-Buylla, A. (2015). Embryonic origin of postnatal neural stem cells. Cell *161*, 1644–1655.

Fujimoto, S., Leiwe, M.N., Sakaguchi, R., Muroyama, Y., Kobayakawa, R., Kobayakawa, K., Saito, T., and Imai, T. (2019). Spontaneous activity generated within the olfactory bulb establishes the discrete wiring of mitral cell dendrites. bioRxiv, 625616.

Fukunaga, I., Berning, M., Kollo, M., Schmaltz, A., and Schaefer, A.T. (2012). Two distinct channels of olfactory bulb output. Neuron *75*, 320–329.

Fukunaga, I., Herb, J.T., Kollo, M., Boyden, E.S., and Schaefer, A.T. (2014). Independent control of gamma and theta activity by distinct interneuron networks in the olfactory bulb. Nature Neuroscience *17*, 1208–1216.

Gadye, L., Das, D., Sanchez, M.A., Street, K., Baudhuin, A., Wagner, A., Cole, M.B., Choi, Y.G., Yosef, N., Purdom, E., *et al.* (2017). Injury activates transient olfactory stem cell states with diverse lineage capacities. Cell Stem Cell *21*, 775–790 e779.

George, N.M., Macklin, W.B., and Restrepo, D. (2021). Excitable axonal domains adapt to sensory deprivation in the olfactory system. bioRxiv, 2021.01.25.428132.

Giessel, A.J., and Datta, S.R. (2014). Olfactory maps, circuits and computations. Current Opinion in Neurobiology *24C*, 120–132.

Gire, D.H., Franks, K.M., Zak, J.D., Tanaka, K.F., Whitesell, J.D., Mulligan, A.A., Hen, R., and Schoppa, N.E. (2012). Mitral cells in the olfactory bulb are mainly excited through a multistep signaling path. Journal of Neuroscience *32*, 2964–2975.

Greer, P.L., Bear, D.M., Lassance, J.M., Bloom, M.L., Tsukahara, T., Pashkovski, S.L., Masuda, F.K., Nowlan, A.C., Kirchner, R., Hoekstra, H.E., *et al.* (2016). A family of non-GPCR chemosensors defines an alternative logic for mammalian olfaction. Cell *165*, 1734–1748.

Grobman, M., Dalal, T., Lavian, H., Shmuel, R., Belelovsky, K., Xu, F.Q., Korngreen, A., and Haddad, R. (2018). A mirror-symmetric excitatory link coordinates odor maps across olfactory bulbs and enables odor perceptual unity. Neuron *99*, 800–813.e6.

Grosmaitre, X., Santarelli, L.C., Tan, J., Luo, M., and Ma, M. (2007). Dual functions of mammalian olfactory sensory neurons as odor detectors and mechanical sensors. Nature Neuroscience *10*, 348–354.

Haddad, R., Lanjuin, A., Madisen, L., Zeng, H., Murthy, V.N., and Uchida, N. (2013). Olfactory cortical neurons read out a relative time code in the olfactory bulb. Nature Neuroscience *16*, 949–957.

Hanchate, N.K., Kondoh, K., Lu, Z.H., Kuang, D.H., Ye, X.L., Qiu, X.J., Pachter, L., Trapnell, C., and Buck, L.B. (2015). Single-cell transcriptomics reveals receptor transformations during olfactory neurogenesis. Science *350*, 1251–1255.

Herrada, G., and Dulac, C. (1997). A novel family of putative pheromone receptors in mammals with a topographically organized and sexually dimorphic distribution. Cell *90*, 763–773.

Hirata, T., Shioi, G., Abe, T., Kiyonari, H., Kato, S., Kobayashi, K., Mori, K., and Kawasaki, T. (2019). A novel birthdate-labeling method reveals segregated parallel projections of mitral and external tufted cells in the main olfactory system. eNeuro *6*, ENEURO.0234-19.2019.

Holland, L. (2020). Invertebrate origins of vertebrate nervous systems. In Evolutionary Neuroscience (Elsevier, Amsterdam), pp. 51–73.

Hong, W., and Luo, L. (2014). Genetic control of wiring specificity in the fly olfactory system. Genetics *196*, 17–29.

Hopfield, J.J. (1995). Pattern-recognition computation using action-potential timing for stimulus representation. Nature *376*, 33–36.

Hu, J., Zhong, C., Ding, C., Chi, Q.Y., Walz, A., Mombaerts, P., Matsunami, H., and Luo, M.M. (2007). Detection of near-atmospheric concentrations of CO_2 by an olfactory subsystem in the mouse. Science *317*, 953–957.

Huberman, A.D., Feller, M.B., and Chapman, B. (2008). Mechanisms underlying development of visual maps and receptive fields. Annual Review of Neuroscience *31*, 479–509.

Igarashi, K.M., Ieki, N., An, M., Yamaguchi, Y., Nagayama, S., Kobayakawa, K., Kobayakawa, R., Tanifuji, M., Sakano, H., Chen, W.R., *et al.* (2012). Parallel mitral and tufted cell pathways route distinct odor information to different targets in the olfactory cortex. Journal of Neuroscience *32*, 7970–7985.

Igarashi, K.M., Lu, L., Colgin, L.L., Moser, M.B., and Moser, E.I. (2014). Coordination of entorhinal-hippocampal ensemble activity during associative learning. Nature *510*, 143–147.

Imai, T. (2014). Construction of functional neuronal circuitry in the olfactory bulb. Seminars in Cell and Developmental Biology *35*, 180–188.

Imai, T. (2020). Odor coding in the olfactory bulb. In The Senses, B. Fritzsch, ed. (Elsevier, Amsterdam), pp. 640–649.

Imai, T., and Sakano, H. (2011). Axon-axon interactions in neuronal circuit assembly: lessons from olfactory map formation. The European Journal of Neuroscience *34*, 1647–1654.

Imai, T., Sakano, H., and Vosshall, L.B. (2010). Topographic mapping–the olfactory system. Cold Spring Harbor Perspectives in Biology *2*, a001776.

Imai, T., Suzuki, M., and Sakano, H. (2006). Odorant receptor-derived cAMP signals direct axonal targeting. Science *314*, 657–661.

Imai, T., Yamazaki, T., Kobayakawa, R., Kobayakawa, K., Abe, T., Suzuki, M., and Sakano, H. (2009). Pre-target axon sorting establishes the neural map topography. Science *325*, 585–590.

Imamura, F., and Rodriguez Gil, D. (2020). Functional architecture of the olfactory bulb. In The Senses, B. Fritzsch, ed. (Elsevier, Amsterdam), pp. 591–609.

Imamura, F., Ayoub, A.E., Rakic, P., and Greer, C.A. (2011). Timing of neurogenesis is a determinant of olfactory circuitry. Nature Neuroscience *14*, 331–337.

Imayoshi, I., Sakamoto, M., Ohtsuka, T., Takao, K., Miyakawa, T., Yamaguchi, M., Mori, K., Ikeda, T., Itohara, S., and Kageyama, R. (2008). Roles of continuous neurogenesis in the structural and functional integrity of the adult forebrain. Nature Neuroscience *11*, 1153–1161.

Inagaki, S., Iwata, R., Iwamoto, M., and Imai, T. (2020). Widespread inhibition, antagonism, and synergy in mouse olfactory sensory neurons in vivo. Cell Reports *31*, 107814.

Inaki, K., Nishimura, S., Nakashiba, T., Itohara, S., and Yoshihara, Y. (2004). Laminar organization of the developing lateral olfactory tract revealed by differential expression of cell recognition molecules. Journal of Comparative Neurology *479*, 243–256.

Inokuchi, K., Imamura, F., Takeuchi, H., Kim, R., Okuno, H., Nishizumi, H., Bito, H., Kikusui, T., and Sakano, H. (2017). Nrp2 is sufficient to instruct circuit formation of mitral-cells to mediate odour-induced attractive social responses. Nature Communications *8*, 15977.

Ishii, T., Hirota, J., and Mombaerts, P. (2003). Combinatorial coexpression of neural and immune multigene families in mouse vomeronasal sensory neurons. Current Biology *13*, 394–400.

Isosaka, T., Matsuo, T., Yamaguchi, T., Funabiki, K., Nakanishi, S., Kobayakawa, R., and Kobayakawa, K. (2015). Htr2a-expressing cells in the central amygdala control the hierarchy between innate and learned fear. Cell *163*, 1153–1164.

Iwata, R., Kiyonari, H., and Imai, T. (2017). Mechanosensory-based phase coding of odor identity in the olfactory bulb. Neuron *96*, 1139–1152.e7.

Jiang, Y., Gong, N.N., Hu, X.S., Ni, M.J., Pasi, R., and Matsunami, H. (2015). Molecular profiling of activated olfactory neurons identifies odorant receptors for odors in vivo. Nature Neuroscience *18*, 1446.

Johnson, B.A., and Leon, M. (2000). Modular representations of odorants in the glomerular layer of the rat olfactory bulb and the effects of stimulus concentration. Journal of Comparative Neurology *422*, 496–509.

Kaji, T., Reimer, J.D., Morov, A.R., Kuratani, S., and Yasui, K. (2016). Amphioxus mouth after dorso-ventral inversion. Zoological Letters *2*, 1–14.

Kaneko-Goto, T., Yoshihara, S., Miyazaki, H., and Yoshihara, Y. (2008). BIG-2 mediates olfactory axon convergence to target glomeruli. Neuron *57*, 834–846.

Kass, M.D., Rosenthal, M.C., Pottackal, J., and McGann, J.P. (2013). Fear learning enhances neural responses to threat-predictive sensory stimuli. Science *342*, 1389–1392.

Kato, H.K., Chu, M.W., Isaacson, J.S., and Komiyama, T. (2012). Dynamic sensory representations in the olfactory bulb: modulation by wakefulness and experience. Neuron *76*, 962–975.

Kawasaki, T., Ito, K., and Hirata, T. (2006). Netrin 1 regulates ventral tangential migration of guidepost neurons in the lateral olfactory tract. Development *133*, 845–853.

Kawauchi, S., Kim, J., Santos, R., Wu, H.H., Lander, A.D., and Calof, A.L. (2009). Foxg1 promotes olfactory neurogenesis by antagonizing Gdf11. Development *136*, 1453–1464.

Ke, M.T., Fujimoto, S., and Imai, T. (2013). SeeDB: a simple and morphology-preserving optical clearing agent for neuronal circuit reconstruction. Nature Neuroscience *16*, 1154–1161.

Kepecs, A., Uchida, N., and Mainen, Z.F. (2006). The sniff as a unit of olfactory processing. Chemical Senses *31*, 167–179.

Kersigo, J., D'Angelo, A., Gray, B.D., Soukup, G.A., and Fritzsch, B. (2011). The role of sensory organs and the forebrain for the development of the craniofacial shape as revealed by Foxg1-cre-mediated microRNA loss. Genesis *49*, 326–341.

Kikuta, S., Fletcher, M.L., Homma, R., Yamasoba, T., and Nagayama, S. (2013). Odorant response properties of individual neurons in an olfactory glomerular module. Neuron *77*, 1122–1135.

Kikuta, S., Sato, K., Kashiwadani, H., Tsunoda, K., Yamasoba, T., and Mori, K. (2010). From the Cover: Neurons in the anterior olfactory nucleus pars externa detect right or left localization of odor sources. Proceedings of the National Academy of Sciences of the United States of America *107*, 12363–12368.

Kimoto, H., Haga, S., Sato, K., and Touhara, K. (2005). Sex-specific peptides from exocrine glands stimulate mouse vomeronasal sensory neurons. Nature *437*, 898–901.

Kishida, T., Thewissen, J.G.M., Hayakawa, T., Imai, H., and Agata, K. (2015). Aquatic adaptation and the evolution of smell and taste in whales. Zoological Letters *1*, 9.

Kobayakawa, K., Kobayakawa, R., Matsumoto, H., Oka, Y., Imai, T., Ikawa, M., Okabe, M., Ikeda, T., Itohara, S., Kikusui, T., *et al.* (2007). Innate versus learned odour processing in the mouse olfactory bulb. Nature *450*, 503–508.

Kondoh, K., Lu, Z.H., Ye, X.L., Olson, D.P., Lowell, B.B., and Buck, L.B. (2016). A specific area of olfactory cortex involved in stress hormone responses to predator odours. Nature *532*, 103–106.

Korsching, S.I. (2020). Taste and smell in zebrafish. In The Senses, B. Fritzsch, ed. (Elsevier, Oxford), pp. 466–492.

Kurian, S.M., Naressi, R.G., Manoel, D., Barwich, A.S., Malnic, B., and Saraiva, L.R. (2021). Odor coding in the mammalian olfactory epithelium. Cell and Tissue Research.

Lee, D., Kume, M., and Holy, T.E. (2019). Sensory coding mechanisms revealed by optical tagging of physiologically defined neuronal types. Science *366*, 1384–1389.

Leinders-Zufall, T., Brennan, P., Widmayer, P., Chandramani, P., Maul-Pavicic, A., Jager, M., Li, X.H., Breer, H., Zufall, F., and Boehm, T. (2004). MHC class I peptides as chemosensory signals in the vomeronasal organ. Science *306*, 1033–1037.

Leinders-Zufall, T., Ishii, T., Mombaerts, P., Zufall, F., and Boehm, T. (2009). Structural requirements for the activation of vomeronasal sensory neurons by MHC peptides. Nature Neuroscience *12*, 1551–U1598.

Leinders-Zufall, T., Lane, A.P., Puche, A.C., Ma, W.D., Novotny, M.V., Shipley, M.T., and Zufall, F. (2000). Ultrasensitive pheromone detection by mammalian vomeronasal neurons. Nature *405*, 792–796.

Li, Q., Korzan, W.J., Ferrero, D.M., Chang, R.B., Roy, D.S., Buchi, M., Lemon, J.K., Kaur, A.W., Stowers, L., Fendt, M., *et al.* (2013). Synchronous evolution of an odor biosynthesis pathway and behavioral response. Current Biology *23*, 11–20.

Li, Y., Xu, J., Liu, Y., Zhu, J., Liu, N., Zeng, W., Huang, N., Rasch, M.J., Jiang, H., Gu, X., *et al.* (2017). A distinct entorhinal cortex to hippocampal CA1 direct circuit for olfactory associative learning. Nature Neuroscience *20*, 559–570.

Liberles, S.D., and Buck, L.B. (2006). A second class of chemosensory receptors in the olfactory epithelium. Nature *442*, 645–650.

Lin, D.M., Wang, F., Lowe, G., Gold, G.H., Axel, R., Ngai, J., and Brunet, L. (2000). Formation of precise connections in the olfactory bulb occurs in the absence of odorant-evoked neuronal activity. Neuron *26*, 69–80.

Lin, D.Y., Zhang, S.Z., Block, E., and Katz, L.C. (2005). Encoding social signals in the mouse main olfactory bulb. Nature *434*, 470–477.

Lledo, P.M., Merkle, F.T., and Alvarez-Buylla, A. (2008). Origin and function of olfactory bulb interneuron diversity. Trends in Neurosciences *31*, 392–400.

Loconto, J., Papes, F., Chang, E., Stowers, L., Jones, E.P., Takada, T., Kumanovics, A., Lindahl, K.F., and Dulac, C. (2003). Functional expression of murine V213 pheromone receptors involves selective association with the M10 and M1 families of MHC class Ib molecules. Cell *112*, 607–618.

Lomvardas, S., Barnea, G., Pisapia, D.J., Mendelsohn, M., Kirkland, J., and Axel, R. (2006). Interchromosornal interactions and olfactory receptor choice. Cell *126*, 403–413.

Lyons, D.B., Allen, W.E., Goh, T., Tsai, L., Barnea, G., and Lomvardas, S. (2013). An Epigenetic Trap Stabilizes Singular Olfactory Receptor Expression. Cell *154*, 325–336.

Ma, L., Qiu, Q., Gradwohl, S., Scott, A., Yu, E.Q., Alexander, R., Wiegraebe, W., and Yu, C.R. (2012). Distributed representation of chemical features and tunotopic organization of glomeruli in the mouse olfactory bulb. Proceedings of the National Academy of Sciences of the United States of America *109*, 5481–5486.

Ma, L.M., Wu, Y.M., Qiu, Q., Scheerer, H., Moran, A., and Yu, C.R. (2014). A developmental switch of axon targeting in the continuously regenerating mouse olfactory system. Science *344*, 194–197.

Magklara, A., Yen, A., Colquitt, B.M., Clowney, E.J., Allen, W., Markenscoff-Papadimitriou, E., Evans, Z.A., Kheradpour, P., Mountoufaris, G., Carey, C., *et al.* (2011). An epigenetic signature for monoallelic olfactory receptor expression. Cell *145*, 555–570.

Mainland, J.D., Keller, A., Li, Y.R., Zhou, T., Trimmer, C., Snyder, L.L., Moberly, A.H., Adipietro, K.A., Liu, W.L.L., Zhuang, H.Y., *et al.* (2014). The missense of smell: functional variability in the human odorant receptor repertoire. Nature Neuroscience *17*, 114–120.

Malnic, B., Hirono, J., Sato, T., and Buck, L.B. (1999). Combinatorial receptor codes for odors. Cell *96*, 713–723.

Malun, D., and Brunjes, P.C. (1996). Development of olfactory glomeruli: Temporal and spatial interactions between olfactory receptor axons and mitral cells in opossums and rats. Journal of Comparative Neurology *368*, 1–16.

Markenscoff-Papadimitriou, E., Allen, W.E., Colquitt, B.M., Goh, T., Murphy, K.K., Monahan, K., Mosley, C.P., Ahituv, N., and Lomvardas, S. (2014). Enhancer interaction networks as a means for singular olfactory receptor expression. Cell *159*, 543–557.

Markopoulos, F., Rokni, D., Gire, D.H., and Murthy, V.N. (2012). Functional properties of cortical feedback projections to the olfactory bulb. Neuron *76*, 1175–1188.

Marks, C.A., Cheng, K., Cummings, D.M., and Belluscio, L. (2006). Activity-dependent plasticity in the olfactory intrabulbar map. Journal of Neuroscience *26*, 11257–11266.

Matsunami, H., and Buck, L.B. (1997). A multigene family encoding a diverse array of putative pheromone receptors in mammals. Cell *90*, 775–784.

Millman, D.J., and Murthy, V.N. (2020). Rapid learning of odor-value association in the olfactory striatum. Journal of Neuroscience *40*, 4335–4347.

Miyamichi, K., Amat, F., Moussavi, F., Wang, C., Wickersham, I., Wall, N.R., Taniguchi, H., Tasic, B., Huang, Z.J., He, Z.G., *et al.* (2011). Cortical representations of olfactory input by trans-synaptic tracing. Nature *472*, 191–196.

Miyamichi, K., Serizawa, S., Kimura, H.M., and Sakano, H. (2005). Continuous and overlapping expression domains of odorant receptor genes in the olfactory epithelium determine the dorsal/ventral positioning of glomeruli in the olfactory bulb. Journal of Neuroscience 25, 3586–3592.

Miyasaka, N., Arganda-Carreras, I., Wakisaka, N., Masuda, M., Sumbul, U., Seung, H.S., and Yoshihara, Y. (2014). Olfactory projectome in the zebrafish forebrain revealed by genetic single-neuron labelling. Nature Communications 5, 3639.

Mombaerts, P., Wang, F., Dulac, C., Chao, S.K., Nemes, A., Mendelsohn, M., Edmondson, J., and Axel, R. (1996). Visualizing an olfactory sensory map. Cell 87, 675–686.

Monahan, K., and Lomvardas, S. (2015). Monoallelic expression of olfactory receptors. Annual Review of Cell and Developmental Biology 31, 721–740.

Moody, S.A., and LaMantia, A.S. (2015). Transcriptional regulation of cranial sensory placode development. Neural Crest and Placodes 111, 301–350.

Mori, K., and Sakano, H. (2011). How is the olfactory map formed and interpreted in the mammalian brain? Annual Review of Neuroscience, Vol 34 34, 467–499.

Mori, K., Takahashi, Y.K., Igarashi, K.M., and Yamaguchi, M. (2006). Maps of odorant molecular features in the mammalian olfactory bulb. Physiological Reviews 86, 409–433.

Munger, S.D., Leinders-Zufall, T., and Zufall, F. (2009). Subsystem organization of the mammalian sense of smell. Annual Review of Physiology 71, 115–140.

Munger, S.D., Leinders-Zufall, T., McDougall, L.M., Cockerham, R.E., Schmid, A., Wandernoth, P., Wennemuth, G., Biel, M., Zufall, F., and Kelliher, K.R. (2010). An olfactory subsystem that detects carbon disulfide and mediates food-related social learning. Current Biology 20, 1438–1444.

Murai, A., Iwata, R., Fujimoto, S., Aihara, S., Tsuboi, A., Muroyama, Y., Saito, T., Nishizaki, K., and Imai, T. (2016). Distorted coarse axon targeting and reduced dendrite connectivity underlie dysosmia after olfactory axon injury. eNeuro 3, ENEURO.0242-16.2016.

Murata, K., Kanno, M., Ieki, N., Mori, K., and Yamaguchi, M. (2015). Mapping of learned odor-induced motivated behaviors in the mouse olfactory tubercle. Journal of Neuroscience 35, 10581–10599.

Nakashima, A., Ihara, N., Shigeta, M., Kiyonari, H., Ikegaya, Y., and Takeuchi, H. (2019). Structured spike series specify gene expression patterns for olfactory circuit formation. Science 365, 46-+.

Nakashima, A., Takeuchi, H., Imai, T., Saito, H., Kiyonari, H., Abe, T., Chen, M., Weinstein, L.S., Yu, C.R., Storm, D.R., et al. (2013). Agonist-independent GPCR activity regulates anterior-posterior targeting of olfactory sensory neurons. Cell 154, 1314–1325.

Nei, M., Niimura, Y., and Nozawa, M. (2008). The evolution of animal chemosensory receptor gene repertoires: roles of chance and necessity. Nature Reviews Genetics 9, 951–963.

Nguyen-Ba-Charvet, K.T., Di Meglio, T., Fouquet, C., and Chédotal, A. (2008). Robos and slits control the pathfinding and targeting of mouse olfactory sensory axons. Journal of Neuroscience 28, 4244–4249.

Niimura, Y. (2009). On the origin and evolution of vertebrate olfactory receptor genes: comparative genome analysis among 23 chordate species. Genome Biology and Evolution 1, 34–44.

Niimura, Y., and Nei, M. (2007). Extensive gains and losses of olfactory receptor genes in mammalian evolution. PLoS One 2, e708.

Niimura, Y., Matsui, A., and Touhara, K. (2014). Extreme expansion of the olfactory receptor gene repertoire in African elephants and evolutionary dynamics of orthologous gene groups in 13 placental mammals. Genome Research 24, 1485–1496.

Nishizumi, H., Miyashita, A., Inoue, N., Inokuchi, K., Aoki, M., and Sakano, H. (2019). Primary dendrites of mitral cells synapse unto neighboring glomeruli independent of their odorant receptor identity. Communications Biology 2, 14.

Norlin, E.M., Alenius, M., Gussing, F., Hagglund, M., Vedin, V., and Bohm, S. (2001). Evidence for gradients of gene expression correlating with zonal topography of the olfactory sensory map. Molecular and Cellular Neuroscience *18*, 283–295.

Oka, Y., Omura, M., Kataoka, H., and Touhara, K. (2004). Olfactory receptor antagonism between odorants. The EMBO Journal *23*, 120–126.

Pacifico, R., Dewan, A., Cawley, D., Guo, C.Y., and Bozza, T. (2012). An Olfactory Subsystem that Mediates High-Sensitivity Detection of Volatile Amines. Cell Rep *2*, 76–88.

Panaliappan, T.K., Wittmann, W., Jidigam, V.K., Mercurio, S., Bertolini, J.A., Sghari, S., Bose, R., Patthey, C., Nicolis, S.K., and Gunhaga, L. (2018). Sox2 is required for olfactory pit formation and olfactory neurogenesis through BMP restriction and Hes5 upregulation. Development *145*, dev153791.

Pashkovski, S.L., Iurilli, G., Brann, D., Chicharro, D., Drummey, K., Franks, K., Panzeri, S., and Datta, S.R. (2020). Structure and flexibility in cortical representations of odour space. Nature *583*, 253-+.

Pfister, P., Smith, B.C., Evans, B.J., Brann, J.H., Trimmer, C., Sheikh, M., Arroyave, R., Reddy, G., Jeong, H.Y., Raps, D.A., *et al.* (2020). Odorant receptor inhibition is fundamental to odor encoding. Current Biology *30*, 2574–2587.

Ressler, K.J., Sullivan, S.L., and Buck, L.B. (1993). A zonal organization of odorant receptor gene expression in the olfactory epithelium. Cell *73*, 597–609.

Riviere, S., Challet, L., Fluegge, D., Spehr, M., and Rodriguez, I. (2009). Formyl peptide receptor-like proteins are a novel family of vomeronasal chemosensors. Nature *459*, 574–577.

Root, C.M., Denny, C.A., Hen, R., and Axel, R. (2014). The participation of cortical amygdala in innate, odour-driven behaviour. Nature *515*, 269–273.

Roper, S.D., and Chaudhari, N. (2017). Taste buds: cells, signals and synapses. Nature Reviews Neuroscience *18*, 485–497.

Rubin, B.D., and Katz, L.C. (1999). Optical imaging of odorant representations in the mammalian olfactory bulb. Neuron *23*, 499–511.

Saito, H., Chi, Q.Y., Zhuang, H.Y., Matsunami, H., and Mainland, J.D. (2009). Odor Coding by a Mammalian Receptor Repertoire. Science Signaling *2*, ra9.

Sakamoto, M., Ieki, N., Miyoshi, G., Mochimaru, D., Miyachi, H., Imura, T., Yamaguchi, M., Fishell, G., Mori, K., Kageyama, R., *et al.* (2014). Continuous postnatal neurogenesis contributes to formation of the olfactory bulb neural circuits and flexible olfactory associative learning. Journal of Neuroscience *34*, 5788–5799.

Sanchez-Guardado, L., and Lois, C. (2019). Lineage does not regulate the sensory synaptic input of projection neurons in the mouse olfactory bulb. Elife *8*, e46675.

Sato, K., Pellegrino, M., Nakagawa, T., Nakagawa, T., Vosshall, L.B., and Touhara, K. (2008). Insect olfactory receptors are heteromeric ligand-gated ion channels. Nature *452*, 1002–U1009.

Sato, Y., Hirata, T., Ogawa, M., and Fujisawa, H. (1998). Requirement for early-generated neurons recognized by monoclonal antibody Lot1 in the formation of lateral olfactory tract. Journal of Neuroscience *18*, 7800–7810.

Sato, Y., Miyasaka, N., and Yoshihara, Y. (2005). Mutually exclusive glomerular innervation by two distinct types of olfactory sensory neurons revealed in transgenic zebrafish. Journal of Neuroscience *25*, 4889–4897.

Sato, Y., Miyasaka, N., and Yoshihara, Y. (2007). Hierarchical regulation of odorant receptor gene choice and subsequent axonal projection of olfactory sensory neurons in zebrafish. Journal of Neuroscience *27*, 1606–1615.

Schwarting, G.A., Kostek, C., Ahmad, N., Dibble, C., Pays, L., and Puschel, A.W. (2000). Semaphorin 3A is required for guidance of olfactory axons in mice. Journal of Neuroscience *20*, 7691–7697.

Schwob, J.E., Costanzo, R.M., and Youngentob, S.L. (2020). Regeneration of the olfactory epithelium. In The Senses, B. Fritzsch, ed. (Elsevier), pp. 565–590.

Schwob, J.E., Jang, W., Holbrook, E.H., Lin, B., Herrick, D.B., Peterson, J.N., and Coleman, J.H. (2017). Stem and progenitor cells of the mammalian olfactory epithelium: Taking poietic license. Journal of Comparative Neurology *525*, 1034–1054.

Serizawa, S., Ishii, T., Nakatani, H., Tsuboi, A., Nagawa, F., Asano, M., Sudo, K., Sakagami, J., Sakano, H., Ijiri, T., *et al.* (2000). Mutually exclusive expression of odorant receptor transgenes. Nature Neuroscience *3*, 687–693.

Serizawa, S., Miyamichi, K., and Sakano, H. (2004). One neuron-one receptor rule in the mouse olfactory system. Trends in Genetics *20*, 648–653.

Serizawa, S., Miyamichi, K., Nakatani, H., Suzuki, M., Saito, M., Yoshihara, Y., and Sakano, H. (2003). Negative feedback regulation ensures the one receptor-one olfactory neuron rule in mouse. Science *302*, 2088–2094.

Serizawa, S., Miyamichi, K., Takeuchi, H., Yamagishi, Y., Suzuki, M., and Sakano, H. (2006). A neuronal identity code for the odorant receptor-specific and activity-dependent axon sorting. Cell *127*, 1057–1069.

Shatz, C.J. (1992). The developing brain. Scientific American *267*, 61–67.

Shepherd, G.M. (2006). Smell images and the flavour system in the human brain. Nature *444*, 316–321.

Shusterman, R., Smear, M.C., Koulakov, A.A., and Rinberg, D. (2011). Precise olfactory responses tile the sniff cycle. Nature Neuroscience *14*, 1039–1044.

Smear, M., Resulaj, A., Zhang, J., Bozza, T., and Rinberg, D. (2013). Multiple perceptible signals from a single olfactory glomerulus. Nature Neuroscience *16*, 1687–1691.

Sosulski, D.L., Bloom, M.L., Cutforth, T., Axel, R., and Datta, S.R. (2011). Distinct representations of olfactory information in different cortical centres. Nature *472*, 213–216.

Spors, H., and Grinvald, A. (2002). Spatio-temporal dynamics of odor representations in the mammalian olfactory bulb. Neuron *34*, 301–315.

St John, J.A., and Key, B. (2003). Axon mis-targeting in the olfactory bulb during regeneration of olfactory neuroepithelium. Chemical Senses *28*, 773–779.

Suryanarayana, S.M., Perez-Fernandez, J., Robertson, B., and Grillner, S. (2021). Olfaction in lamprey pallium revisited-dual projections of mitral and tufted cells. Cell Reports *34*, 108596.

Takeuchi, H., Inokuchi, K., Aoki, M., Suto, F., Tsuboi, A., Matsuda, I., Suzuki, M., Aiba, A., Serizawa, S., Yoshihara, Y., *et al.* (2010). Sequential arrival and graded secretion of Sema3F by olfactory neuron axons specify map topography at the bulb. Cell *141*, 1056–1067.

Treloar, H.B., Miller, A.M., Ray, A., and Greer, C.A. (2010). Development of the Olfactory System. In The Neurobiology of Olfaction, A. Menini, ed. (CRC Press/Taylor & Francis, Boca Raton, FL).

Tsuboi, A., Miyazaki, T., Imai, T., and Sakano, H. (2006). Olfactory sensory neurons expressing class I odorant receptors converge their axons on an antero-dorsal domain of the olfactory bulb in the mouse. European Journal of Neuroscience *23*, 1436–1444.

Tucker, E.S., Lehtinen, M.K., Maynard, T., Zirlinger, M., Dulac, C., Rawson, N., Pevny, L., and LaMantia, A.-S. (2010). Proliferative and transcriptional identity of distinct classes of neural precursors in the mammalian olfactory epithelium. Development *137*, 2471–2481.

Uchida, N., Takahashi, Y.K., Tanifuji, M., and Mori, K. (2000). Odor maps in the mammalian olfactory bulb: domain organization and odorant structural features. Nature Neuroscience *3*, 1035–1043.

Vassalli, A., Rothman, A., Feinstein, P., Zapotocky, M., and Mombaerts, P. (2002). Minigenes impart odorant receptor-specific axon guidance in the olfactory bulb. Neuron *35*, 681–696.

Vassar, R., Ngai, J., and Axel, R. (1993). Spatial segregation of odorant receptor expression in the mammalian olfactory epithelium. Cell *74*, 309–318.

Veeman, M.T., Newman-Smith, E., El-Nachef, D., and Smith, W.C. (2010). The ascidian mouth opening is derived from the anterior neuropore: reassessing the mouth/neural tube relationship in chordate evolution. Developmental Biology *344*, 138–149.

von der Weid, B., Rossier, D., Lindup, M., Tuberosa, J., Widmer, A., Dal Col, J., Kan, C.D., Carleton, A., and Rodriguez, I. (2015). Large-scale transcriptional profiling of chemosensory neurons identifies receptor-ligand pairs in vivo. Nature Neuroscience *18*, 1455–1463.

Vosshall, L.B., Amrein, H., Morozov, P.S., Rzhetsky, A., and Axel, R. (1999). A spatial map of olfactory receptor expression in the *Drosophila antenna*. Cell *96*, 725–736.

Wachowiak, M., and Cohen, L.B. (2001). Representation of odorants by receptor neuron input to the mouse olfactory bulb. Neuron *32*, 723–735.

Wang, F., Nemes, A., Mendelsohn, M., and Axel, R. (1998). Odorant receptors govern the formation of a precise topographic map. Cell *93*, 47–60.

Wang, P.Y., Boboila, C., Chin, M., Higashi-Howard, A., Shamash, P., Wu, Z., Stein, N.P., Abbott, L.F., and Axel, R. (2020). Transient and persistent representations of odor value in prefrontal cortex. Neuron *108*, 209–224.e6.

Wanner, A.A., and Friedrich, R.W. (2020). Whitening of odor representations by the wiring diagram of the olfactory bulb. Nature Neuroscience *23*, 433–442.

Wierman, M.E., Kiseljak-Vassiliades, K., and Tobet, S. (2011). Gonadotropin-releasing hormone (GnRH) neuron migration: Initiation, maintenance and cessation as critical steps to ensure normal reproductive function. Frontiers in Neuroendocrinology *32*, 43–52.

Wilson, C.D., Serrano, G.O., Koulakov, A.A., and Rinberg, D. (2017). A primacy code for odor identity. Nature Communications *8*, 1477.

Wilson, D.A., and Sullivan, R.M. (2011). Cortical processing of odor objects. Neuron *72*, 506–519.

Wilson, R.I., and Mainen, Z.F. (2006). Early events in olfactory processing. Annual Review of Neuroscience *29*, 163–201.

Wray, S. (2010). From nose to brain: development of gonadotrophin-releasing hormone-1 neurones. Journal of Neuroendocrinology *22*, 743–753.

Wu, A., Yu, B., and Komiyama, T. (2020b). Plasticity in olfactory bulb circuits. Current Opinion in Neurobiology *64*, 17–23.

Wu, A., Yu, B., Chen, Q.Y., Matthews, G.A., Lu, C., Campbell, E., Tye, K.M., and Komiyama, T. (2020a). Context-dependent plasticity of adult-born neurons regulated by cortical feedback. Science Advances *6*, eabc8319.

Wu, Y.M., Ma, L.M., Duyck, K., Long, C.C., Moran, A., Scheerer, H., Blanck, J., Peak, A., Box, A., Perera, A., *et al.* (2018). A population of navigator neurons is essential for olfactory map formation during the critical period. Neuron *100*, 1066–1082.

Xu, L., Li, W.Z., Voleti, V., Zou, D.J., Hillman, E.M.C., and Firestein, S. (2020). Widespread receptor-driven modulation in peripheral olfactory coding. Science *368*, 154.

Yamada, Y., Bhaukaurally, K., Madarasz, T.J., Pouget, A., Rodriguez, I., and Carleton, A. (2017). Context- and output layer-dependent long-term ensemble plasticity in a sensory circuit. Neuron *93*, 1198-+.

Yan, Z.Q., Tan, J., Qin, C., Lu, Y., Ding, C., and Luo, M.M. (2008). Precise circuitry links bilaterally symmetric olfactory maps. Neuron *58*, 613–624.

Yokoi, M., Mori, K., and Nakanishi, S. (1995). Refinement of odor molecule tuning by dendrodendritic synaptic inhibition in the olfactory bulb. Proceedings of the National Academy of Sciences of the United States of America *92*, 3371–3375.

Yoon, H.Y., Enquist, L.W., and Dulac, C. (2005). Olfactory inputs to hypothalamic neurons controlling reproduction and fertility. Cell *123*, 669–682.

Yu, C.R., Power, J., Barnea, G., O'Donnell, S., Brown, H.E., Osborne, J., Axel, R., and Gogos, J.A. (2004). Spontaneous neural activity is required for the establishment and maintenance of the olfactory sensory map. Neuron *42*, 553–566.

Zak, J.D., Reddy, G., Vergassola, M., and Murthy, V.N. (2020). Antagonistic odor interactions in olfactory sensory neurons are widespread in freely breathing mice. Nature Communications *11*, 3350.

Zeppilli, S., Ackels, T., Attey, R., Klimpert, N., Ritola, K.D., Boeing, S., Crombach, A., Schaefer, A.T., and Fleischmann, A. (2020). Molecular characterization of projection neuron subtypes in the mouse olfactory bulb. Elife, *10*, e65445.

Zhang, X.M., and Firestein, S. (2002). The olfactory receptor gene superfamily of the mouse. Nature Neuroscience *5*, 124–133.

Zhang, X.X., Yan, W.J., Wang, W.L., Fan, H.M., Hou, R.Q., Chen, Y.L., Chen, Z.Q., Ge, C.F., Duan, S.M., Compte, A., *et al.* (2019). Active information maintenance in working memory by a sensory cortex. Elife *8*, e43191.

Zhao, H., and Reed, R.R. (2001). X Inactivation of the OCNC1 channel gene reveals a role for activity-dependent competition in the olfactory system. Cell *104*, 651–660.

Zhao, S.H., Tian, H.K., Ma, L.M., Yuan, Y., Yu, C.R., and Ma, M.H. (2013). Activity-dependent modulation of odorant receptor gene expression in the mouse olfactory epithelium. PLoS ONE *8*, e69862.

Zheng, C., Feinstein, P., Bozza, T., Rodriguez, I., and Mombaerts, P. (2000). Peripheral olfactory projections are differentially affected in mice deficient in a cyclic nucleotide-gated channel subunit. Neuron *26*, 81–91.

Zou, D.J., Chesler, A.T., Le Pichon, C.E., Kuznetsov, A., Pei, X., Hwang, E.L., and Firestein, S. (2007). Absence of adenylyl cyclase 3 perturbs peripheral olfactory projections in mice. Journal of Neuroscience *27*, 6675–6683.

Zou, D.J., Feinstein, P., Rivers, A.L., Mathews, G.A., Kim, A., Greer, C.A., Mombaerts, P., and Firestein, S. (2004). Postnatal refinement of peripheral olfactory projections. Science *304*, 1976–1979.

3 Vision and Retina Information Processing
From Opsins to the Visual Cortex

Paul R. Martin, Bernd Fritzsch

CONTENTS

DOI: 10.1201/9781003092810-3

3.1   OPSIN, EYES AND CENTRAL PROJECTIONS
ARE THE BASIS OF VERTEBRATE VISION

Opsin is central to phototransduction and vision (Arendt et al., 2016; Lamb et al., 2007; Land and Nilsson, 2012; Nilsson, 2020; Pisani et al., 2020). There are different families of opsins (Lamb, 2013; Pisani et al., 2020; Valencia et al., 2019), which are all G-protein-coupled receptors (GPCRs). It is now known that from the earliest stages, opsins evolved together with the regulatory gene *Pax6*, together with *Eya/Six*, *Sox2*, and *Mitf*, among others. An independent control mechanism comprises retinoic acids (*RAs*) and their synthesizing Raldh enzymes (Feuda et al., 2012; Luo et al., 2006; Neitz et al., 2020; Wagner et al., 2002). In addition to retinal ganglion cells, development needs to generate, in sequence, horizontal, cone, amacrine, rod, bipolar, and Müller cells. A second, equally important aspect of development requires the transformation of vertebrate eyes with evagination and provides the interaction of the lens placode to develop a retina and central projections from retinal ganglion cells (Fuhrmann, 2010; Lamb, 2013; Sanes et al., 2011; Wässle, 2020). Finally, additional molecular, development, and neuronal control mechanisms work to transform some retinal ganglion cells, driven by *Atoh7*, to yield a melanopsin-based phototransduction cell family with unique properties and brain connections (Hattar et al., 2002; Lucas et al., 2020; Provencio et al., 1998). In this chapter, we give an overview of the ciliary photoreceptors, the opsins and the development and evolution of photoreceptors and ganglion cells. We will then detail the retinal ganglion cell projections to the midbrain or *tectum optimum* and the lateral geniculate body in dorsal thalamus, which provides the input to cortical visual pathways. Before these details we will provide an overview of retina evolution and development, based on molecular biology of *Opsins, Pax6, Pax2, Eya/Six, Shh*, and *Retinaldehyde (RA)*.

3.2   OPSINS DRIVE EVOLUTION OF VISION OF CHORDATES

3.2.1   AMPHIOXUS (LANCELET)

Four different photoreceptive cells are known among lancelets that are either associated with the neuropore or are distinct and physically separated from the neuropore: These are the frontal eye next to the neuropore, the lamellar body, the Joseph cells, and the dorsal ocelli (Holland, 2020; Lacalli et al., 1994). The frontal eye has certain similarities to the retina, with specific gene expression domains pointing to counterparts in the vertebrate retina (Kozmik et al., 2003; Pergner and Kozmik, 2017; Vopalensky et al., 2012). To date, the ciliary cells of the frontal eye and lamellar body have not been demonstrated to be light-responsive (Lamb, 2013). Most importantly, the 'frontal eye' does not conform to the ciliary morphology as it lacks other features of ciliary neurons. In contrast, the Joseph cells and the dorsal ocelli express the enzymatic machinery of the transduction cascade and have established photoreceptor function with excitation maximum near 470 nm (del Pilar Gomez et al., 2009). Near the neuropore are pigmented cells that express *Pax2/5/8*, *Otx*, *Mitf*, and melanin synthesis genes (Figure 3.1a). Further from the pigmented cells

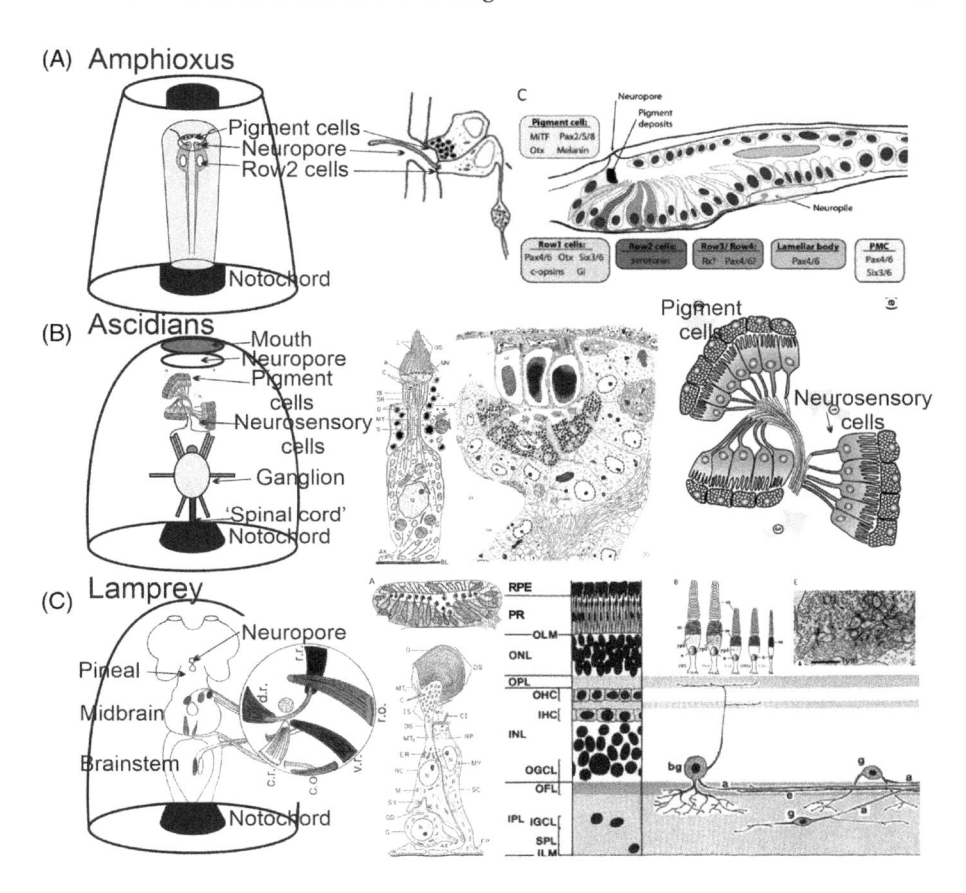

FIGURE 3.1 Comparing lancelet (A), ascidians (B), and lamprey (C) eyes. Lancelet has cilia without lamellae in a 'frontal eye' that protrudes through the neuropore and has neurosensory cells that project rostrally. Note that the notochord is extending beyond the rostrum (section view) containing pigment cells. Ascidians have a short (or absent) notochord and a neuropore, which will open later through the mouth and possess single or complex cilia with an outer segment of lamellae. The outer segments are next to or separate from pigment cells and project to the nearby ganglion that may or may not exhibit a 'spinal cord' in derived ascidians. Lampreys have asymmetric pineal dorsal photoreceptor cells that connect with ganglion cells and form synapses. A transient neuropore opens anterior to the pineal and expands into a bilateral retina containing rods and cones that interconnect with bipolar, horizontal, amacrine, and ganglion cells, some of which extend to the outer plexiform layer. (Modified from Barnes, 1971; Braun and Stach, 2017, 2019; Collin et al., 2009; Eakin, 1973; Eakin and Kuda, 1970; Fritzsch and Collin, 1990; Fritzsch et al., 1990; Lacalli et al., 1994.)

are cells positive for *Pax4/6* as well as *Otx*, *Six3/6*, and c-Opsins (row 1 cells; Vopalensky et al., 2012). The sensory cells in the lancelet have a ciliary cilium that extends to the neuropore and has a central projection of its own axon to reach a central target (Lacalli et al., 1994; Pergner and Kozmik, 2017). In contrast, the central projection of vertebrate retina is from ganglion cells that depend on *Atoh7*, which is a bHLH gene that is not found in the lancelet (Holland, 2020; Lamb, 2013; Nilsson, 2020). It is important to point out that ciliary sensory cells of vertebrates end in a synaptic ribbon, which contacts secondary sensory cells and is under control of *Atoh7*, whereas in the lancelet, sensory neurons lack a bHLH gene and project directly to the ventral neuropil.

Molecular fingerprinting of lancelets has identified a total of 21 opsin genes (Pergner and Kozmik, 2017), which show highly variable tripeptide sequences of opsins, comparable to vertebrates. Like vertebrates, lancelets have genes for melanopsin that may terminate in the same ventral neuropil and could be comparable to vertebrate melanopsin (Lucas et al., 2020; Pantzartzi et al., 2017; Provencio et al., 1998). Current data suggest that all-*trans*-retinal as well 11-*cis*-retinal in lancelets is different from that in vertebrates (Albalat et al., 2012; Poliakov et al., 2012). Furthermore, lancelets do not possess the retinal pigmented epithelium that is ubiquitous in vertebrates (Crandall et al., 2011). A duplication and change in chromophore binding site (from position 181 to 113) also distinguishes lancelet from vertebrate opsin (Lamb, 2013).

In summary, beyond the early steps of a 'frontal eye' under control of certain primordial genes (*Pax2/5/6, Pax4/6, Six3/6, Otx, c-opsins*) in lancelets, the evolution of vertebrate eyes required major steps to transform a ciliated sensory neuron (lancelet) into the split arrangement of a ciliated sensory cell and a central projection neuron (vertebrates). We suggest two or more additional genes evolved in vertebrates to convert a single neurosensory cell into a ciliary sensory cell connected with a ganglion neuron (Fritzsch and Elliott, 2017; Lamb, 2013)

3.2.2 TUNICATES

Several of over 3,000 species of tunicates (ascidians) have been characterized (Braun et al., 2020; Winkley et al., 2020). Ciliated sensory cells are present for nearly all ascidians (at least in earliest steps of development) and are characterized by pigment cells adjacent to them. The sensory cells have an outer segment with multiple lamellae, comparable to three out of four tunicate sensory cell types and are different to the 'frontal eye' of the lancelet (Figure 3.1). Each sensory cell of ascidians has the central projection from the nearby neuronal ganglion (Braun and Stach, 2019; Konno et al., 2010; Lacalli and Holland, 1998; Lacalli et al., 1994; Ryan et al., 2018). Detailed analysis shows the origin from the neuropore in the sea squirt *Ciona* (Veeman and Reeves, 2015) and confirmed the independent connection of the neuropores with the opening of the gut (Veeman et al., 2010). In a unique arrangement, the neuropore is fused with the mouth opening, something that is prevented by the extended notochord of lancelets (Figure 3.1). In the free-floating salp *Thalia*, oocytes develop into a blastozooid stage within a few hours (Winkley et al., 2020). The oozooid stage of *Thalia* has a single large pigmented set of cells that are adjacent to sensory cells (Lacalli and Holland, 1998). In the blastozooid stage, receptor cells form three distinct pigment cups, each one pointing in a different direction (Braun and Stach, 2017). The cups have no lens associated with them, but have transparent transient 'lens cells' (Barnes, 1971) and have three discrete 'sensory cells' of which only one is associated with melanin (Konno et al., 2010). The simple brain of all ascidians is connected by multiple branches to the 'eyes' which have not been analyzed in terms of individual terminals (Braun and Stach, 2019; Ryan et al., 2018).

Unlike the lancelet, *Ciona* expresses an opsin-1 gene, which is closely related to the vertebrate non-visual opsins (Pisani et al., 2020; Terakita, 2005). As stated above for the lancelet, the tunicates are unable to transform all-*trans* to 11-*cis* reginal (Albalat et al., 2012; Poliakov et al., 2012). Interestingly, certain species do not bind RA, while others use conventional RA signaling (Fujiwara and Cañestro, 2018). Molecular experiments showed expression of neural crest-like and placode-like structures (Bassham and Postlethwait, 2005; Cao et al., 2019) suggesting that certain steps toward eye development are evident in the ascidians. Interestingly, the ascidian arrestin binds to opsin-based pigments, a function similar to the vertebrate non-visual arrestin, β-*arrestin* (Kawano-Yamashita et al., 2011)

Although certain similarities are clear with respect to opsins in the lancelet, ascidians, and vertebrates, we lack a model of how to transform the simple frontal eye of lancelet to the evaginated ascidian eye. To date, the relationship of ascidian to vertebrate eyes is unclear (Lamb, 2013) as certain genes (*Pax4/6, Eya/Six, Mitf*) are required at the level of vertebrates. Interestingly enough, *RA* in lancelet, ascidians, and vertebrates affects rostro-caudal development, suggesting a different influence of *RA* on sensory cells in chordates (Campo-Paysaa et al., 2008; Glover et al., 2006; Hinman and Degnan, 2000), which clearly depend on *RA* for adult retina development (Crandall et al., 2011; Manns and Fritzsch, 1991).

In summary, ascidians have neurosensory cells that are associated with pigment cells and likely depend on *Mitf* signals from the mesenchyme as in lancelets and vertebrates. We suggest that a split of a ciliary neurosensory cell into the formation of ganglion neurons and sensor cells must have happened in vertebrates to allow the formation of these two distinct cell types (sensory cells and ganglion neurons).

3.2.3 Neuropore, Retina, and Eye Morphology Require a Molecular Cascade in Vertebrates

Obviously, eye evolution requires several developmental steps to transform from chordates such as lancelet and ascidians (with a single neurosensory cell type) into a vertebrate (with distinct receptor cells and projection ganglion neurons). Beyond a few genes that are identified across chordates and are associated with earliest gene expression requires a basic set of innovations, including the novel duplication of sensory neuron development using *Atoh7* (retina ganglion neurons) and *Neurod1* (sensory cells). Unfortunately, critical data with respect to eye development in cyclostomes such as lampreys are scarce (Lamb, 2013; Pombal and Megías, 2019; Suzuki and Grillner, 2018) and we are also far behind in understanding hagfish eye development (Ota and Kuratani, 2006; Wicht, 1996). We here endeavor to develop a combined perspective on neuropores and retina across different chordates.

3.2.3.1 Neuropore

Lamprey neuropore starts as a keel that will form into the neural tube in later development (Richardson et al., 2010; Sauka-Spengler and Bronner-Fraser, 2008). Overall, this early step is comparable to the keel of zebrafish neuropore formation and similar factors may influence closure of the neuropore (Shinotsuka et al., 2018). A neuropore is described for hagfish but not much information beyond that is available due to limited knowledge of hagfish development (Higuchi et al., 2019; Ota and Kuratani, 2006; Wicht, 1996). Similarly, more details are needed to understand anterior neuropore defects in mice and man (Copp and Greene, 2010; Copp et al., 2003). Formation of the neural tube starts in the hindbrain and fusion progresses anteriorly, leaving the neuropore at the last step of closure. *Wnt/Frizzled* signaling involves a set of the planar cell-polarity genes (*Vang, Scrb1, Crash, Disheveled*) that will cause the craniorachischisis in mutant mice, implying these signal pathways in closure of the neuropore (Rolo et al., 2019). The gene *Opb* (open brain) and *Zic2* interact with anti-*Shh* to regulate this process and reduced eye development is one consequence of failure of neural tube closure (Eggenschwiler et al., 2001; Gunther et al., 1994; Guo et al., 2006; Rolo et al., 2019). In part, the absence of *Msx1* and *Wnt3a* underlies the effects seen in *Opb* and *Zic2* null mice, in which the brain remains open without any closure (Copp and Greene, 2010). Downstream to *Opb/Zic2* are *Fgfs, Wnts*, retinoic acid (*RA*), *Nodal*, and *BMPs* provide the signals that will, eventually, close the neuropore (Wilson and Houart, 2004) and are known to involve *Shh* in mutants (Eggenschwiler et al., 2001). Triazole interferes with *RA* signaling and produces numerous defects in retinal and anterior neuropore formation as well as otic defects (Shinotsuka et al., 2018). In summary, there are clear common factors in neural tube development in mice and man that can affect the pineal and retina in extreme cases.

3.2.3.2 Retina and Morphology

Correlated with the previous outline of chordate neuropore development is a set of human genes that are associated with anophthalmia (Table 3.1). These defects include lack or retina formation in the extreme cases. The earliest step is formation of the eye field before the gastrulation has completed (Plaisancié et al., 2019). During gastrulation, the eye evaginates laterally, splitting the eye field into a right and left optic vesicle. Among the various human genes involved are *SOX2, OTX2, RAX, VSX2, STRA6, RARB, VAX1, BMP4/7, SIX6, PTCH1*, and *GDF3/6* (Plaisancié et al., 2019). In most case of anophthalmia, the retina develops to some extent, but there are also coloboma defects that affect the optic nerve leaving it incomplete.

3.2.3.3 Development of the Retina

A detailed description of retinal development is provided by Fuhrmann (2010), highlighting multiple steps involved in retina expansion and interaction with the lens (Figure 3.2). Starting with the eye field in the neural plate, the eye field splits into two parts and interacts with a cascade of genes to form the anlage, which then protrude and interacts with the lens placode and forms the stalk and optic fissure that rolls up to form the retina. Initial retina expansion induces development of

TABLE 3.1
Genes Affecting Eye Development

Gene	Anoph-thalmia	Colo-bomas
SOX2 (SRY [sex-determine region Y]-box2)	+	+/−
OTX2 (orthodenticle homeobox 2)	+	+/−
RAX (retina and anterior neural fold homeobox)	+	+/−
VSC2 (visual system homeobox 2)	+	+/−
RARB (retinoic acid receptor [β] beta)[a]	+	+
MALB2 male abnormal gene family 21)	+	+/−
VAX1 (ventral anterior homeobox 1)	+	−
PAX6 (paired box 6)	+/−	+/−
PAX2 (paired box 2)	−	+
BMP4 (bone morphogenetic protein 3)	+/−	+/−
BMP7 (bone morphogenetic protein 7)	+	+/−
GDF3 (growth differentiation factor 3)	+	+/−
GDF6 (growth differentiation factor 7)	+	+/−
SIX6 (sine oculis homeobox, Drosophila, homolog of, 6)	+/−	−
PTCH1 (patched, Drosophila homolog of, 1)	+/−	+

[a] Incomplete of RA associated mutants.

retinal pigment epithelium (RPE). A next step is driven by mesenchymal induction of *Mitf, Pax6, Rax, Otx2, Six3*, and *Lhx2*. In addition, expression of *Fgf* is secreted from the ectoderm and upregulates *pERK* genes as well as *Vsx2* and *Sox2* to be suppressed by *Mitf*. Expression of *Sox2* defines the neuronal part of the retina development. Under expression of *Six3*, it regulates *Pax6* and *Sox2* in the lens

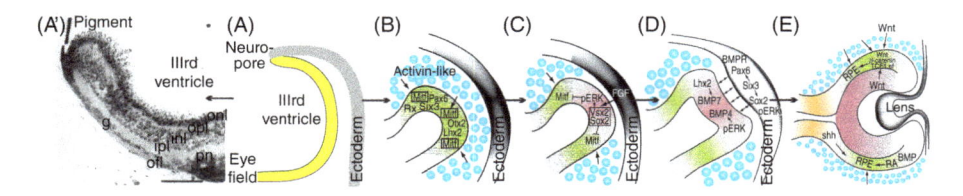

FIGURE 3.2 Retina formation starts ventral to the neuropore, expressing the IIIrd ventricle and shows expression of the eye field (A). Right panels display sequential upregulation of *Mitf* and *Rax* to interact with *Pax6, Six1, Otx2*, and *Lhx2* (B) followed by *Fgf* expression collaborating with *Sox2* that suppresses *Mitf* (C). Several genes interact to upregulate the lens placode during eye formation (D). *Shh* defines the optic stalk, *Mitf* develops pigments, whereas the retina turns through *Sox2* and *Pax6* into the neuronal precursors (E). (A') shows the effect of high doses of *RA* causing the evagination to be suppressed but a normal retina develops adjacent to the pigment. (Modified from Buono and Martinez-Morales, 2020; Fuhrmann, 2010; Manns and Fritzsch, 1991.)

placode and interacts with *BMP4/BMP7*, fine-tuned by *Lhx2* and *pERK* during morphogenesis of the retina.

Once the lens begins forming, three distinct areas (RPE, neural retina, optic stalk) require additional interactions under *Wnt* signal pathways to retain RPE under *Mitf*, *BMP*, *RA*, and β-*catenin* control, and interaction with *Shh* to define the optic stalk (Buono and Martinez-Morales, 2020; Fuhrmann, 2010). These complex interactions have been detailed in *Xenopus* with emphasis on the role of *noggin* as an antagonist for *BMPs* involving *tbx3, pax6, pax2, vax2, shh*, and *ra* (Sasagawa et al., 2002). Comparable information exists for early ear development in the zebrafish for which interesting effects of the same controlling genes can be studied in anophthalmia, retinitis pigmentosa, cyclopia, and coloboma (Cavodeassi et al., 2005; Richardson et al., 2017), pointing to similarities across jawed vertebrates. Unfortunately, much work on cyclostomes is still required to fill in the gaps in our understanding.

In summary, a set of intrinsic and extracellular factors control eye organogenesis, but there are gaps in knowledge of these interactions, in particularly control of evagination of the retina and its control by *RA* and other genes (Fuhrmann, 2010; Lamb, 2013; Manns and Fritzsch, 1991).

3.3 LAMPREYS AND HAGFISH COMPARED TO JAWED FISH: MAKING RETINA DEVELOPMENT

Eye development starts with the neuropore that is the basis of retina formation, followed by the retina development, and expanded by central projections in the two extant cyclostomes, that evolved ~482 million years ago into 35 lamprey and 79 hagfish species.

3.3.1 LAMPREYS

Compared to retina development in jawed vertebrates, knowledge of the molecular basis of retina development in lamprey is very incomplete. It is known that lampreys have a unique developmental trajectory. For example, retina formation begins in the late embryos with a few receptors and retinal ganglion cells (RGCs) that receive a prominent efferent input from the brain (Anadon et al., 1998; Barreiro-Iglesias et al., 2017; De Miguel et al., 1989; Fain, 2019; Fritzsch, 1991; Fritzsch and Collin, 1990; Lamb, 2013; Pombal and Megías, 2019; Suzuki and Grillner, 2018; Villar-Cheda et al., 2006). Opsin-expressing photoreceptor cells are reportedly present early in larvae and make contact with RGCs and bipolar cells (Suzuki and Grillner, 2018). If substantiated, it means the developmental sequence of RGC > horizontal > cone > amacrine > rod > bipolar in jawed vertebrates is different in lampreys, with simultaneous formation of rods > RGC > bipolar followed later by horizontal and amacrine (Suzuki and Grillner, 2018). Most data suggest the presence of at least two cone types in lampreys and they can have up to five different cone-like cells (Collin et al., 2009). Moreover, the synaptic terminals resemble cone terminals (Lamb, 2013; Lamb et al., 2007) and the cone morphology resembles that of jawed vertebrates. Some cells have features like cones but express rhodopsin and show rod-like transduction (Lamb, 2013) yet saturate in bright light comparable to cone-like cells (Fain, 2019; Morshedian and Fain, 2017). The majority (~74%) of RGC somas are in the inner nuclear layer (IPL), compared to 1% in the inner plexiform layer (Fritzsch and Collin, 1990), that receives afferent and efferent fibers next to the INL. In addition to the different organization of RGCs (which form a distinct ganglion cell layer [GCL] in gnathostomes), the optic fibers run sclerad to the inner plexiform layer (Fritzsch and Collin, 1990). Further interesting features are biplexiform and o-RGCs that make direct contact with photoreceptors: A situation otherwise unprecedented in gnathostomes (Fain, 2019; Jones et al., 2009). The different position in RGCs (IPL versus GCL) combined with branches that reach photoreceptors may indicate a unique and direct input of cones to RGCs in lampreys (Fritzsch and Collin, 1990; Jones et al., 2009; Suzuki and Grillner, 2018).

3.3.2 PINEAL

An additional sensory system is the pineal and parapineal of lampreys and gnathostomes, also known as parietal or pineal eyes (de Vlaming and Olcese, 2020; Koyanagi et al., 2004; Pu and Dowling, 1981; Reiter, 1980; Roberts, 1978; Simonneaux and Ribelayga, 2003; Vigh-Teichmann et al., 1983). Pineal eyes are absent in hagfish (Lamb, 2013) and have different connections in mammals (Reiter, 1980; Simonneaux and Ribelayga, 2003). The pineal forms anterior to the neuropore and forms a dorsal expansion of the diencephalon. Ciliated sensory cells connect to ganglion cells with ribbon synapses (Lamb, 2013). The direct connection of ganglion cells thus represents the output of ciliated photoreceptors at ribbons (Pu and Dowling, 1981). Pineal receptors have rod-like function and use rhodopsin (Lamb, 2013). A mediator of clathrin-mediated GPCR internalization is β-*arrestin*. Granules are generated by light-dependent absorption by β-*arrestin*-mediated internalization of parapinopsins from the outer segments. This β-*arrestin*-mediated internalization is responsible for eliminating the stable photoproduct and restoring cell conditions to the original dark state (Kawano-Yamashita et al., 2011). The presence of two cell types (ciliated sensory and ganglion cells) sets it apart from neurosensory output in lancelet and ascidians (neurosensory cells), suggesting an early split in pineal eyes that had to that time provided a direct ciliated sensory connection in lateral eyes with RGCs (Fain, 2019; Jones et al., 2009). Left and right pineal and parapineal in lampreys and several gnathostomes have different sizes (de Vlaming and Olcese, 2020; Pu and Dowling, 1981; Smith et al., 2018). The pineal is connected with the right whereas the parapineal is connected to the left, forming an asymmetric habenula (Anadon et al., 1998; Puzdrowski and Northcutt, 1989).

3.3.3 HAGFISH

Compared to lamprey, hagfish differ in certain features. To start with, unlike lamprey eye, the hagfish eye does not have a lens (Collin and Fritzsch, 1993; Gustafsson et al., 2008; Baden et al., 2020; Song and Kim, 2020; Dong and Allison, 2021), which is now known to play a major role in eye development from the earliest stages of retina formation (Fuhrmann, 2010). If confirmed, it would the only eye without a lens. The ciliated sensory cells require rhodopsin and are likely rod cells (Lamb, 2013): A suggestion consistent with recent evidence for true rods and cones in lampreys (Fain, 2019). The retinal ganglion cells are not organized into one or more layers (as in jawed vertebrates), and the RGC disposition resembles that of lampreys (Fritzsch and Collin, 1990; Suzuki and Grillner, 2018), likely correlated with the absence of myelin (Fritzsch and Glover, 2007). The neuronal makeup is unclear with respect to bipolar, horizontal, and amacrine neurons, suggesting a direct input from ciliated sensory input to RGCs via ribbons (Holmberg, 1971, 1977; Locket and Jørgensen, 1998).

3.3.4 GNATHOSTOME GENE EXPRESSION

The developmental sequence of different cell types has been established following the initial insight of proliferation of a column of retinal cells (Cepko et al., 1996). The earliest cell stage is the formation of RGCs and depends on *Atoh7* in all vertebrates. The next stage is generation of cone and amacrine followed by horizontal neurons. Rods and cones depend on *Neurod1* (Dennis et al., 2019; Neitz et al., 2020) that interacts with other genes to develop as distinct sensory cells (Brzezinski and Reh, 2015). There is a noticeable delay in bipolar and rod formation (Arendt et al., 2016; Lamb, 2013). For example, the number of different distinct bipolar neurons is currently at around 10 subtypes that depend on *Otx2* to generate these distinct types (Cepko, 2014; Chan et al., 2020).

3.3.5 SUMMARY AND OUTLOOK

A new hypothesis of retina formation is proposed: building on previous work describing the formation of two distinct cell types through gene duplication (Fritzsch and Elliott, 2017; Holland and Daza, 2018), we can now include eye development through retina multiplication (Figure 3.3). Neurosensory cells have bifunctions that have the ciliary process with its own axon in lancelet and ascidians and form three distinct clusters in *Thalia*. We propose two different sized clusters in *Thalia* and *Ciona* evolved into pineal/parapineal and the third, largest, cluster will be the basis of the lateral eyes of vertebrates. Consistent with a split into two sensory cells and ganglion cells is the presence in lampreys of the pineal/parapineal and also the direct neuropil projections in certain lampreys and hagfish. Formation of two kinds of ciliated sensory cells (*Neurod1*) and a set of neurons (*Atoh7*) is proposed to be similar to gnathostomes with minor adjustment of developing many more basic neuron types (Lamb, 2013). The vertebrate genome duplicated twice prior to other chordates: this whole-genome duplication permitted the splitting of sensory cells from ganglion neurons (Holland and Daza, 2018; Smith et al., 2013).

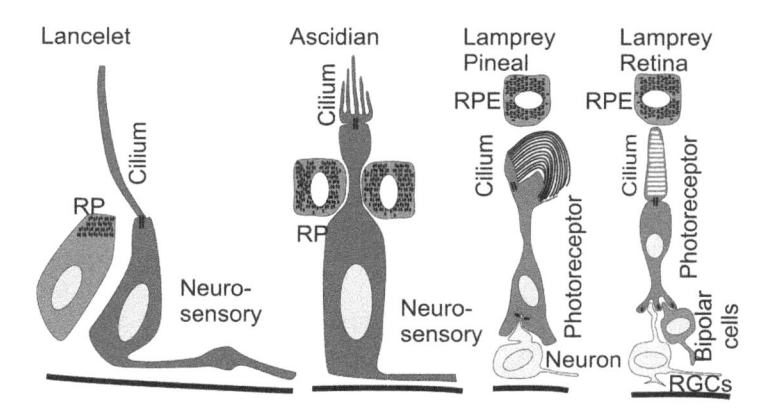

FIGURE 3.3 Lancelet has a simple cilium (double bar) that is adjacent to pigment cells and projects to the neuropil. In contrast, ascidians have cilium branches, tightly associated pigment cells, and project to the neuropil. Lampreys have two photoreceptors, in pineal/parapineal and in the retina. Both terminate in ribbons, comprising either a single neuron (pineal) or retinal ganglion cells (RGCs; in retina) that together with bipolar cells, horizontal, and amacrine cells process the visual input and transmit it to the brain.

3.4 RETINA NEURONS: DIVERSITY OF RETINAE AND RECEPTORS

Retina formation can be envisioned as a progressive restriction in the potentiality of the precursor cells. Like marbles rolling down through Waddington's epigenetic landscape, progenitors will progress from a pool of uncommitted precursors (those comprising the eye morphogenetic field) to a complex multicellular organ. Each developmental decision entails branching the genetic program once to specify the seven basic cell types comprising the mature retina (Buono and Martinez-Morales, 2020; Sanchez et al., 2020; Wässle, 2020). Overall, the role of *Sox2* and *Pax6* for neuroblast formation is confirmed with scRNA-seq and ended with *Otx2* to define the last steps of photoreceptors (Sridhar et al., 2020). The first cell types that develop are the retinal ganglion cells (RGC), which depends on *Atoh7* for their development (Cepko et al., 1996; Dennis et al., 2019; Lamb, 2013; Mao et al., 2008). The developing cells are considered to have the following sequence of proliferation: ***RGC > horizontal > cone > amacrine > rod > bipolar > Müller cells*** (Brzezinski and Reh, 2015; Lamba and Reh, 2020; Sridhar et al., 2020; Wässle, 2020). These steps are controlled mostly by bHLH genes such as *Atoh7, Neurod1, Neurod2, Ascl1, bHLHb5*, and *Ptf1a*, and analysis suggests a complex cross-regulatory interaction (Mao et al., 2013). For example, the loss of *Neurod1* leads to the loss of cone cells, whereas *Ptf1a* defines horizontal and amacrine cells. The potential can switch from photoreceptors to form horizontal, amacrine, and bipolar cells. Misexpression of *Atoh7* can substitute by *Neurod1* (Mao et al., 2008), suggesting that a switch is not possible but follows the RGC developmental direction. In contrast, the replacement of *Neurod1* by *Atoh7* can cause differentiation into amacrine and photoreceptor cells, instead of RGCs (Mao et al., 2013).

Retinal progenitors are multipotent and their potential change over development mediates the sequence of cell types that correlates with differential proliferation and fate outcomes (Brzezinski and Reh, 2015; Lamba and Reh, 2020). How can bistable states define a given fate (Chan et al., 2020)? Recent human scRNA-seq confirms the role of *Atoh7* but shows a different cluster of fate determination in organoid models; with one cluster of RGC, amacrine and horizontal neurons and a separate cluster of photoreceptors and bipolar neurons (Sridhar et al., 2020). It is the final step of *Hes1* and *Rax* expression that drives *Notch1* and regulates the *Hes5/Sox8/Sox9* genes driving the formation of Müller cells (Buono and Martinez-Morales, 2020), consistent with a role of *Sox2* upstream of all neuronal precursor formation (Imayoshi and Kageyama, 2014).

In summary, a defined sequence of gene expression is characteristic among vertebrates. *Atoh7* generates retinal ganglion neurons, *Neurod1* depends on rods and cones.

3.5 CENTRAL PROJECTIONS OF THE RETINA: FINDING THE RIGHT CONNECTIONS

As noted earlier retinal ganglion cells (RGCs; retinal ganglion cells) are the first retinal neurons to develop (Cepko et al., 1996; Lamb, 2013). RGCs depend on *Atoh7, Pou4f2, Isl1,* and others (Buono and Martinez-Morales, 2020; Clause et al., 2014; Roberts and Vetter, 2018; Sanchez et al., 2021) to project to the central targets, the diencephalon and midbrain. Upon reaching the chiasm, the RGC axons sort out to reach different targets, partially bilateral, and largely contralateral to reach about 30 distinct targets of RGCs in mammals (Sanes and Masland, 2015; Varadarajan and Huberman, 2018; Wässle, 2020). Based on Sperry's and Gierer's work (Gierer, 1987; Sperry, 1971), the chemoaffinity map hypothesis was developed, with the proposal of orthogonal diffusion gradients of ephrin ligands and receptors in the retina and matching expression gradients in the midbrain. These gradients ensure that a given RGC projects to a matching area of a retinotopic map, including suprachiasmatic nucleus (SCN) mapping of melanopsin inputs (Liu et al., 1997; Lucas et al., 2020), the basal optic tract (BOT) terminating next to the oculomotor neurons (Brecha et al., 1980; Fritzsch, 1980; Lilley et al., 2019), the lateral geniculate body (LGB) (Suryanarayana et al., 2017) and the topographical projections in the midbrain [tectum opticum or superior colliculus of mammals; (Fritzsch et al., 2019; Triplett, 2014)].

3.5.1 SUPRACHIASMATIC NUCLEUS (SCN)

The input from RGCs projections to the suprachiasmatic nuclei of the hypothalamus (SCN) is bilateral and derives from the retinohypothalamic tract [RHT; (Abrahamson and Moore, 2001)]. The SCN is a circadian clock that is responsible

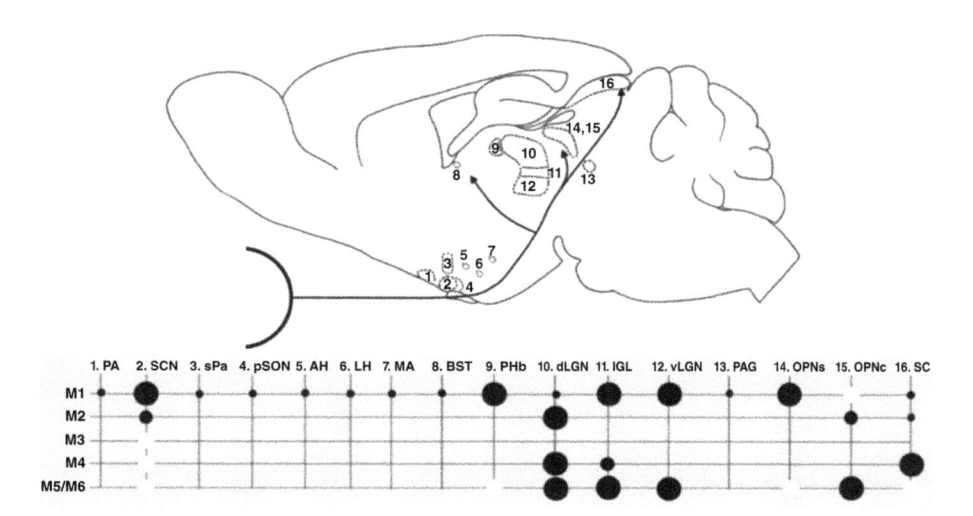

FIGURE 3.4 A sample of ipRGC brain targets is depicted in a quasi-sagittal schematic of the mouse brain. Below is a plot of innervation densities across ipRGC types. Each dot indicates the approximate density of innervation by its size. AH, anterior hypothalamus; BST, bed nucleus of the stria terminalis; dLGN, dorsal lateral geniculate nucleus; IGL, intergeniculate leaflet; LH, lateral hypothalamus; MA, medial amygdala; OPN, olivary pretectal nucleus (with shell, s, and core, c, regions); PA, preoptic area, which includes the VLPO (ventrolateral preoptic area); PAG, periaqueductal gray; PHb, perihabenular zone; pSON, peri-supraoptic nucleus; SC, superior colliculus; SCN, suprachiasmatic nucleus; sPa, subparaventricular zone; and vLGN, ventral lateral geniculate nucleus. (Adapted from Do, 2019.)

for entraining the solar cycle for the temporal regulation of behavioral rhythms. The molecular basis has been elucidated *(Per, Cry)* (Hastings et al., 2003; Takahashi, 2017). The retina and multiple other sources (Abrahamson and Moore, 2001) innervate about 10,000 neurons of each side (Hastings et al., 2018). The precision and robustness of the SCN molecular clockwork can sustain stable circadian time-keeping at periods spanning 16-48 hours. Furthermore, the rhythms of individual cells are tightly synchronized across the SCN at different phases (Hastings et al., 2018). Several subnuclei have been identified; each is part of the SCN complex (Figure 3.4). Pacemakers are embedded in circuits with a defined topology to control the oscillatory behavior of excitable tissues and heart and respiration rhythms.

More recently, the SCN was shown to receive specific input from intrinsic photosensitive retinal ganglion cells (ipRGCs; Hattar et al., 2002; Provencio et al., 1998) that project specifically to the SCN (Qiu et al., 2005) with additional subtypes that spread to other central targets (Do, 2019; Hastings et al., 2018; Lucas et al., 2020). Among the widespread distribution of melanopsin are cells in the lancelet that may be the functional equivalent to igRGCs (del Pilar Gomez et al., 2009). In contrast, available data suggest no evidence in lamprey or hagfish

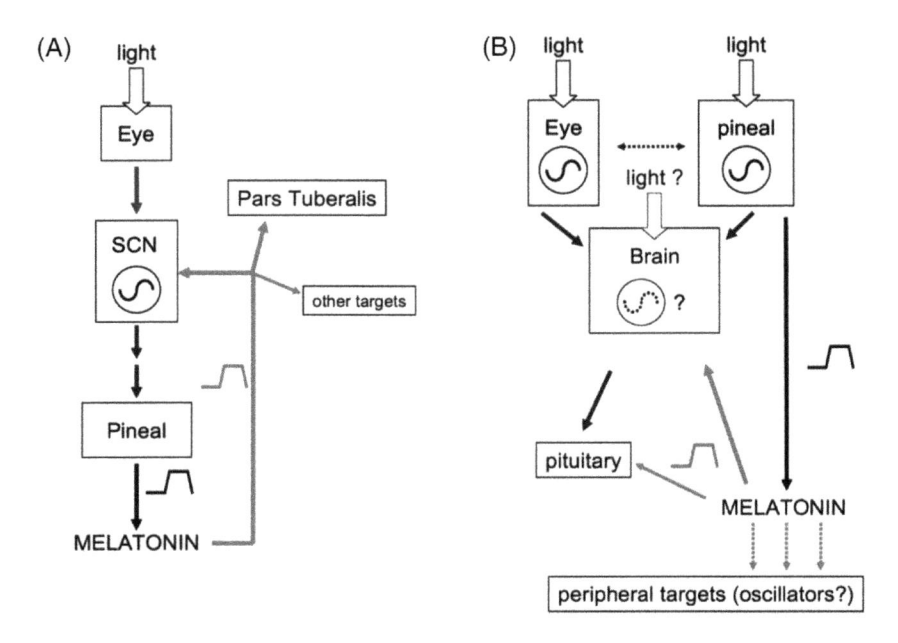

FIGURE 3.5 Mammals (A) have a linear connection from the light (eye) through the SCN and reaches the pineal to produce melatonin. Control through multisynaptic pathways provides feedback to modulate seasonal function. (B) Other vertebrates have a network of three light-sensitive oscillators in the retina, the pineal, and (possibly) the brain to affect the pituitary and peripheral targets. Pineal melatonin drives various targets for oscillator variations. (Modified from Falcon et al., 2010.)

of fibers projecting to the brainstem (Sun et al., 2014). Further work is needed to understand the fine tuning of diverse input from a set of at least six distinct populations.

In mammals, photic information is received by the eye and projects via the retinohypothalamic tract (RHT) to reach the SCN and (eventually via several pathways) the pineal gland (Cipolla-Neto and Amaral, 2018; Falcon et al., 2010). In most other vertebrates, the pineal gland is exposed directly to light (Figure 3.5). Ultimately, melatonin is produced in the pineal through a cascade from tryptophan via intermediates of serotonin to generate melatonin.

Melatonin reaches the pituitary via receptors and has complex interactions within at least four components: the rostral preoptic area (POA), the SCN, the lateral tuberal nucleus (LTN), and the ventromedial thalamic nucleus (VTN). In addition, all four of these structures receive hormone signals and may also receive retinal input via SCN and pineal (Cipolla-Neto and Amaral, 2018; Feuda et al., 2012; Vernadakis et al., 1998). What is clear is that pineal and melatonin receptors and expression are linked to neuroendocrine regulation, in particular the brain-pituitary-gonadal axis that is affected by GNRH (de Crapona et al., 1986; Falcon et al., 2010; Münz et al., 1982).

3.5.2 Basal Optic Tract (BOT) or Accessory Optic System (AOS)

This projection is present in all vertebrates (Brecha et al., 1980; Fritzsch, 1980; Fritzsch and Collin, 1990; Wang et al., 2020) with the exception of the BOT in hagfish (Wicht and Northcutt, 1990), which lacks all oculomotor input. The BOT/ AOS is composed of a series of terminal nuclei and receives input from the retina that is characterized by large receptive fields (Giolli et al., 2006). In addition, inputs are received from pretectal projections (nucleus of the optic tract; NOT) and the ventral lateral geniculate nucleus (vLGN). The AOS system is connected to the oculomotor system and influences optokinetic nystagmus, among other functions. Among mammals, the AOS subdivides into the medial terminal nucleus (MTN), the dorsal terminal nucleus (DTN), and lateral terminal nucleus (LTN) that receives input from the nucleus of the optic tract (NOT). Each nucleus communicates with the others and is part of the optokinetic control system (Liu et al., 2016). More recently, ipRGCs have been demonstrated to reach the AOS (Delwig et al., 2016). Further, recent data suggest a unique population of RGCs reaches the AOS through a *Sema6A* targeting mechanism (Lilley et al., 2019) and reports ventral-directed image slip. Interestingly enough, these RGC neurons are among the few neurons that can regrow after a nerve crush and re-establish connection to central targets; a property that is shared with the SCN-projecting RGCS (the M1 type of RGC). With modern techniques will likely come new approaches going beyond the currently limited knowledge of the BOT and its comparison to the sophistication of the AOS.

3.5.3 Lateral Geniculate Nucleus (LGN) and Pulvinar

The LGN is the major diencephalic relay to the visual cortex (Hubel and Wiesel, 1962; Karten, 2013; Lean et al., 2019; McLaughlin et al., 2000; Sherman, 2020; Stellwagen and Shatz, 2002). The presence of LGN-cortex connections was considered a novel expansion in gnathostomes, but a homologous thalamic connection to visual pallium is now recognized in lampreys, (Suryanarayana et al., 2020; Suryanarayana et al., 2017), suggesting a basic similarity between lamprey and gnathostomes (Suzuki and Grillner, 2018). No such connection from the retina through to the cortex has been identified in hagfish (Wicht, 1996; Wicht and Northcutt, 1998).

The LGN (Figure 3.6) receives both uncrossed and crossed retinal projections that typically occupy distinct regions of the nucleus (Kalish et al., 2018; Tassinari et al., 1997). Developmental data demonstrate that eye-specific domains in the dorsal LGN (dLGN) become segregated through spontaneous electrical activity (left and right) prior to photic-driven activity (Guido, 2018; Huberman et al., 2008). Initially, ipsilateral and contralateral LGN fibers overlap broadly before being segregated by activity (Dhande and Huberman, 2014; Sherman, 2020). This is a competitive process, as seen by the expansion of LGN input if the retina projection is removed from one eye (Huberman et al., 2008). *BDNF* must be involved in the segregation, because *BDNF* inhibition during development causes retraction of terminals from the blocked retina

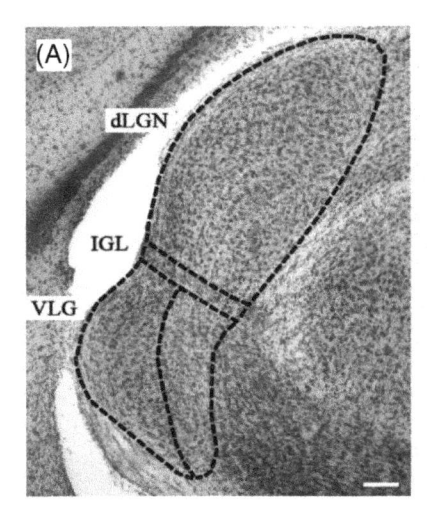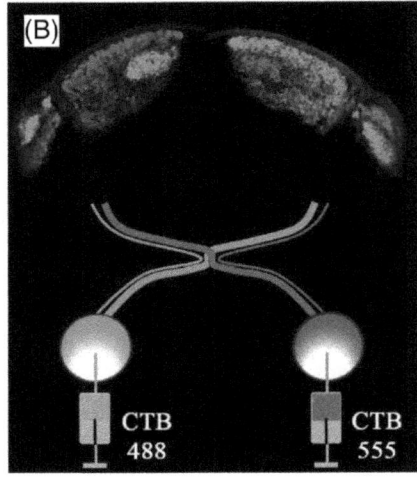

FIGURE 3.6 Segregated inputs from the left and right eyes are demonstrated with double labeling (CTB 488, CTB 555). dLGN and vLGN show a core and shell organization (A) sublayers formed in vLGN (left and right), whereas the retinal input to dLGN (B) comes as an integrated segregation without overt sublayer structure. (Modified from Guido, 2018)

within the dLGN (Menna et al., 2003). The corticothalamic pathway specifically targets the dLGN, avoiding the vLGN (Guido, 2018). Detailed analysis in mammals shows that distinct functional types of relay neurons can also be segregated within the dLGN: for example primates have distinct layers of magnocellular and parvocellular cells. Such overt segregation is not present in mammals such as the mouse (Kalish et al., 2018).

A molecular update for LGN was recently published and defines several genes involved in synapse and circuit development. These genes are highly expressed and dynamically regulated in both neurons and non-neuronal cells (glial, astrocytes). This data set shows that the assembly and refinement of neural connectivity during postnatal development involves collaboration between the majority of cell types in the LGN and dynamic changes of expression levels between P5-P21 day old mice (Kalish et al., 2018).

The lateral geniculate nucleus and pulvinar are examples of two different types of relay: the lateral geniculate nucleus is a first order relay, transmitting information from a subcortical source (retina), while the pulvinar is mostly a higher order relay, transmitting information from layer 5 of one cortical area to another. Higher order relays seem to play a key role in cortical function via a cortico-thalamo-cortical route. Other examples of first and higher order relays also represent the majority of thalamic volume. Thus, the thalamus not only provides a behaviorally relevant, dynamic control over the nature of information relayed, it also plays a key role in basic cortico-cortical communication (Sherman, 2020).

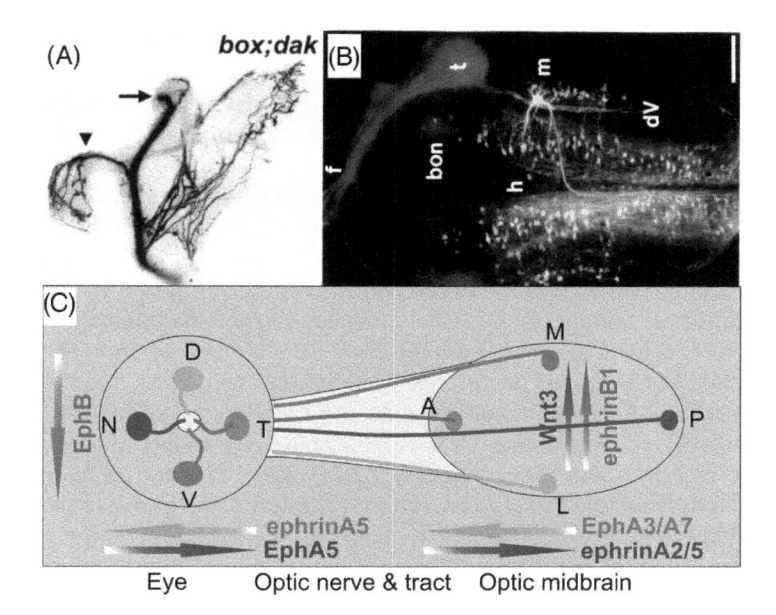

FIGURE 3.7   The retinotectal system maps a 2D surface (the retina ganglion cells) onto another 2D surface (the midbrain roof or tectum opticum) via highly ordered optic nerve/tract fiber pathways (C). Within the midbrain, the presorted fibers are further guided by molecular gradients matching retinal gradients of ligand/receptor distributions, highlighted in certain expression distribution. Certain double mutants [*dak/box*; (A)] show disorganized projections. Retinal projection can expand to reach the hindbrain directly after high levels of RA treatment (B). (Modified from Fritzsch et al., 2019; Lee et al., 2004; Manns and Fritzsch, 1991.)

3.5.4   SUPERIOR COLLICULUS (SC) OR TECTUM OPTICUM

The major retinal projection to the superior colliculus (SC) comes from the contralateral eye (Huberman et al., 2008; Triplett, 2014). Graded expression of several molecules and receptors redundantly defines how the surface of the retina is mapped via targeted projection of RGCs onto the midbrain (Figure 3.7). *Ephrin-A/EphA* has naso-temporal and *ephrin-B/EphB* dorso-ventral concentration-dependent attractive and repellent effects. The orthogonal gradients define a narrow region in which terminal arbors of a given RGC can form. Eliminating multiple ligand/receptor pairs causes expanded distribution of RGC axons (Kuwajima et al., 2017; Sitko et al., 2018; Triplett, 2014; Wu et al., 2018). Some very crude topology nevertheless remains even after the main ligand/receptor pairings have been deleted. Multigene knockouts combined with removal of activity result in diffuse and broad innervation. Additional molecular gradients are provided by a *Wnt3* gradient that defines, redundant to the *ephrin-B/EphB* gradient, the mediolateral slope in the midbrain for dorso-ventral RGN axonal sorting (Schmitt et al., 2006). The midbrain *Wnt3* gradient is translated into differential projections using *Ryk* gradients on RGN axons to modify, via repellent actions, the attraction mediated by *Fzd* receptor activity. Additional redundancy is provided by other secreted factors such as *En-2* (Zhang et al., 2017). Activity of axons is not needed to define the overall projection, but axonal arbors in the midbrain become less confined without activity. If both molecular map and activity are disrupted in combined mutants, the resulting maps of individual RGNs can cover large areas of the midbrain (Figure 3.7). This result demonstrates that neuronal activity combines with molecular specificity to sharpen the retinotopic map and provides the basis for ocular stripe formation in three-eyed frogs (Constantine-Paton and Law, 1978). Additional finding of optic tract guidance showed the role of specific gene expression (dackel, *dak*; boxer, *box*) for accurate fiber navigation to the SC (Lee et al., 2004). The genes encode glycosyltransferases, suggesting the role of heparan sulfates. *dak/box* double null mutants show additional defects that resemble the effects of *Robo2* knockout. Interestingly enough, contralateral projections are dominant in vertebrates (Vigouroux et al., 2021) except for nearly bilateral projection in amphibians that have a reduced and limited retina projections (Fritzsch et al., 1985).

In summary, the projections of retinal ganglion cells are dispersed across multiple thalamic target areas. Among targets are the suprachiasmatic nucleus (SCN), the basal optic tract (BOT), the lateral geniculate body (LGB), and the topographical projections in the midbrain (tectum opticum or superior colliculus).

3.6 PROJECTIONS TO THE FOREBRAIN SHOWS SIMILARITIES ACROSS MOST VERTEBRATES

Early suggestions that the cortically projecting LGN fibers are a novel feature unique to mammals (Striedter and Northcutt, 2019) have been revised significantly through discovery of homologous pathways in birds, reptiles, fish, and now more recent findings in lampreys (Suryanarayana et al., 2020; Suryanarayana et al., 2017). Basically, the revised insights into lampreys changes our perspective by demonstrating the presence of visual retinotopic information and expand previous suggestions concerning the evolution of vertebrate brains (Briscoe and Ragsdale, 2018; Karten, 2013; Tosches and Laurent, 2019). In addition, the work of Suryanarayana et al. (2020) demonstrates pallial somatosensory and motor areas, essentially showing a high degree of similarity across vertebrates. Demonstration of homologous visual pathways in birds, reptiles, teleosts, and sharks means that retina output in cyclostomes reaches sensory cortex. This pathway is lacking in hagfish, likely associated with secondary reduction of the visual system (Wicht and Northcutt, 1998).

While these broad similarities are now established across vertebrates, this does not mean that their visual pathways show similarities in detail. The six-layered cytoarchitecture of mammalian neocortex is not present among non-mammals, which have three layers or fewer (Tosches and Laurent, 2019). An equivalent of the LGN projection is known among all mammals but there are still gaps in our detailed understanding of non-mammalian retinal inputs to LGN-equivalent structures

(Suryanarayana et al., 2020). Further work generates highly specific neurons that respond to complex conjunctions of features and the dynamic formation of coherent assemblies of cells, whereby each cell codes only for particular aspects of the perceptual object (Singer, 2020). Evidence suggests that the presence of the respective feature constellation is signaled by an increase in discharge rate while the relations with other neurons are established by precise synchronization of action potentials. The oscillatory patterning of neuronal responses that occurs in different frequency bands and in most structures in the brain is proposed to serve this adjustment and to provide the basis for the generation and read-out of temporal codes (Singer, 2020).

3.7 SUMMARY AND CONCLUSION

Opsins are the start of all thinking about vision and show surprisingly similarities across chordates. Central projection in chordates has a single neurosensory origin (lancelet, ascidians) that is diversified through a unique split of a sensory ciliated cell connecting via ribbons of rods/cones to the *Atoh7*-regulated cell families that send the information to the brain (vertebrates). Outputs of melanopsins can be identified in the earliest chordates and is now identified with RGC types that drive the suprachiasmatic and circadian rhythms in structures receiving bilateral input. It is possible to suggest that the three different branches of vertebrate eyes (retina, pineal, parapineal) could have originated in the three retinal outputs of ascidians. The total of ~30 RGC types is different in their projections, showing a diversity of inputs, and shows various susceptibility to nerve crush. Topology is well understood among retina projections under control of distinct genes that are now characterized with respect to melanopsins and other outputs. Improved molecular understanding is also helping to clarify how diverse RGC populations accurately innervate multiple brain targets. Finally, we show that the retina reaches out to generate a topographic map in the superior colliculi and extend, via the LGN, to the cortex, showing that all vertebrates have a generally similar forebrain input.

REFERENCES

Abrahamson, E.E., and Moore, R.Y. (2001). Suprachiasmatic nucleus in the mouse: retinal innervation, intrinsic organization and efferent projections. Brain Research *916*, 172–191.

Albalat, R., Martí-Solans, J., and Cañestro, C. (2012). DNA methylation in amphioxus: from ancestral functions to new roles in vertebrates. Briefings in Functional Genomics *11*, 142–155.

Anadon, R., Melendez-Ferro, M., Perez-Costas, E., Pombal, M.A., and Rodicio, M.C. (1998). Centrifugal fibers are the only GABAergic structures of the retina of the larval sea lamprey: an immunocytochemical study. Brain Research *782*, 297–302.

Arendt, D., Musser, J.M., Baker, C.V., Bergman, A., Cepko, C., Erwin, D.H., Pavlicev, M., Schlosser, G., Widder, S., and Laubichler, M.D. (2016). The origin and evolution of cell types. Nature Reviews Genetics *17*, 744–757.

Baden, T., Euler, T. and Berens, P. (2020). Understanding the retinal basis of vision across species. Nature Reviews Neuroscience 21(1), 5–20.

Barnes, S.N. (1971). Fine structure of the photoreceptor and cerebral ganglion of the tadpole larva of Amaroucium constellatum (Verrill) (Subphylum: Urochordata; Class: Ascidiacea). Zeitschrift für Zellforschung und mikroskopische Anatomie *117*, 1–16.

Barreiro-Iglesias, A., Fernández-López, B., Sobrido-Cameán, D., and Anadón, R. (2017). Organization of alpha-transducin immunoreactive system in the brain and retina of larval and young adult Sea Lamprey (*Petromyzon marinus*), and their relationship with other neural systems. Journal of Comparative Neurology *525*, 3683–3704.

Bassham, S., and Postlethwait, J.H. (2005). The evolutionary history of placodes: a molecular genetic investigation of the larvacean urochordate *Oikopleura dioica*. Development *132*, 4259–4272.

Braun, K., and Stach, T. (2017). Structure and ultrastructure of eyes and brains of Thalia democratica (Thaliacea, Tunicata, Chordata). Journal of Morphology *278*, 1421–1437.

Braun, K., and Stach, T. (2019). Morphology and evolution of the central nervous system in adult tunicates. Journal of Zoological Systematics and Evolutionary Research *57*, 323–344.

Braun, K., Leubner, F., and Stach, T. (2020). Phylogenetic analysis of phenotypic characters of Tunicata supports basal Appendicularia and monophyletic Ascidiacea. Cladistics *36*, 259–300.

Brecha, N., Karten, H., and Hunt, S. (1980). Projections of the nucleus of the basal optic root in the pigeon: an autoradiographic and horseradish peroxidase study. Journal of Comparative Neurology *189*, 615–670.

Briscoe, S.D., and Ragsdale, C.W. (2018). Homology, neocortex, and the evolution of developmental mechanisms. Science *362*, 190–193.

Brzezinski, J.A., and Reh, T.A. (2015). Photoreceptor cell fate specification in vertebrates. Development *142*, 3263–3273.

Buono, L., and Martinez-Morales, J.R. (2020). Retina development in vertebrates: systems biology approaches to understanding genetic programs: on the contribution of next-generation sequencing methods to the characterization of the regulatory networks controlling vertebrate eye development. BioEssays *42*, 1900187.

Campo-Paysaa, F., Marlétaz, F., Laudet, V., and Schubert, M. (2008). Retinoic acid signaling in development: tissue-specific functions and evolutionary origins. Genesis *46*, 640–656.

Cao, C., Lemaire, L.A., Wang, W., Yoon, P.H., Choi, Y.A., Parsons, L.R., Matese, J.C., Levine, M., and Chen, K. (2019). Comprehensive single-cell transcriptome lineages of a proto-vertebrate. Nature *571*, 349–354.

Cavodeassi, F., Carreira-Barbosa, F., Young, R.M., Concha, M.L., Allende, M.L., Houart, C., Tada, M., and Wilson, S.W. (2005). Early stages of zebrafish eye formation require the coordinated activity of Wnt11, Fz5, and the Wnt/β-catenin pathway. Neuron *47*, 43–56.

Cepko, C. (2014). Intrinsically different retinal progenitor cells produce specific types of progeny. Nature Reviews Neuroscience *15*, 615–627.

Cepko, C.L., Austin, C.P., Yang, X., Alexiades, M., and Ezzeddine, D. (1996). Cell fate determination in the vertebrate retina. Proceedings of the National Academy of Sciences *93*, 589–595.

Chan, C.S., Lonfat, N., Zhao, R., Davis, A.E., Li, L., Wu, M.-R., Lin, C.-H., Ji, Z., Cepko, C.L., and Wang, S. (2020). Cell type-and stage-specific expression of Otx2 is regulated by multiple transcription factors and cis-regulatory modules in the retina. Development *147*.

Cipolla-Neto, J., and Amaral, F.G.d. (2018). Melatonin as a hormone: new physiological and clinical insights. Endocrine Reviews *39*, 990–1028.

Clause, A., Kim, G., Sonntag, M., Weisz, C.J., Vetter, D.E., Rűbsamen, R., and Kandler, K. (2014). The precise temporal pattern of prehearing spontaneous activity is necessary for tonotopic map refinement. Neuron *82*, 822–835.

Collin, S.P., and Fritzsch, B. (1993). Observations on the shape of the lens in the eye of the silver lamprey, Ichthyomyzon unicuspis. Canadian Journal of Zoology *71*, 34–41.

Collin, S.P., Davies, W.L., Hart, N.S., and Hunt, D.M. (2009). The evolution of early vertebrate photoreceptors. Philosophical Transactions of the Royal Society B: Biological Sciences *364*, 2925-2940.

Constantine-Paton, M., and Law, M.I. (1978). Eye-specific termination bands in tecta of three-eyed frogs. Science *202*, 639–641.

Copp, A.J., and Greene, N.D. (2010). Genetics and development of neural tube defects. The Journal of Pathology: A Journal of the Pathological Society of Great Britain and Ireland *220*, 217–230.

Copp, A.J., Greene, N.D., and Murdoch, J.N. (2003). The genetic basis of mammalian neurulation. Nature Reviews Genetics *4*, 784–793.

Crandall, J.E., Goodman, T., McCarthy, D.M., Duester, G., Bhide, P.G., Dräger, U.C., and McCaffery, P. (2011). Retinoic acid influences neuronal migration from the ganglionic eminence to the cerebral cortex. Journal of Neurochemistry *119*, 723–735.

de Crapona, M.C., Münz, H., Fritzsch, B., and Claas, B. (1986). Structure and development of the LHRH-immunoreactive nucleus olfacto-retinalis in the cichlid fish brain. In Ontogeny of Olfaction, W. Breipol ed. (Heidelberg: Springer), pp. 211–223.

De Miguel, E., Rodicio, M.C., and Anadón, R. (1989). Ganglion cells and retinopetal fibers of the larval lamprey retina: an HRP ultrastructural study. Neuroscience Letters *106*, 1–6.

de Vlaming, V., and Olcese, J. (2020). The pineal and reproduction in fish, amphibians, and reptiles. In The Pineal Gland, R.J. Reiter ed. (Boca Raton: CRC Press), pp. 1–29.

del Pilar Gomez, M., Angueyra, J.M., and Nasi, E. (2009). Light-transduction in melanopsin-expressing photoreceptors of Amphioxus. Proceedings of the National Academy of Sciences *106*, 9081–9086.

Delwig, A., Larsen, D.D., Yasumura, D., Yang, C.F., Shah, N.M., and Copenhagen, D.R. (2016). Retinofugal projections from melanopsin-expressing retinal ganglion cells revealed by intraocular injections of Cre-dependent virus. PLoS One *11*, e0149501.

Dennis, D.J., Han, S., and Schuurmans, C. (2019). bHLH transcription factors in neural development, disease, and reprogramming. Brain Research *1705*, 48–65.

Dhande, O.S., and Huberman, A.D. (2014). Retinal ganglion cell maps in the brain: implications for visual processing. Current Opinion in Neurobiology *24*, 133–142.

Do, M.T.H. (2019). Melanopsin and the intrinsically photosensitive retinal ganglion cells: biophysics to behavior. Neuron *104*, 205–226.

Dong, E.M. and Allison, W.T. (2021). Vertebrate features revealed in the rudimentary eye of the Pacific hagfish (*Eptatretus stoutii*). Proceedings of the Royal Society B, 288(1942), 20202187.

Eakin, R.M. (1973). The Third Eye (Berkeley: University of California Press).

Eakin, R.M., and Kuda, A. (1970). Ultrastructure of sensory receptors in ascidian tadpoles. Zeitschrift für Zellforschung und mikroskopische Anatomie *112*, 287–312.

Eggenschwiler, J.T., Espinoza, E., and Anderson, K.V. (2001). Rab23 is an essential negative regulator of the mouse Sonic hedgehog signalling pathway. Nature *412*, 194–198.

Fain, G.L. (2019). Lamprey vision: photoreceptors and organization of the retina. Seminars in Cell and Developmental Biology 106, 5–11.

Falcon, J., Migaud, H., Munoz-Cueto, J.A., and Carrillo, M. (2010). Current knowledge on the melatonin system in teleost fish. General and Comparative Endocrinology *165*, 469–482.

Feuda, R., Hamilton, S.C., McInerney, J.O., and Pisani, D. (2012). Metazoan opsin evolution reveals a simple route to animal vision. Proceedings of the National Academy of Sciences of the United States of America *109*, 18868–18872.

Fritzsch, B. (1980). Retinal projections in European salamandridae. Cell and Tissue Research *213*, 325–341.

Fritzsch, B. (1991). Ontogenetic clues to the phylogeny of the visual system. In The Changing Visual System, P. Bagnoli, W. Hodos eds. (Boston: Springer), pp. 33–49.

Fritzsch, B., and Collin, S.P. (1990). Dendritic distribution of two populations of ganglion cells and the retinopetal fibers in the retina of the silver lamprey (*Ichthyomyzon unicuspis*). Visual Neuroscience *4*, 533–545.

Fritzsch, B., and Elliott, K.L. (2017). Gene, cell, and organ multiplication drives inner ear evolution. Developmental Biology *431*, 3–15.

Fritzsch, B., and Glover, J.C. (2007). 2.01-Evolution of the Deuterostome Central Nervous System: An Intercalation of Developmental Patterning Processes with Cellular Specification Processes. In Evolution of Nervous Systems. In J.H. Kaas, ed, New York, pp. 1–24.

Fritzsch, B., Elliott, K.L., and Pavlinkova, G. (2019). Primary sensory map formations reflect unique needs and molecular cues specific to each sensory system. F1000Research *8*.

Fritzsch, B., Himstedt, W., and de Caprona, M.C. (1985). Visual projections in larval Ichthyophis kohtaoensis (Amphibia: Gymnophiona). Developmental Brain Research *23*, 201–210.

Fritzsch, B., Sonntag, R., Dubuc, R., Ohta, Y., and Grillner, S. (1990). Organization of the six motor nuclei innervating the ocular muscles in lamprey. Journal of Comparative Neurology *294*, 491–506.

Fuhrmann, S. (2010). Eye morphogenesis and patterning of the optic vesicle. In Current Topics in Developmental Biology, R.L. Cagan, T. Reh, eds. (Elsevier), pp. 61–84.

Fujiwara, S., and Cañestro, C. (2018). Reporter Analyses Reveal Redundant Enhancers that Confer Robustness on Cis-Regulatory Mechanisms. In Transgenic Ascidians, Y. Sasakura, ed. (Singapore: Springer), pp. 69–79.

Gierer, A. (1987). Directional cues for growing axons forming the retinotectal projection. Development *101*, 479–489.

Giolli, R.A., Blanks, R.H., and Lui, F. (2006). The accessory optic system: basic organization with an update on connectivity, neurochemistry, and function. Progress in Brain Research *151*, 407–440.

Glover, J.C., Renaud, J.S., and Rijli, F.M. (2006). Retinoic acid and hindbrain patterning. Journal of Neurobiology *66*, 705–725.

Guido, W. (2018). Development, form, and function of the mouse visual thalamus. Journal of Neurophysiology *120*, 211–225.

Gunther, T., Struwe, M., Aguzzi, A., and Schughart, K. (1994). Open brain, a new mouse mutant with severe neural tube defects, shows altered gene expression patterns in the developing spinal cord. Development *120*, 3119–3130.

Guo, A., Wang, T., Ng, E.L., Aulia, S., Chong, K.H., Teng, F.Y.H., Wang, Y., and Tang, B.L. (2006). Open brain gene product Rab23: expression pattern in the adult mouse brain and functional characterization. Journal of Neuroscience Research *83*, 1118–1127.

Gustafsson, O., Collin, S., and Kröger, R. (2008). Early evolution of multifocal optics for well-focused colour vision in vertebrates. Journal of Experimental Biology *211*, 1559–1564.

Hastings, M.H., Maywood, E.S., and Brancaccio, M. (2018). Generation of circadian rhythms in the suprachiasmatic nucleus. Nature Reviews Neuroscience *19*, 453–469.

Hastings, M.H., Reddy, A.B., and Maywood, E.S. (2003). A clockwork web: circadian timing in brain and periphery, in health and disease. Nature Reviews Neuroscience *4*, 649–661.

Hattar, S., Liao, H.-W., Takao, M., Berson, D.M., and Yau, K.-W. (2002). Melanopsin-containing retinal ganglion cells: architecture, projections, and intrinsic photosensitivity. Science *295*, 1065–1070.

Higuchi, S., Sugahara, F., Pascual-Anaya, J., Takagi, W., Oisi, Y., and Kuratani, S. (2019). Inner ear development in cyclostomes and evolution of the vertebrate semicircular canals. Nature *565*, 347–350.

Hinman, V.F., and Degnan, B.M. (2000). Retinoic acid perturbs Otx gene expression in the ascidian pharynx. Development Genes and Evolution *210*, 129–139.

Holland, L. (2020). Invertebrate origins of vertebrate nervous systems. In Evolutionary Neuroscience, J.H. Kaas, ed. (New York: Elsevier), pp. 51–73.

Holland, L.Z., and Daza, D.O. (2018). A new look at an old question: when did the second whole genome duplication occur in vertebrate evolution? Genome Biology *19*, 1–4.

Holmberg, K. (1971). The hagfish retina: electron microscopic study comparing receptor and epithelial cells in the *Pacific hagfish, Polistotrema stouti*, with those in the *Atlantic hagfish, Myxine glutinosa*. Zeitschrift für Zellforschung und Mikroskopische Anatomie *121*, 249–269.

Holmberg, K. (1977). The cyclostome retina. In The Visual System in Vertebrates, F. Crescitelli, ed. (Berlin: Springer), pp. 47–66.

Hubel, D.H., and Wiesel, T.N. (1962). Receptive fields, binocular interaction and functional architecture in the cat's visual cortex. The Journal of Physiology *160*, 106.

Huberman, A.D., Feller, M.B., and Chapman, B. (2008). Mechanisms underlying development of visual maps and receptive fields. Annual Review of Neuroscience *31*, 479–509.

Imayoshi, I., and Kageyama, R. (2014). bHLH Factors in self-renewal, multipotency, and fate choice of neural progenitor cells. Neuron *82*, 9–23.

Jones, M.R., Grillner, S., and Robertson, B. (2009). Selective projection patterns from subtypes of retinal ganglion cells to tectum and pretectum: distribution and relation to behavior. Journal of Comparative Neurology *517*, 257–275.

Kalish, B.T., Cheadle, L., Hrvatin, S., Nagy, M.A., Rivera, S., Crow, M., Gillis, J., Kirchner, R., and Greenberg, M.E. (2018). Single-cell transcriptomics of the developing lateral geniculate nucleus reveals insights into circuit assembly and refinement. Proceedings of the National Academy of Sciences *115*, E1051–E1060.

Karten, H.J. (2013). Neocortical evolution: neuronal circuits arise independently of lamination. Current Biology *23*, R12–R15.

Kawano-Yamashita, E., Koyanagi, M., Shichida, Y., Oishi, T., Tamotsu, S., and Terakita, A. (2011). Beta-arrestin functionally regulates the non-bleaching pigment parapinopsin in lamprey pineal. PloS One *6*, e16402.

Konno, A., Kaizu, M., Hotta, K., Horie, T., Sasakura, Y., Ikeo, K., and Inaba, K. (2010). Distribution and structural diversity of cilia in tadpole larvae of the ascidian Ciona intestinalis. Developmental Biology *337*, 42–62.

Koyanagi, M., Kawano, E., Kinugawa, Y., Oishi, T., Shichida, Y., Tamotsu, S., and Terakita, A. (2004). Bistable UV pigment in the lamprey pineal. Proceedings of the National Academy of Sciences *101*, 6687–6691.

Kozmik, Z., Daube, M., Frei, E., Norman, B., Kos, L., Dishaw, L.J., Noll, M., and Piatigorsky, J. (2003). Role of Pax genes in eye evolution: a cnidarian PaxB gene uniting Pax2 and Pax6 functions. Developmental Cell *5*, 773–785.

Kuwajima, T., Soares, C.A., Sitko, A.A., Lefebvre, V., and Mason, C. (2017). SoxC transcription factors promote contralateral retinal ganglion cell differentiation and axon guidance in the mouse visual system. Neuron *93*, 1110–1125. e1115.

Lacalli, T., and Holland, L. (1998). The developing dorsal ganglion of the salp *Thalia* democratica, and the nature of the ancestral chordate brain. Philosophical Transactions of the Royal Society of London Series B: Biological Sciences *353*, 1943–1967.

Lacalli, T.C., Holland, N., and West, J. (1994). Landmarks in the anterior central nervous system of amphioxus larvae. Philosophical Transactions of the Royal Society of London Series B: Biological Sciences *344*, 165–185.

Lamb, T.D. (2013). Evolution of phototransduction, vertebrate photoreceptors and retina. Progress in Retinal and Eye Research *36*, 52–119.

Lamb, T.D., Collin, S.P., and Pugh, E.N. (2007). Evolution of the vertebrate eye: opsins, photoreceptors, retina and eye cup. Nature Reviews Neuroscience *8*, 960–976.

Lamba, D.A., and Reh, T.A. (2020). Disease and repair approaches in vision. In The Senses, 2nd Edition, B. Fritzsch, P.R. Martin, eds. (Oxford: Elsevier), pp. 54–65.

Land, M.F., and Nilsson, D.-E. (2012). Animal eyes (Oxford: Oxford University Press).

Lean, G.A., Liu, Y.J., and Lyon, D.C. (2019). Cell type specific tracing of the subcortical input to primary visual cortex from the basal forebrain. Journal of Comparative Neurology *527*, 589–599.

Lee, J.-S., Von Der Hardt, S., Rusch, M.A., Stringer, S.E., Stickney, H.L., Talbot, W.S., Geisler, R., Nüsslein-Volhard, C., Selleck, S.B., and Chien, C.-B. (2004). Axon sorting in the optic tract requires HSPG synthesis by ext2 (dackel) and extl3 (boxer). Neuron *44*, 947–960.

Lilley, B.N., Sabbah, S., Hunyara, J.L., Gribble, K.D., Al-Khindi, T., Xiong, J., Wu, Z., Berson, D.M., and Kolodkin, A.L. (2019). Genetic access to neurons in the accessory optic system reveals a role for Sema6A in midbrain circuitry mediating motion perception. Journal of Comparative Neurology *527*, 282–296.

Liu, B.-h., Huberman, A.D., and Scanziani, M. (2016). Cortico-fugal output from visual cortex promotes plasticity of innate motor behaviour. Nature *538*, 383–387.

Liu, C., Weaver, D.R., Jin, X., Shearman, L.P., Pieschl, R.L., Gribkoff, V.K., and Reppert, S.M. (1997). Molecular dissection of two distinct actions of melatonin on the suprachiasmatic circadian clock. Neuron *19*, 91–102.

Locket, N.A., and Jørgensen, J.M. (1998). The eyes of hagfishes. In The Biology of Hagfishes, J.M. Jørgensen et al., eds (Berlin: Springer), pp. 541–556.

Lucas, R.J., Allen, A.E., Milosavljevic, N., Storchi, R., and Woelders, T. (2020). Can We See with Melanopsin? Annual Review of Vision Science *6*.

Luo, T., Sakai, Y., Wagner, E., and Dräger, U.C. (2006). Retinoids, eye development, and maturation of visual function. Journal of Neurobiology *66*, 677–686.

Manns, M., and Fritzsch, B. (1991). The eye in the brain: retinoic acid effects morphogenesis of the eye and pathway selection of axons but not the differentiation of the retina in Xenopus laevis. Neuroscience Letters *127*, 150–154.

Mao, C.-A., Cho, J.-H., Wang, J., Gao, Z., Pan, P., Tsai, W.-W., Frishman, L.J., and Klein, W.H. (2013). Reprogramming amacrine and photoreceptor progenitors into retinal ganglion cells by replacing Neurod1 with Atoh7. Development *140*, 541–551.

Mao, C.-A., Wang, S.W., Pan, P., and Klein, W.H. (2008). Rewiring the retinal ganglion cell gene regulatory network: Neurod1 promotes retinal ganglion cell fate in the absence of Math5. Development *135*, 3379–3388.

McLaughlin, D., Shapley, R., Shelley, M., and Wielaard, D.J. (2000). A neuronal network model of macaque primary visual cortex (V1): Orientation selectivity and dynamics in the input layer 4Cα. Proceedings of the National Academy of Sciences *97*, 8087–8092.

Menna, E., Cenni, M.C., Naska, S., and Maffei, L. (2003). The anterogradely transported BDNF promotes retinal axon remodeling during eye specific segregation within the LGN. Molecular and Cellular Neuroscience *24*, 972–983.

Morshedian, A., and Fain, G.L. (2017). Light adaptation and the evolution of vertebrate photoreceptors. The Journal of Physiology *595*, 4947–4960.

Münz, H., Claas, B., Stumpf, W., and Jennes, L. (1982). Centrifugal innervation of the retina by luteinizing hormone releasing hormone (LHRH)-immunoreactive telencephalic neurons in teleostean fishes. Cell and Tissue Research *222*, 313–323.

Neitz, M., Patterson, S.S., and Neitz, J. (2020). The genetics of cone opsin based vision disorders. In The Senses, B. Fritzsch, ed. (Elsevier), pp. 493–507.

Nilsson, D.-E. (2020). Eye evolution in animals. In The Senses, 2nd Edition, B. Fritzsch, P.R. Martin, eds. (Oxford: Elsevier), pp. 96–121.

Ota, K.G., and Kuratani, S. (2006). The history of scientific endeavors towards understanding hagfish embryology. Zoological Science *23*, 403–418.

Pantzartzi, C.N., Pergner, J., Kozmikova, I., and Kozmik, Z. (2017). The opsin repertoire of the European lancelet: a window into light detection in a basal chordate. International Journal of Developmental Biology *61*, 763–772.

Pergner, J., and Kozmik, Z. (2017). Amphioxus photoreceptors-insights into the evolution of vertebrate opsins, vision and circadian rhythmicity. International Journal of Developmental Biology *61*, 665–681.

Pisani, D., Rota-Stabelli, O., and Feuda, R. (2020). Sensory Neuroscience: A Taste for Light and the Origin of Animal Vision. Current Biology: CB *30*, R773–r775.

Plaisancié, J., Ceroni, F., Holt, R., Seco, C.Z., Calvas, P., Chassaing, N., and Ragge, N.K. (2019). Genetics of anophthalmia and microphthalmia. Part 1: Non-syndromic anophthalmia/microphthalmia. Human Genetics, 1–32.

Poliakov, E., Gubin, A.N., Stearn, O., Li, Y., Campos, M.M., Gentleman, S., Rogozin, I.B., and Redmond, T.M. (2012). Origin and evolution of retinoid isomerization machinery in vertebrate visual cycle: hint from jawless vertebrates. PLoS One *7*, e49975.

Pombal, M.A., and Megías, M. (2019). Development and functional organization of the cranial nerves in lampreys. The Anatomical Record *302*, 512–539.

Provencio, I., Jiang, G., Willem, J., Hayes, W.P., and Rollag, M.D. (1998). Melanopsin: An opsin in melanophores, brain, and eye. Proceedings of the National Academy of Sciences *95*, 340–345.

Pu, G.A., and Dowling, J.E. (1981). Anatomical and physiological characteristics of pineal photoreceptor cell in the larval lamprey, Petromyzon marinus. Journal of Neurophysiology *46*, 1018–1038.

Puzdrowski, R.L., and Northcutt, R.G. (1989). Central projections of the pineal complex in the silver lamprey Ichthyomyzon unicuspis. Cell and Tissue Research *255*, 269–274.

Qiu, X., Kumbalasiri, T., Carlson, S.M., Wong, K.Y., Krishna, V., Provencio, I., and Berson, D.M. (2005). Induction of photosensitivity by heterologous expression of melanopsin. Nature *433*, 745–749.

Reiter, R.J. (1980). The pineal and its hormones in the control of reproduction in mammals. Endocrine Reviews *1*, 109–131.

Richardson, M.K., Admiraal, J., and Wright, G.M. (2010). Developmental anatomy of lampreys. Biological Reviews *85*, 1–33.

Richardson, R., Tracey-White, D., Webster, A., and Moosajee, M. (2017). The zebrafish eye—a paradigm for investigating human ocular genetics. Eye *31*, 68–86.

Roberts, A. (1978). Pineal eye and behaviour in Xenopus tadpoles. Nature *273*, 774–775.

Roberts, J.M., and Vetter, M.L. (2018). From Retina to Stem Cell and Back Again: Memories of a Chromatin Journey. Cell Reports *22*, 2519–2520.

Rolo, A., Galea, G.L., Savery, D., Greene, N.D., and Copp, A.J. (2019). Novel mouse model of encephalocele: post-neurulation origin and relationship to open neural tube defects. Disease Models & Mechanisms *12*, 1–10.

Ryan, K., Lu, Z., and Meinertzhagen, I.A. (2018). The peripheral nervous system of the ascidian tadpole larva: Types of neurons and their synaptic networks. Journal of Comparative Neurology *526*, 583–608.

Sanchez, A.N., Alitto, H.J., and Usrey, W.M. (2020). Eye to brain; parallel visual pathways. In The Senses, 2nd Edition, B. Fritzsch, P.R. Martin, eds. (Oxford: Elsevier), pp. 362–368.

Sanes, D.H., Reh, T.A., and Harris, W.A. (2011). Development of the Nervous System (Sanes: Academic Press).

Sanes, J.R., and Masland, R.H. (2015). The types of retinal ganglion cells: current status and implications for neuronal classification. Annual Review of Neuroscience *38*, 221–246.

Sasagawa, S., Takabatake, T., Takabatake, Y., Muramatsu, T., and Takeshima, K. (2002). Axes establishment during eye morphogenesis in Xenopus by coordinate and antagonistic actions of BMP4, Shh, and RA. Genesis *33*, 86–96.

Sauka-Spengler, T., and Bronner-Fraser, M. (2008). Insights from a sea lamprey into the evolution of neural crest gene regulatory network. The Biological Bulletin *214*, 303–314.

Schmitt, A.M., Shi, J., Wolf, A.M., Lu, C.-C., King, L.A., and Zou, Y. (2006). Wnt–Ryk signalling mediates medial–lateral retinotectal topographic mapping. Nature *439*, 31.

Sherman, S.M. (2020). The lateral geniculate nucleus and pulvinar. In The Senses, 2nd Edition, B. Fritzsch, P.R. Martin, eds. (Oxford: Elsevier), pp. 369–391.

Shinotsuka, N., Yamaguchi, Y., Nakazato, K., Matsumoto, Y., Mochizuki, A., and Miura, M. (2018). Caspases and matrix metalloproteases facilitate collective behavior of non-neural ectoderm after hindbrain neuropore closure. BMC Developmental Biology *18*, 1–11.

Simonneaux, V., and Ribelayga, C. (2003). Generation of the melatonin endocrine message in mammals: a review of the complex regulation of melatonin synthesis by norepinephrine, peptides, and other pineal transmitters. Pharmacological Reviews *55*, 325–395.

Singer, W. (2020). Temporal coherence: a versatile code for the definition of relations. In The Senses, 2nd Edition, B. Fritzsch, P.R. Martin, eds. (Oxford: Elsevier), pp. 480–486.

Sitko, A.A., Kuwajima, T., and Mason, C.A. (2018). Eye-specific segregation and differential fasciculation of developing retinal ganglion cell axons in the mouse visual pathway. Journal of Comparative Neurology *526*, 1077–1096.

Smith, J.J., Kuraku, S., Holt, C., Sauka-Spengler, T., Jiang, N., Campbell, M.S., Yandell, M.D., Manousaki, T., Meyer, A., and Bloom, O.E. (2013). Sequencing of the sea lamprey (Petromyzon marinus) genome provides insights into vertebrate evolution. Nature Genetics *45*, 415–421.

Smith, K.T., Bhullar, B.-A.S., Köhler, G., and Habersetzer, J. (2018). The only known jawed vertebrate with four eyes and the Bauplan of the pineal complex. Current Biology *28*, 1101–1107.e1102.

Song, Y.S. and Kim, J.K. (2020). Molecular phylogeny and classification of the family Myxinidae (Cyclostomata: Myxiniformes) using the supermatrix method. Journal of Asia-Pacific Biodiversity, 13(4), 533–538.

Sperry, R. (1971). How a developing brain gets itself properly wired for adaptive function. In The Biopsychology of Development (New York: Academic Press), pp. 27–44.

Sridhar, A., Hoshino, A., Finkbeiner, C.R., Chitsazan, A., Dai, L., Haugan, A.K., Eschenbacher, K.M., Jackson, D.L., Trapnell, C., and Bermingham-McDonogh, O. (2020). Single-cell transcriptomic comparison of human fetal retina, hPSC-derived retinal organoids, and long-term retinal cultures. Cell Reports *30*, 1644–1659. e1644.

Stellwagen, D., and Shatz, C. (2002). An instructive role for retinal waves in the development of retinogeniculate connectivity. Neuron *33*, 357–367.

Striedter, G.F., and Northcutt, R.G. (2019). Brains through Time: A Natural History of Vertebrates (Oxford: Oxford University Press).

Sun, L., Kawano-Yamashita, E., Nagata, T., Tsukamoto, H., Furutani, Y., Koyanagi, M., and Terakita, A. (2014). Distribution of mammalian-like melanopsin in cyclostome retinas exhibiting a different extent of visual functions. PLoS One *9*, e108209.

Suryanarayana, S.M., Pérez-Fernández, J., Robertson, B., and Grillner, S. (2020). The evolutionary origin of visual and somatosensory representation in the vertebrate pallium. Nature Ecology & Evolution *4*, 639–651.

Suryanarayana, S.M., Robertson, B., Wallén, P., and Grillner, S. (2017). The lamprey pallium provides a blueprint of the mammalian layered cortex. Current Biology *27*, 3264–3277. e3265.

Suzuki, D.G., and Grillner, S. (2018). The stepwise development of the lamprey visual system and its evolutionary implications. Biological Reviews *93*, 1461–1477.

Takahashi, J.S. (2017). Transcriptional architecture of the mammalian circadian clock. Nature Reviews Genetics *18*, 164.

Tassinari, G., Bentivoglio, M., Chen, S., and Campara, D. (1997). Overlapping ipsilateral and contralateral retinal projections to the lateral geniculate nucleus and superior colliculus in the cat: a retrograde triple labelling study. Brain Research Bulletin *43*, 127–139.

Terakita, A. (2005). The opsins. Genome biology *6*, 213.

Tosches, M.A., and Laurent, G. (2019). Evolution of neuronal identity in the cerebral cortex. Current Opinion in Neurobiology *56*, 199–208.

Triplett, J.W. (2014). Molecular guidance of retinotopic map development in the midbrain. Current Opinion in Neurobiology *24*, 7–12.

Valencia, J.E., Feuda, R., Mellott, D.O., Burke, R.D., and Peter, I.S. (2019). Ciliary photoreceptors in sea urchin larvae indicate pan-deuterostome cell type conservation. Bioarxiv, doi.org/10.1101/683318.

Varadarajan, S.G., and Huberman, A.D. (2018). Assembly and repair of eye-to-brain connections. Current Opinion in Neurobiology *53*, 198–209.

Veeman, M., and Reeves, W. (2015). Quantitative and in toto imaging in ascidians: Working toward an image-centric systems biology of chordate morphogenesis. Genesis *53*, 143–159.

Veeman, M.T., Newman-Smith, E., El-Nachef, D., and Smith, W.C. (2010). The ascidian mouth opening is derived from the anterior neuropore: reassessing the mouth/neural tube relationship in chordate evolution. Developmental Biology *344*, 138–149.

Vernadakis, A.J., Bemis, W.E., and Bittman, E.L. (1998). Localization and partial characterization of melatonin receptors in amphioxus, hagfish, lamprey, and skate. General and Comparative Endocrinology *110*, 67–78.

Vigh-Teichmann, I., Korf, H.-W., Nürnberger, F., Oksche, A., Vigh, B., and Olsson, R. (1983). Opsin-immunoreactive outer segments in the pineal and parapineal organs of the lamprey (*Lampetra fluviatilis*), the eel (*Anguilla anguilla*), and the rainbow trout (*Salmo gairdneri*). Cell and Tissue Research *230*, 289–307.

Vigouroux, R.J., Duroure, K., Vougny, J., Albadri, S., Kozulin, P., Herrera, E., Nguyen-Ba-Charvet, K., Braasch, I., Suárez, R., and Del Bene, F. (2021). Bilateral visual projections exist in non-teleost bony fish and predate the emergence of tetrapods. Science *372*, 150–156.

Villar-Cheda, B., Abalo, X.M., Anadón, R., and Rodicio, M.C. (2006). Calbindin and calretinin immunoreactivity in the retina of adult and larval sea lamprey. Brain Research *1068*, 118–130.

Vopalensky, P., Pergner, J., Liegertova, M., Benito-Gutierrez, E., Arendt, D., and Kozmik, Z. (2012). Molecular analysis of the amphioxus frontal eye unravels the evolutionary origin of the retina and pigment cells of the vertebrate eye. Proceedings of the National Academy of Sciences *109*, 15383–15388.

Wagner, E., Luo, T., and Dräger, U.C. (2002). Retinoic acid synthesis in the postnatal mouse brain marks distinct developmental stages and functional systems. Cerebral Cortex *12*, 1244–1253.

Wang, K., Hinz, J., Zhang, Y., Thiele, T.R., and Arrenberg, A.B. (2020). Parallel channels for motion feature extraction in the pretectum and tectum of larval zebrafish. Cell Reports *30*, 442–453. e446.

Wässle, H. (2020). Decomposing a cone's output. In The Senses, Bioarxiv, doi.org/10.1101/683318 (Elsevier), pp. 212–233.

Wicht, H. (1996). The brains of lampreys and hagfishes: Characteristics, characters, and comparisons. Brain, Behavior and Evolution *48*, 248–261.

Wicht, H., and Northcutt, R.G. (1990). Retinofugal and retinopetal projections in the pacific hagfish, Eptatretus stouti (Myxinoidea)(Part 1 of 2). Brain, Behavior and Evolution *36*, 315–321.

Wicht, H., and Northcutt, R.G. (1998). Telencephalic connections in the Pacific hagfish (Eptatretus stouti), with special reference to the thalamopallial system. Journal of Comparative Neurology *395*, 245–260.

Wilson, S.W., and Houart, C. (2004). Early steps in the development of the forebrain. Developmental Cell *6*, 167–181.

Winkley, K.M., Kourakis, M.J., DeTomaso, A.W., Veeman, M.T., and Smith, W.C. (2020). Tunicate gastrulation. In Current Topics in Developmental Biology, L. Solnica-Krezel, ed. (New York: Elsevier), pp. 219–242.

Wu, S., Chang, K.-C., and Goldberg, J.L. (2018). Retinal cell fate specification. Trends in Neurosciences *41*, 165–167.

Zhang, C., Kolodkin, A.L., Wong, R.O., and James, R.E. (2017). Establishing wiring specificity in visual system circuits: from the retina to the brain. Annual Review of Neuroscience *40*, 395–424.

4 Trigeminal and Related Spinal Projections

How to Cross or Not the Multisensory Projections

Bernd Fritzsch

CONTENTS

4.1 INTRODUCTION TO THE TRIGEMINAL SYSTEM

The somatic sensory input is essential for vertebrates and provides touch sensors, including proprioception, pain, and thermoreception (Goodman and Bensmaia, 2020); it reaches trigeminal neurons, crosses (mostly) to the contralateral thalamus and continues to the somatosensory cortex (Erzurumlu, 2020; Kratochwil and Rijli, 2020; Zilles and Palomero-Gallagher, 2020). The trigeminal central sensory projection starts at rhombomere 2 (r2; Kratochwil et al., 2017) and blends with the spinal cord past r11 (Gray, 2013; Nieuwenhuys and Puelles, 2015). Its major sensory neurons are derived from *Neurog1* (Ma et al., 1998), which provide the three branches of the trigeminal areas (ophthalmic, maxillary, mandibular branches; (Erzurumlu et al., 2010; Noden and Trainor, 2005). A large set of different cutaneous mechanoreceptors (Merkel, Ruffini, Meissner, Pacinian) and distinct lanceolate endings are combined with pain, proprioception and thermoreception to provide the touch and related sensations (Abraira and Ginty, 2013; Abraira et al., 2017; Goodman and Bensmaia, 2020). In mice, we can distinguish among 'barrelettes' (brainstem), 'barreloids' (thalamus), and 'barrels' (cortex; Erzurumlu, 2021; Erzurumlu and Gaspar, 2020) that start from the whiskers and continue to

DOI: 10.1201/9781003092810-4

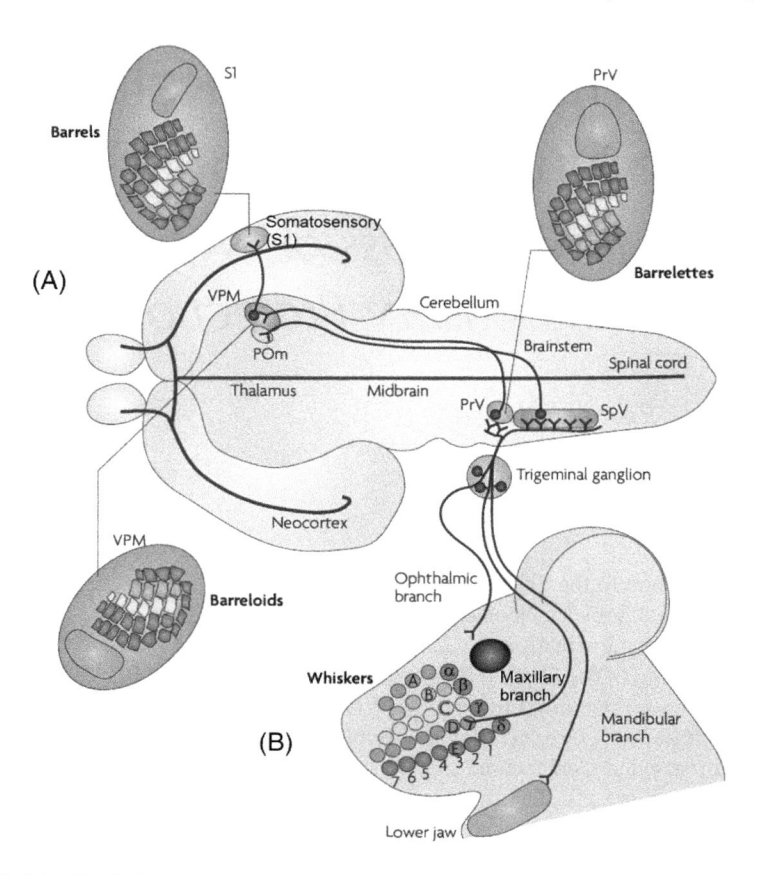

FIGURE 4.1 Depicting the mouse skin that gives rise to whiskers (B) and shows the central projections (A) with three different levels of the trigeminal presentation (barrels, barreloids, barrelettes). Note the different whiskers are five and four additional terminal branches. The trigeminal reaches from the mandibular branch to continue as the principal (PRV) to reach the thalamus. In contrast, the mandibular branch is extending via the spinal (SpV) to reach the cortical innervation. POm, posteromedial nucleus; VPM, ventral posteromedial nucleus. (Modified from Erzurumlu et al., 2010.)

the somatosensory cortex (Figure 4.1). In contrast, different mammals lack whiskers but have a profound human lip size and sensation whereas that of elephants has a unique proboscis that is extremely enlarged, likewise for the star-nosed mole (Leitch and Gaede, 2020). All peripheral and central projections are highly organized to present a somatotopic representation from the trigeminal neurons, brainstem, thalamus, and cortex (Erzurumlu et al., 2010; Zilles and Palomero-Gallagher, 2020). A unique placode adds to the neural crest derived trigeminal neurons and depends on *Eya1/Six1, Pax3, Sox2, Fgf8, Tbx1/3, Onecut1/2, Hmx1, BMP4, Megf8,* and *Isl1* for earliest genetic formation (Buzzi et al., 2019; Engelhard et al., 2013; Moody and LaMantia, 2015). The distribution of the trigeminal neurons derive, in part, from a placode (Fode et al., 1998; Ma et al., 1998) and, in part, from the neural crest (Ziermann et al., 2018). Additional minor contributions from the facial, glossopharyngeal, and vagal neurons are originated from placodes (Fritzsch et al., 1997a; O'Neill et al., 2012) and proximal neural crest neurons (Vermeiren et al., 2020; Ziermann et al., 2018). The geniculate, petrosal and nodose are derived from placodes and receive mixed (geniculate) or a separate (superior, jugular) neural crest input (Ziermann et al., 2018). Downstream of the earliest placode and neural crest, it depends on bHLH genes such as *Neurog1, Neurog2 and Neurod1* (Dennis et al., 2019; Fode et al., 1998; Kim et al., 2001; Ma et al., 1998) that are driving the genes for neuronal differentiation, including *Pou4f1* and other genes (Huang et al., 2001; Lassiter et al., 2014). In addition, the role of *BDNF* and *Ntf3* are essential for viability of neurons and require *Semaphorin*, *Slit, Robo*, and *Npr2* (López-Bendito et al., 2007; O'Connor and Tessier-Lavigne, 1999; Ter-Avetisyan et al., 2018) that is needed for proper projections (Figure 4.1). Several genes are upregulated in trigeminal projections, among them is CGRP, *Nos*, and others (Messlinger et al., 2020).

Central projections extend the bHLH gene *Ascl1*, which defines the trigeminal and spinal cord (Hernandez-Miranda et al., 2017; Iskusnykh et al., 2016; Lai et al., 2016). In addition, a set of different bHLH genes are expressed in *Ptf1a, Olig2, Neurog1*, and *Neurog2* that are upstream of several neuron defining genes (*Pou4f1, Lmx1a, Phox2b, Lbx1, Pax2*, among others). Downstream to the various degree of migration will be followed by different inhibitory and excitatory genes (Lai et al., 2016; Nothwang, 2016). A separate input to the trigeminal fibers has a novel addition from the mesencephalic trigeminal ganglion (MTN, MesV) that is a unique transformation of dorsal cells in lampreys and hagfish (Fritzsch and Northcutt, 1993; Lipovsek et al., 2017; Mastick et al., 1996). The MTN neurons branch, like dorsal cells in cyclostomes and Rohon-Beard cells in the spinal cord, have a central projection that bifurcates, regulated by *Nrp2* (Ter-Avetisyan et al., 2018) to branch off into a proximal and a distal projection.

Finally, we are defining three levels of the trigeminal organization:

- Primary neurosensory cells from placode/neural crest to reach trigeminal, facial, glossopharyngeal, and vagal projections.
- Second-order sensory neurons receive trigeminal descending fibers that reach the rhombencephalon.
- Third input projects from the somatosensory system via the trigeminal lemniscus to reach the thalamus and somatosensory cortex (Erzurumlu et al., 2010).

I will provide an overview of the molecular basis of the trigeminal origin and its diversity followed by the brainstem projections and describe in detail the branches to add to the rhombomeres and the trigeminal lemniscus to reach the VPM at the thalamus. Finally, I will provide an overview of the somatosensory cortex.

4.2 TRIGEMINAL SENSORY NEURONS DEPEND ON PLACODE AND NEURAL CREST FORMATION

The trigeminal ganglion starts as a preplacodal ectoderm (PPE) adjacent to the neural crest (NC; Buzzi et al., 2019; Moody and LaMantia, 2015) that is next to the neural plate (NP) forming the brainstem (Figure 4.2). Initially, either PPE or NC can differentiate and is in an unstable state that is progressively segregated in the trigeminal future genes (Thiery et al., 2020). The border between *Otx2* rostral and *Gbx2* caudal is overlapping in the domain for the NP (Glover et al., 2018; Moody and LaMantia, 2015) that will be reduced into a clear separation of somites in mice and chicken (Thiery et al., 2020). *Sox2* and *Zic* genes are initially upregulated in the NP which will be downregulated later (Aruga and Hatayama, 2018; Lee et al., 2014). At later NP stages, *Six1, Eya1,* and *Irx1* genes are expressed in the PPE followed by *Msx1, Pax3, Dlx, Irx, Tbx,* and *Foxi.* The NC specifies and defines *Msx1, Pax3, Foxd3, Sox9/10, Dlx3, Tfap2,* and *Snail2* (Moody and LaMantia, 2015). Underlying is *Wnt* to drive NC that inhibits PPE to be driven by *Fgfs* and, with delay for PPE, *BMPs* (Riddiford and Schlosser, 2016; Thiery et al., 2020). Expression of *Pax3*

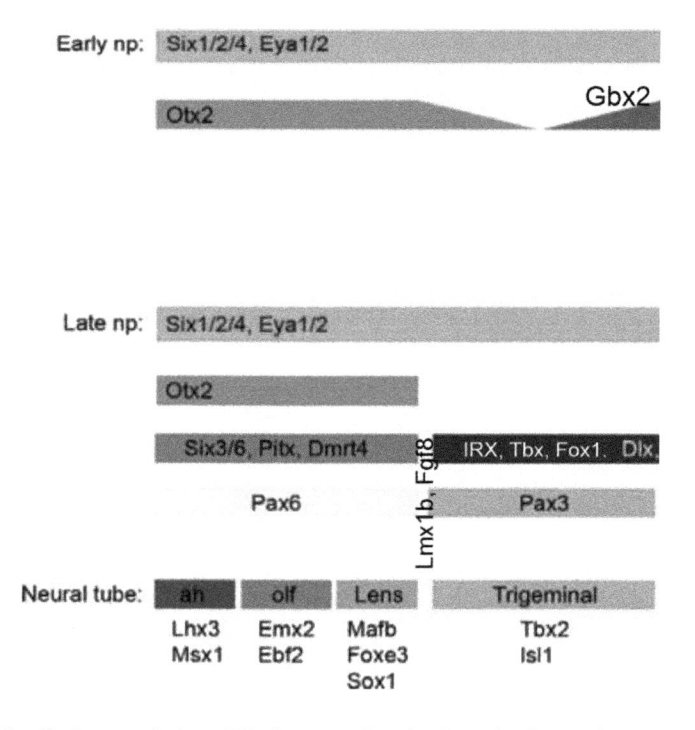

FIGURE 4.2 Early neural plate (NP) is separating *Otx2* caudal from *Gbx2* rostral that will continue to stay positive for *Six/Eya* (top). At later stages, it has developed a gap, free of either *Otx2* or *Gbx2*. An expression of *Lmx1b* is upstream of *Fgf8* that combines with *Pax3* that drives *Tbx, Irx, Fox1,* and *Dlx* for its progressive expression. (Modified from Glover et al., 2018; Moody and LaMantia, 2015; Watson et al., 2017)

represses *Pax6* (rostral) and *Pax2/8* (caudal) and is forming a restricted upregulation of the trigeminal placode/neural crest (Thiery et al., 2020). The ophthalmic and maxilla-mandibular parts of the *Pax3* are forming from placodes, which are generated, with delay, largely from ventral NC (Lassiter et al., 2014). *Fgf8* is expressed in the r0 isthmic region (Lee et al., 1997; Watson et al., 2017) to drive the formation of *Pax3* expression in the ophthalmic and maxillary development that limits the rostral level to r1 (Iwasato and Erzurumlu, 2018). *Lmx1b* is necessary for the r0 expression that regulates *Fgf8* and *Wnt1* (Glover et al., 2018) which defines the trochlear motoneurons (Jahan et al., 2021).

All neurons of the trigeminal development (and all other central and peripheral neurons) require *Sox2* expression for their development (Dvorakova et al., 2020; Imayoshi and Kageyama, 2014; Kageyama et al., 2019; Reiprich and Wegner, 2015). *Sox2* (Figure 4.2) functions as a pioneer factor that pre-patterns neural regulatory genes (Sarlak and Vincent, 2016). Downstream of *Sox2* genes require the expression of the bHLH gene, *Neurog1*, in mammals (Ma et al., 1998). Data suggest that *Delta/Notch* expression disappears in *Neurog2* in chicken (Lassiter et al., 2014) that is also an upstream of *Neurod1* in mice and chicken (Kim et al., 2001; Ma et al., 1998). In addition, *Fgfr4* is co-expressed with *Neurog2* in chicken (Lassiter et al., 2014). *Wnt1* and *Wnt3a* are part of the expression of *Isl1*, a gene that interacts with *Neurod1* (Dvorakova et al., 2016; Macova et al., 2019). Interestingly enough, using *Wnt-cre* mediated deletion of *Sox10* results not only in the loss of all Schwann cells but reduces the trigeminal neurons (Mao et al., 2014). *Pou4f1* is an essential gene driving the neuronal trigeminal differentiation (Eng et al., 2001; Huang et al., 2001); that projection is extended through the entire spinal cord (Eng et al., 2007). Interacting with *Pou4f1* is *Isl1* (Cai et al., 2003) and *Runx* (Kramer et al., 2006) that combine to regulate different projections (Yoshikawa et al., 2016). Differentiation of discrete trigeminal central projection is driven by three branches: The first differentiating projections develop from the ophthalmic ganglion that depends on *Tbx1/3* which overlaps with the maxillary ganglion. Subdivisions are driven by *Oc2* (maxillary and mandibular ganglion), *Oc1* and *Hmx1* (mandibular ganglion) that depend on *BMP4* expression for normal development (Erzurumlu et al., 2010). The three branches of the ganglion are projecting distinctly, starting with the most ventral (ophthalmic ganglion) followed dorsal by the maxillary ganglion and ends with the most dorsal central projection provided by the mandibular branch. Even more dorsal is the projection of the facial, glossopharyngeal and vagal projections that add to the most dorsal fibers, generating a reversal distribution of r2/3 input fibers (Erzurumlu, 2020; Fritzsch et al., 2019).

An overlapping dependency of *Ntf3* and *BDNF* is distributed through the trigeminal ganglion (Bothwell, 2016; Huang and Reichardt, 2003; Sharma et al., 2020). In contrast to the overlap of both neurotrophins, the distal part of facial, glossopharyngeal and vagal depend on *BDNF* and *Ntrk2*, whereas the proximal ganglion is depending on *Ntf3* and *Ntrk3* (Fritzsch et al., 1997b; Hellard et al., 2004). Neurotrophins are playing a role for neuronal development but are reduced in the trigeminal neurons in both *BDNF* and *Ntf3* double null mice (O'Connor and Tessier-Lavigne, 1999). Additional neurotrophins provide trophic support by other neurons such as *GDNF* and related genes (Bothwell, 2016). The complex interaction is driven by distinct and partial overlap of *Ntf3* and *BDNF* that is interacting with distinct neurotrophins that depend on *Ntrk2* (*TrkB*) and *Ntrk3* (*TrkC*; Lessmann et al., 2003; Sharma et al., 2020). A set of regulatory genes are identified that allow the correct central projections, such a *Semaphorin, Neuropilin, Slit, Robo,* and others (Erzurumlu, 2020).

In summary, the impact of distinct expression of specific earliest genes depends on *Eya1/Six1, Pax3, Fgf8,* and *BMP4.* The expression of neuronal differentiation following *Sox2, Isl1, Neurog1, Neurod1,* and *Pou4f1* for neuronal differentiation is targeting the trigeminal projections. I briefly describe the role of neurotrophins for viability of postnatal neurons. The expression guidance of genes projects to distinct fascicles of the trigeminal projection to the rhombomeres.

4.3 ORGANIZING OF THE CENTRAL SENSORY INPUTS TO PROVIDE TOUCH

Touch, in combination of proprioceptors, thermoreceptors, pain and lanceolate endings, are a broad perception of warm breeze, raindrops, gentle touch of the skin of the mouth, head and the rest of the skin (Goodman and Bensmaia, 2020). Tactile perception is the largest end organ of the skin in any vertebrate (Figure 4.3). All sensory input is innervated by four sensory inputs from the cranial (trigeminal, facial, glossopharyngeal, vagal) and spinal dorsal root ganglia (DRG). The exteroceptive somatosensory system provides a sensory world map to decipher the various tactile stimuli. It is convenient to distinguish between low-threshold and high-threshold

Associated fiber (conduction velocity[1])	Skin type	End organ/ending type	Location	Optimal Stimulus[4]	Response properties
Aβ (16-96m/s)	Glabrous	Merkel cell	Basal Layer of epidermis	Indentation	
	Hairy	Merkel cell (touch dome)	Around Guard hair follicles		
Aβ (20-100m/s)	Glabrous	Ruffini[2]	Dermis[3]	Stretch	
	Hairy	unclear	unclear		
Aβ (26-91m/s)	Glabrous	Meissner corpuscle	Dermal papillae	Skin movement	
	Hairy	Longitudinal lanceolate ending	Guard/Awl-Auchene hair follicles	Hair follicle deflection	
Aβ (30-90m/s)	Glabrous	Pacinian corpuscle	Deep dermis	Vibration	
Aδ (5-30m/s)	Hairy	Longitudinal lanceolate ending	Awl-Auchene/ Zigzag hair follicles	Hair follicle deflection	
C (0.2-2m/s)	Hairy	Longitudinal lanceolate ending	Awl-Auchene/ Zigzag hair follicles	Hair follicle deflection	
Aβ/Aδ/C (0.5-100m/s)	Glabrous Hairy	Free nerve ending	Epidermis/Dermis	Noxious mechanical	

FIGURE 4.3 The different types of the various cutaneous sensory neurons are presented. We can distinguish the four Aβ cells that are larger neurons that conduct faster compared to the Aδ smaller neurons that conduct slower where the smallest C fiber is the slowest conducting fibers. Note the distinct glabrous and hairy innervation that sets aside the noxious innervation. Modified after (Abraira and Ginty, 2013)

perception to define innocuous from harmful stimulation (Abraira and Ginty, 2013). Sensory modalities are anatomically and physiologically discrete and provide a single line that faithfully conveys information.

Cutaneous sensory neurons are classified as Aβ, Aδ, or C fibers (Figure 4.3). Their difference will be in cell body size, axon diameter, level of myelination and axon conduction velocities. C type is the smallest and slowest (range 0.2–2 m/s). Aδ is medium in size and speed (range 5–30 m/s). Aβ are larger but are variable in speed (range 16–96 m/s). Aδ and C fibers are nociceptors whereas Aβ fibers are light-touch receptors of various kinds. Skin can be distinguished as glabrous (non-hairy) and (non-glabrous) skin that provides Pacinian corpuscles, Ruffini endings, Meissner corpuscles, and Merkel discs (Figure 4.4). Hairy skin is characterized by various lanceolate endings around the follicles (Figure 4.4). Within hair innervation, three major types are known: zigzag, awl/auchene, and guard.

Among many cutaneous sensors are Merkle cells: they depend on *Atoh1*. Downstream of *Atoh1* are *Pou4f3* and other factors (Maricich et al., 2009) that can be induced to proliferation and transformation (Shuda et al., 2015). In contrast to a unique dependence of *Atoh1*, all innervation depends on various neurotrophic supports to develop into sensory-specific organs (Huang and Reichardt, 2003). For example, the hair follicle afferents are complex in form and function and end into three different kind of hair follicle innervation: about 76% of zigzag, 23% of awl/ auchene, and 1% of guard hairs (Abraira and Ginty, 2013). Depending on their

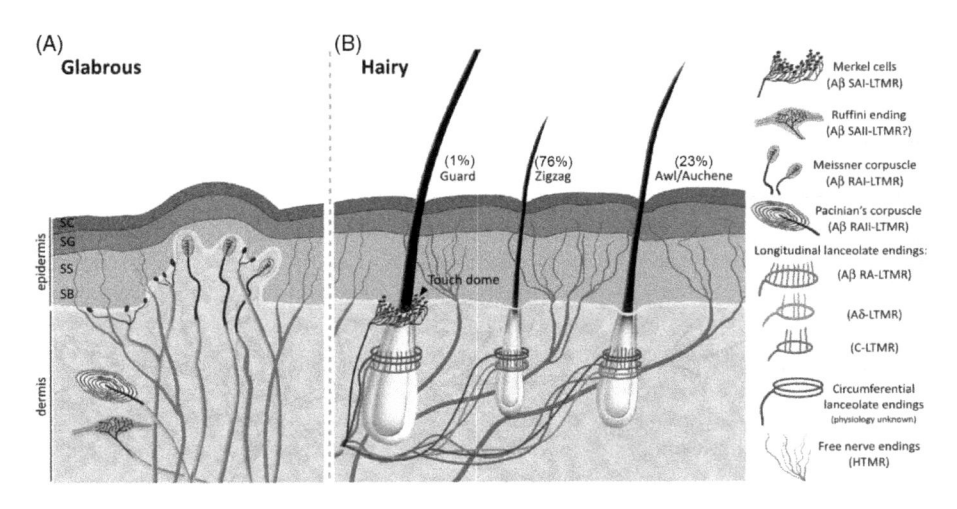

FIGURE 4.4 Cutaneous mechanoreceptors in the skin show distinct distribution of glabrous (A) and hairy innervation (B). Merkel cells are ubiquitous but have distinct organization of the hairy cells that is unique to the 1% of guard hairy innervation. The glabrous innervation has three unique innervations (Ruffini, Meissner, Pacinian) in addition to Merkel cells. In hairy innervation, neurons are reaching three types of lanceolate innervations that spread across fast (Aβ), medium (Aδ), and slow (C) with distinct distribution that is common for the circumferential innervation. (Taken from Abraira and Ginty, 2013.)

activity, poke, stroke, and breeze exit their maximum response of the skin (Figure 4.4). The spinal cord is restricted to a single terminal projection but can expand bilaterally to several segments that can be activated by Aβ terminals. A detailed description is given of the dorsal horn of the spinal cord and its distinct outputs that is partially characterized based on the molecular organization (Abraira et al., 2017; Lai et al., 2016; Sharma et al., 2020).

Molecular sensory innervation is unique to *Piezo1/2* (Cox et al., 2017) with a three propeller helical organization that is essential for its function (Zhao et al., 2018). Piezo encodes a 2500 amino acid protein that aggregates into an unusual trimer with propeller-like extensions. These 'propellers' seem to be essential for the high mechanosensitivity of the untethered channel (Zhao et al., 2018). It is noteworthy that *Piezo* exceeds any other mechanosensitive channel examined to date in terms of 'lipid-from-force' sensitivity (Cox et al., 2017): *Piezo* proteins form the most sensitive untethered mechanosensory channel (Fritzsch et al., 2020). Evidence has been demonstrated to play roles in cellular development, volume regulation, cellular migration, proliferation, and elongation (Bagriantsev et al., 2014). The pore resembles the acid-sensing channel 1 (ASIC1). Each trimeric subunit has 13 parallel transmembrane helix pairs while other units form a blade that is responsible for the high sensitivity to stretch along membrane (Zhao et al., 2018). *Piezo* channels are inherently mechanosensitive and can respond to the stretch of the membrane (Cox et al., 2016; Ridone et al., 2019). Human *Piezo1* exhibits a dimension change of around 8nm during the opening, comparable to dimensional changes of bacterial mechanosensitive channels, supporting that *Piezo1* is sensing bilayer lipid tension (force-from-lipid) similar to bacterial *Msc* channels. *Piezo1* seems to be gated by tensions in the range of 1–3 mN/m and responds to membrane curvature changes upon both positive and negative pressure applied to a patch pipette (Cox et al., 2016). Despite being mechanosensors, *Piezo* channels can be activated by chemical stimuli (Saotome et al., 2018; Syeda et al., 2015; Zhao et al., 2018) and play a role in pain sensation (Szczot et al., 2018).

Piezo molecules are evolutionarily conserved and also found in all multicellular organisms where they play a role in mechanosensation as well as being modulated by voltage (Hu et al., 2019; Moroni et al., 2018; Murthy et al., 2018; Poole et al., 2015) but also play a role in nociception (Kim et al., 2012). Piezo channels show limited multiplication with at the most three *Piezo* genes known in each species. Piezo channels are found in all metazoans, plants, fungi as well as many single-celled organisms, clearly supporting that lipid force measurement with or without extra- or intracellular tethers is an ancient feature of eukaryotes, predating metazoans (Cox et al., 2017). Most interesting in the context of metazoan evolution is the Piezo involvement in stem cell renewal in certain epithelial cells through sensing of lipid tension that regulates proliferation (He et al., 2018). Given *Piezo's* ability to sense membrane tension and curvature (Cox et al., 2016) it could have evolved as an early safety valve in eukaryotes, paralleling the evolution of MscL channels in bacteria (Cox et al., 2017; Kung et al., 2010).

An interesting pattern of innervation is the rodent terminal of the whisker innervation (Erzurumlu and Gaspar, 2020). A specified and consistent pattern of the maxillary branch is innervating the trigeminal hair cells by five whiskers. The central organization depends on *Hoxa2, Lmx1b, Epha4/7, Slit2,3, Robo1/2, Neuropolin1*, and *Sema3a* and they are also needed for normal innervation of r2 (mandibular) and r3 (maxillary) input. Five longitudinal rows are characterized in mice; they form the whiskers. Interestingly enough, the central maxillary projection is formed as 'barrelettes' that are unique to the r3 terminal fibers (Iwasato and Erzurumlu, 2018). Detailed analysis described the various molecular basis in several mutants that change the details of the maxillary projection of the r3 terminal (Erzurumlu and Gaspar, 2020; Iwasato and Erzurumlu, 2018; Kratochwil and Rijli, 2020). Upstream of this organization of whiskers, retinoic acid (RA) plays a role as it affects the detailed patterning of innervation downstream of *Hoxa2* (Glover et al., 2006; Oury et al., 2006).

Comparable to rodents, but distinct in innervation, is the star-nosed mole that has 22 appendages of approximately a centimeter in diameter when splayed out, slightly smaller than the average human fingertip (Catania, 2011) to form in connection to the Eimer's organ (Catania, 2000). Merkel cell-neurite complex, depending on Piezo innervation, is above a layer of lamellated corpuscles and a series of nociceptive free nerve endings that run the length of the column to radiate out in a ring at the top of the column (Leitch and Gaede, 2020). Their innervation is solely from the trigeminal nerve which is greatly expanded in star-nosed moles (~80,000 myelinated fibers), compared to the trigeminal nerve of other terrestrial moles (~30,000 myelinated fibers) and their sister group, the shrew (Leitch and Gaede, 2020). More detailed characterization of Eimer's organ response properties showed populations with rapidly adapting properties, with response of Pacinian corpuscles with maximal sensitivity to 250 Hz sinusoidal stimulation. These responses are presumably related to the anatomically similar lamellated corpuscles found at the base of each Eimer's organ. Other responses were slowly-adapting cells, consistent with recording associated with Merkel cell-neurite complexes. Furthermore, directional sensitivity was found in a majority of the tactile receptive fields, suggesting that the ring of free nerve endings at the tip of the Eimer's organ may confer timing differences that related to direction specific stimuli (Leitch and Gaede, 2020).

In summary, the large number of different sensory receptors that contribute to the skin of the face is part of the sensory innervation from head to toes, providing the largest sensory organ of all. The lips in human, the whiskers of mice and the nose of the star-nosed mole are extreme expansions that are a special reorganization from general skin innervation that depends on Piezo input.

4.4 DEFINING A TRIGEMINAL PROJECTION: WHERE ARE WE NOW?

All genes of the hindbrain and spinal cord are a set of gene expressions that are downstream of *Sox2*. The two dorsal expressions (*Atoh1, Neurog1/2*) are distinct for the ear projection. However, from dA3 through dB4 is the expression of trigeminal neurons that continues to the spinal cord (Hernandez-Miranda et al., 2017; Lai et al., 2016). *Ascl1* is defining the development of the trigeminal neurons. bHLH neuronal genes are building an oscillation that interacts *at Ascl1/Hes1*

(Kageyama et al., 2019; Lai et al., 2016). Superimposed is the reciprocal inhibition of *Neurog1/2* with *Ascl1* that interacts with *Ptf1a*. Upstream of neurons all alar plates depend on *BMPs* and *Wnts* in the roof plate that is dorsal and is counteracted with *Shh* that defines the floor plate (Glover et al., 2018). Downstream of bHLH genes is a specific function of *Lbx1* which drives somatosensory development which counteracts the viscero-sensory system, driven by *Phox2b* (Hernandez-Miranda et al., 2017). Pain is a central set of information that comes in with the trigeminal input with various speeds. One gene, *Lbx1*, is enriched in lamina II of the dorsal horn that requires sensory mechanical but not thermal pain. Ablation of *Lbx1* and these excitatory neurons eliminate mechanical sensation (Duan et al., 2014). Excitatory and inhibitory second-order sensory neurons are distinguished by transient expression of *Pou4f1/Tlx3/Lmx1b* (excitatory) and *Ptf1a/Pax2/Lhx1* (inhibitory) fate (Hernandez-Miranda et al., 2017).

Currently, mature neurons and particular neurotransmitters, receptors, axons and dendrites are clarified; we begin to understand the circuits, mostly in the spinal cord dorsal horn (Lai et al., 2016). Several layers of spinal cord are distinguished between the excitatory *Vglut* positive neurons that contrast with nine different inhibitory neuronal subtypes that are driven by *Gly* (Nothwang, 2016).

As highlighted before, the *Lmx1b* is upstream to *Fgf8* (Figure 4.5) that is upstream of *Pax3* expression (Mishima et al., 2009; Thiery et al., 2020; Watson et al., 2017) and sets up *Hoxa2* to limit the expression of r2 (Erzurumlu et al., 2010). A uniform organization from r2 through the spinal cord is consistent with following the different rhombomeres (Nieuwenhuys and Puelles, 2015) that is identical to lampreys (Glover et al., 2018; Pombal and Megías, 2019). Genetic fate mapping revealed that the mouse PrV is mainly derived from r2 and r3 postmitotic nuclei, demonstrated by *Hoxa2* mutations (Oury et al., 2006). The map of the lower jaw and lips projects from the mandibular ganglia to reach the most dorsal fibers starting in r2 whereas the maxillary fibers reach through r3 to form the 'barrelette' organization (Erzurumlu et al., 2010).

The principal (oralis; PrV) and spinal (interpolaris, caudalis; SpV) projections are organized inverted to the central projection with respect to the ophthalmic, maxillary, and mandibular fibers to generate a map: from most ventral (ophthalmic) neurons extend to most dorsal projections (mandibular) fibers. Unique to rodents, the PrV fibers are organized into 'barrelettes' to mimic the well-known barrel fields of the cortex (Erzurumlu and Gaspar, 2020). In addition, subnuclei are important for subdivisions, the interpolaris and the caudalis part of the SpV. The ascending branches of second order trigeminal neurons give rise to project to the thalamus from the PrV and SpV to form the lemniscal set of fibers; these fibers reach the thalamus on the contralateral trigeminal neurons, except that the principal spinal connection reaches ipsilateral to the thalamus (Erzurumlu et al., 2010). Recent data demonstrate the same central projection for lamprey (Suryanarayana et al., 2020), mimicking the lemniscal fibers in all vertebrates (Briscoe and Ragsdale, 2019; Erzurumlu, 2020).

All trigeminal and other somatosensory projections start in r2 first and extend through the cervical spinal cord where they end (Figure 4.6). Later stages reach mandibular branches from r2 that is selectively innervated by the r3 branches from

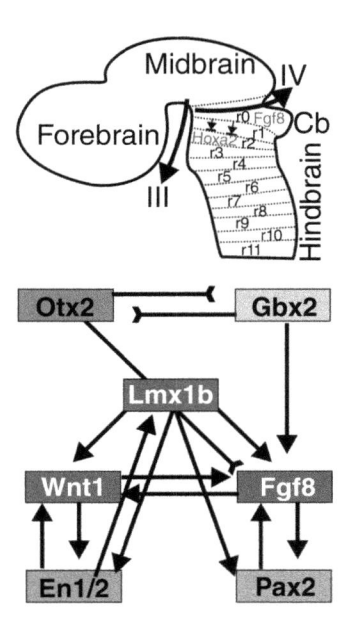

FIGURE 4.5 The organization starts from r2 (*Hoxa2*) that is suppressed by *Fgf8* that is expressed in the isthmic region now referred as r0. The interaction defines the reciprocal inhibition of *Otx2* and *Gbx2*. Upstream is the expression of *Lmx1b* that drives the formation of *Wnt1* and *Fgf8* that are positively reinforcing in r1 and suppresses of *Hoxa2*. Downstream of *En1/2* and *Pax2*, *Hoxa2* depends on a positive feedback from *Wnt1* and *Fgf8*, respectively. Cb, cerebellum; III, oculotmor nerve; IV, trochlear nerve. (Modified from Glover et al., 2018; Jahan et al., 2021; Mishima et al., 2009; Watson et al., 2017)

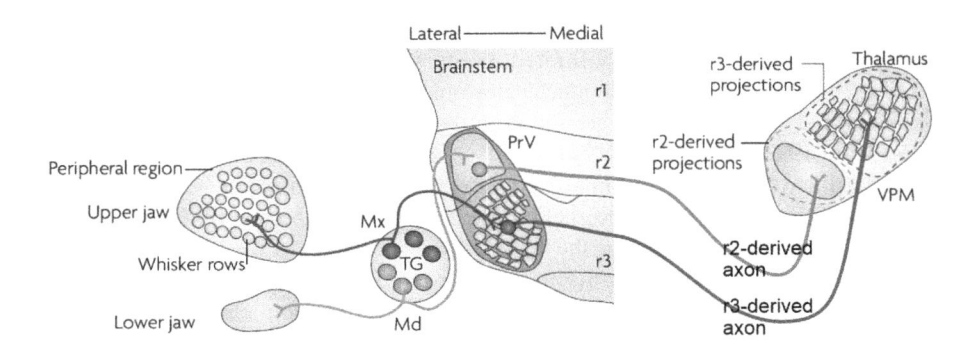

FIGURE 4.6 Connection of whiskers and thalamus show that the maxillary fibers (left) end up in r3 of the brainstem (middle) to reach the VPM of the thalamus (right). Mandibular (left) projects to r2 (middle) and stay in a distinct thalamus (right) to end in its own terminal field. Md, mandibular; Mx, maxillary; PrV, principal trigeminus; Tg, trigeminus; VPM, ventral posterior medial nucleus. (Modified from Erzurumlu et al., 2010.)

the maxillary fibers. Current evidence suggests an altered central projection that expands from r2 dorsally to cross fibers following the *Lmx1a/b* deletion (Chizhikov et al., 2021). Since *Lmx1b* depends on *Fgf8* and *Wnt1* (Mishima et al., 2009) that is limited to r1 *Hox* expression (Erzurumlu et al., 2010) we suggest that the massive expansion of trigeminal projection is due to alteration of rostral rhombomeres, the expansion of *Hoxa2* modification, that allow fibers to extend to the cerebellum (Oury et al., 2006). Several mutants have been identified that order the PrV projections (*Drg11, Nmdar1, Lmx1b)* that lack the formation of the barrelette in these three knockout mice (Erzurumlu et al., 2010).

Projections from somatosensory thalamus innervation start out to reach the cortex before lemniscal fibers from the trigeminal and spinal cord (Iwasato and Erzurumlu, 2018). Connecting fibers continue to the thalamus to end up in the VPM to form the 'barrelettes' (r3); from here the fibers reach the 'barreloids' of the VPM (Figure 4.7). Thalamocortical axons to reach specific cortical targets are controlled by multiple thalamic and subpallial molecular guidance cues. For examples, whisker-related patterning follows the topographic projection zone of the primary sensory cortex (Iwasato and Erzurumlu, 2018). *Fgf8* expression from the rostromedial neocortical primordium acts as an organizer that can generate a duplication of somatosensory cortical maps (Toyoda et al., 2010). A distinct expression of *Wnt* and *Fgf* like systems is uniform among hemichordates and vertebrates. Partial or incomplete similar expressions are known in amphioxus and ascidians (Glover et al., 2018; Holland, 2020; Pani et al., 2012). Most importantly (Figure 4.4), the co-expression of *Wnt/Fgf* expression that is downstream of *Lmx1b* (Glover et al., 2018; Mishima et al., 2009) is unique among hemichordates and vertebrates, shared by unique expression of miRs (Pierce et al., 2008; Banks et al., 2020).

An additional branch of the trigeminal fiber is known as the mesencephalic trigeminal ganglion (MTN) also known as mesencephalic fibers (MesV). The development of MTN expands from the midbrain to the trigeminal branches, starting at r1 (Lipovsek et al., 2017; Mastick et al., 1996). The central projection bifurcates to branch off into a proximal and a distal projection, regulated by *Nrp2* (Ter-Avetisyan et al., 2018). If the bifurcation is not expanding to reach muscle fibers of mandibular branches, it will result in a limited ability to bite with strength (Ter-Avetisyan et al., 2018). Interestingly enough, cyclostomes do not have the MTN fibers but have instead dorsal cells that are found in the trigeminal projections (Fritzsch and Northcutt, 1993; Pombal and Megías, 2019). These dorsal cells are found in Rohon-Beard cells in the spinal cord in all vertebrates except for amniotes.

In summary, generating the specific input requires a cross inhibition of *Hoxa2* by *Fgf8* in r0/isthmus to define the input of three branches of the trigeminal fibers. Central projections from a lemniscus which extends to the second-order trigeminal neurons project to reach the thalamus via ipsilateral (principle) and contralateral (principle and spinal) projections to reach the thalamus.

4.5   SOMATOSENSORY ORGANIZATION AND EVOLUTION

The forebrain gives rise to all sensory input to reach the topology of distinct areas of the mammalian cortex. Following extensive investigation, Brodmann (Brodmann, 1909) defines the cortical somatosensory cortex that was later discovered by

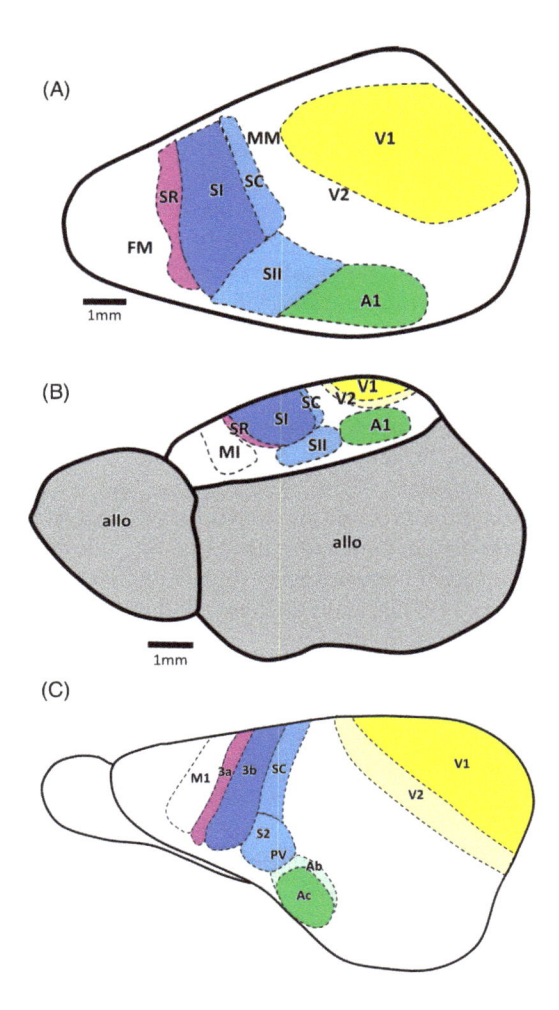

FIGURE 4.7 Comparing three cortical organization of a marsupial the gray short-tailed opossum (A; *Monodelphis*), and two mammals the tenrec (B; *Echinops*) and the tupaia (C; *Tupaia*). Two distinct somatosensory areas (S1, S2) are adjacent to the auditory area. Note that the additional anterior and posterior divisions are common among mammals. The distinct areas of the visual field (V1, V2) vary among the size of the cortex. Note that different nomenclature is depicted. A1 (Ab,Ac), auditory cortex; allo, allocortex; FM, frontal area; M1, primary motor cortex; MM, multimodal cortex; SI (3a, 3b) primary somatosensory cortex; SII (S2, PV) secondary somatosensory cortex; SC, caudal somatosensory field, SR, rostral somatosensory field; V1, V2 primary and secondary visual cortex. (Modified from Zilles and Palomero-Gallagher, 2020.)

Penfield to show the distorted human-like figures, referred to as 'homunculus' (Penfield and Boldrey, 1937). Clearly, some 80 years ago the basic organization of the mammalian somatosensory input was explained. Meanwhile, somatosensory inputs are now known beyond mammals and are now described among lampreys (Suryanarayana et al., 2020). Sensory inputs from somatosensory and visceral input are a common organization of all vertebrates (Briscoe and Ragsdale, 2019; Pani et al., 2012); their basic organization is now refined. Future functional connectivity of the thalamo-cortical system has begun to reveal how the development and age related modifications in humans are affected by functional connectivity (Steiner et al., 2020).

The marsupials' branch (*Monodelphis domestica*) diverged from the common eutherian mammals 180 million years ago (Wong and Kaas, 2009). The same organization is following the topography of neocortical areas with the visual cortex occipitally, the auditory cortex ventrally and two major somatosensory areas rostrally (Figure 4.7). Obviously, the most small-brained placental mammals (Catania et al., 2000) show nevertheless segregation comparable to the larger-brained mammals.

There is no separate motor cortex in marsupials that makes a distinct difference to the motor output in lampreys (Suryanarayana et al., 2020) and gnathostomes (Briscoe and Ragsdale, 2019). The regional and laminar distribution of transmitter receptors in the brain of the marsupial *Monodelphis domestica* (Gray short-tailed opossum) differs noticeably from that of eutherian brains, since area S1 does not stand out by a conspicuously higher or lower density of any of the examined receptor types (Zilles and Palomero-Gallagher, 2020) which also differ (5-HT, AMPA, Gaba, Kainate).

The tenrec (Figure. 4.7B) belongs to an early diverging group of mammals. Despite its basal phylogenetic position, the tenrec somatosensory system displays a topographical and functional organization comparable to that found in most small-brained mammals, with two somatotopically organized fields within the primary somatosensory cortex, and a ventrally adjacent secondary somatosensory region (Krubitzer et al., 1997). The tenrec somatosensory cortex is characterized by a clearly higher packing density of myelinated fibers than surrounding cortical areas (Zilles and Palomero-Gallagher, 2020).

Tree shrews are small mammals initially thought to be rodents (Figure. 4.7C), then classified as primitive prosimians, and currently included in an independent clade encompassing the *Tupaiidae* family of the order *Scandentia* (Emmons, 2000). Five areas have been identified within the *Tupaia* somatosensory cortex, primary somatosensory area 3b, somatosensory areas 3a and SC, and secondary somatosensory areas S2 and PV (Zilles and Palomero-Gallagher, 2020).

Woolsey and van der Loos (1970) first described a novel terminal field of rodents, which was coined the 'barrel field' (Erzurumlu and Gaspar, 2020). This area is unique (Figure 4.8) in that it is a major system that can be manipulated by loss and gain of single or complex whiskers (Erzurumlu and Gaspar, 2020; Welker et al., 1996). In addition, the cortical organization can be affected by genetics which can change the pattern of the central projections (Kratochwil et al., 2017; Moreno-Juan et al., 2017). The study of the barrel field has been driven by the unique organization of rodents (Liao et al., 2020); it became a major investigation of plasticity (Erzurumlu and Gaspar, 2020; Qi et al., 2021).

Injury to the sensory receptors in the whiskers after birth leads to predictable pattern alterations at all levels, from the brainstem (barrelettes) to the thalamus (barreloids) to the cortex (barrels). Currently, investigation of molecular and genetic

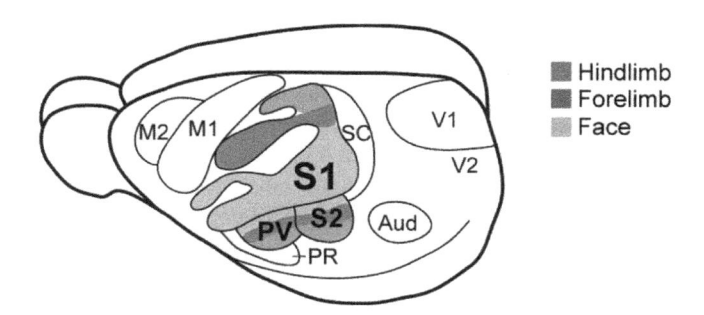

FIGURE 4.8 The location of primary somatosensory cortex (S1), secondary somatosensory cortex (S2) and parietal ventral cortex (PV) of the rat. The subdivisions of S1 represent the face, forelimb, and hindlimb, mirror images of PV and S2 areas. Aud, auditory cortex; M1, M2, primary and secondary cortex; PR, parietal cortex; SC, caudal somatosensory cortex; V1, V2, primary, and secondary visual cortex. (Modified from Liao et al., 2020.)

manipulations are driving the neurobiological model to study the somatotopic and patterning of plasticity at both morphological and physiological levels (Erzurumlu and Gaspar, 2020). For example, a calcium signal between the thalamus and barrel cortex is starting before sensory input is present. Blocking the calcium signals results in hyperexcitability and does not build up the barrel organization (Antón-Bolaños et al., 2019). Dendritic organization is controlled by specific transcription factors; it appears to form cell specification in the barrel cortex (Antón-Bolaños et al., 2018). A self-organized protomap drives the functional assembly of murine thalamocortical sensory circuits.

In fact, future work is aiming to reveal the thalamo-cortical networks and cognition that were associated with functional connectivity between several thalamic nuclei (Kolb et al., 2012; Steiner et al., 2020). In addition, cortical structure and function can be altered by modifying the source or pattern of activity in thalamocortical afferents. Studies of cross-modal plasticity have shown that a sensory cortical area can substitute for another after a switch of input modality during development. Afferent inputs might therefore direct the formation of their own processing circuitry, a possibility that has important implications for brain development, plasticity and evolution or after a surgical removal of part of the cortex (Pallas, 2001).

In summary, an overview of the somatosensory development and organization across certain mammals is provided. I highlight the role of certain innervations that have been defined by the molecular and juvenile studies in the central cortex, when used excessively for the barrel field organizations.

4.6 SUMMARY AND CONCLUSION

The organization of the trigeminal projection has been studied in development of mostly mammals to analyze the innervation from the sensory input through the somatosensory cortex. The very large sensory field is divided into two areas with respect to innervation, the hairy (lanceolate) and glabrous (Ruffini, Meissner, Pacinian) skin innervation. Many unique sensory systems are found in either of these two areas that have a common innervation, the Merkel cells. Molecularly organization depends on a set of genes to define the trigeminal placode and neural crest, both of which are downstream of *Pax3* and *Eya1/Six1*. A set of bHLH genes are playing a role for Merkel cells (*Atoh1*), trigeminal neurons (*Neurog1*) and trigeminal nuclei (*Ascl1, Olig3, Ptf1a*). The central organization will depend on various rhombomere expressions, starting with *Hoxa2* at *r2*. From these sensory inputs follow the somatotopic central projections to the hindbrain providing the three major branches, the ophthalmic, maxillary and mandibular branch of the trigeminal nerve. Second order neurons in the hindbrain have unique innervation that provides mandibular branches to r2 whereas the maxillary fibers redirect to innervate the maxillary branch from r3. The ascending projection requires a set of genes that drive the somatotopic thalamus and cortical organizations. Somatosensory neurons provide a somatotopic projection to the thalamus and eventually continue to the somatosensory cortex. Most recently, a model system has been developed to study axon-pathfinding issues and tactile sensory circuit formation in mice, paving the way to future insights.

ACKNOWLEDGEMENT

The fine-tuning of this chapter was helped by Dr. D. Crapon de Caprona.

REFERENCES

Abraira, V.E., and Ginty, D.D. (2013). The sensory neurons of touch. Neuron *79*, 618–639.

Abraira, V.E., Kuehn, E.D., Chirila, A.M., Springel, M.W., Toliver, A.A., Zimmerman, A.L., Orefice, L.L., Boyle, K.A., Bai, L., and Song, B.J. (2017). The cellular and synaptic architecture of the mechanosensory dorsal horn. Cell *168*, 295–310. e219.

Antón-Bolaños, N., Espinosa, A., and López-Bendito, G. (2018). Developmental interactions between thalamus and cortex: a true love reciprocal story. Current Opinion in Neurobiology *52*, 33–41.

Antón-Bolaños, N., Sempere-Ferràndez, A., Guillamón-Vivancos, T., Martini, F.J., Pérez-Saiz, L., Gezelius, H., Filipchuk, A., Valdeolmillos, M., and López-Bendito, G. (2019). Prenatal activity from thalamic neurons governs the emergence of functional cortical maps in mice. Science *364*, 987–990.

Aruga, J., and Hatayama, M. (2018). Comparative genomics of the Zic family genes. In Zic Family (Springer), pp. 3–26.

Bagriantsev, S.N., Gracheva, E.O., and Gallagher, P.G. (2014). Piezo proteins: regulators of mechanosensation and other cellular processes. Journal of Biological Chemistry *289*, 31673–31681.

Banks, S.A., Pierce, M.L. and Soukup, G.A. (2020). Sensational MicroRNAs: neurosensory roles of the MicroRNA-183 family. Molecular Neurobiology 57(1), 358–371.

Bothwell, M. (2016). Recent advances in understanding neurotrophin signaling. F1000 Research *5*.

Briscoe, S.D., and Ragsdale, C.W. (2019). Evolution of the chordate telencephalon. Current Biology *29*, R647–R662.

Brodmann, K. (1909). Vergleichende lokalisationslehre der grosshirnrinde. In Vergleichende lokalisationslehre der grobhirnrinde (Vancouver), pp. 324–324.

Buzzi, A.L., Hintze, M.S., and Streit, A. (2019). Development of neurogenic placodes in vertebrates. eLS, 1–14.

Cai, C.-L., Liang, X., Shi, Y., Chu, P.-H., Pfaff, S.L., Chen, J., and Evans, S. (2003). Isl1 identifies a cardiac progenitor population that proliferates prior to differentiation and contributes a majority of cells to the heart. Developmental Cell 5, 877–889.

Catania, K.C. (2000). Epidermal sensory organs of moles, shrew moles, and desmans: a study of the family talpidae with comments on the function and evolution of Eimer's organ. Brain, Behavior and Evolution 56, 146–174.

Catania, K.C. (2011). The sense of touch in the star-nosed mole: from mechanoreceptors to the brain. Philosophical Transactions of the Royal Society B 366, 3016–3025.

Catania, K.C., Jain, N., Franca, J.G., Volchan, E., and Kaas, J.H. (2000). The organization of somatosensory cortex in the short-tailed opossum (*Monodelphis domestica*). Somatosensory & Motor Research 17, 39–51.

Chizhikov, V.V., Iskusnykh, I.Y., Fattakhov, N., and Fritzsch, B. (2021). Lmx1a and Lmx1b are redundantly required for the development of multiple components of the mammalian auditory system. Neuroscience 452, 247–264.

Cox, C.D., Bae, C., Ziegler, L., Hartley, S., Nikolova-Krstevski, V., Rohde, P.R., Ng, C.-A., Sachs, F., Gottlieb, P.A., and Martinac, B. (2016). Removal of the mechanoprotective influence of the cytoskeleton reveals PIEZO1 is gated by bilayer tension. Nature Communications 7, 10366.

Cox, C.D., Bavi, N., and Martinac, B. (2017). Origin of the force: the force-from-lipids principle applied to piezo channels. Current Topics in Membranes 79, 59–96.

Dennis, D.J., Han, S., and Schuurmans, C. (2019). bHLH transcription factors in neural development, disease, and reprogramming. Brain Research 1705, 48–65.

Duan, B., Cheng, L., Bourane, S., Britz, O., Padilla, C., Garcia-Campmany, L., Krashes, M., Knowlton, W., Velasquez, T., and Ren, X. (2014). Identification of spinal circuits transmitting and gating mechanical pain. Cell 159, 1417–1432.

Dvorakova, M., Jahan, I., Macova, I., Chumak, T., Bohuslavova, R., Syka, J., Fritzsch, B., and Pavlinkova, G. (2016). Incomplete and delayed Sox2 deletion defines residual ear neurosensory development and maintenance. Scientific Reports 6, 1–16.

Dvorakova, M., Macova, I., Bohuslavova, R., Anderova, M., Fritzsch, B., and Pavlinkova, G. (2020). Early ear neuronal development, but not olfactory or lens development, can proceed without SOX2. Developmental Biology 457, 43–56.

Emmons, L.H. (2000). Tupai: A Field Study of Bornean Tree Shrews (Berkeley: University of California Press).

Eng, S.R., Dykes, I.M., Lanier, J., Fedtsova, N., and Turner, E.E. (2007). POU-domain factor Brn3a regulates both distinct and common programs of gene expression in the spinal and trigeminal sensory ganglia. Neural Development 2, 3.

Eng, S.R., Gratwick, K., Rhee, J.M., Fedtsova, N., Gan, L., and Turner, E.E. (2001). Defects in sensory axon growth precede neuronal death in Brn3a-deficient mice. Journal of Neuroscience 21, 541–549.

Engelhard, C., Sarsfield, S., Merte, J., Wang, Q., Li, P., Beppu, H., Kolodkin, A.L., Sucov, H.M., and Ginty, D.D. (2013). MEGF8 is a modifier of BMP signaling in trigeminal sensory neurons. Elife 2, e01160.

Erzurumlu, R.S. (2020). Development of the somatosensory cortex and patterning of afferent projections. In The Senses, B. Fritzsch, ed. (Elsevier), pp. 372–381.

Erzurumlu, R.S., and Gaspar, P. (2020). How the barrel cortex became a working model for developmental plasticity: a historical perspective. Journal of Neuroscience 40, 6460–6473.

Erzurumlu, R.S., Murakami, Y., and Rijli, F.M. (2010). Mapping the face in the somatosensory brainstem. Nature Reviews Neuroscience 11, 252–263.

Fode, C., Gradwohl, G., Morin, X., Dierich, A., LeMeur, M., Goridis, C., and Guillemot, F. (1998). The bHLH protein NEUROGENIN 2 is a determination factor for epibranchial placode–derived sensory neurons. Neuron *20*, 483–494.

Fritzsch, B., Elliott, K.L., and Pavlinkova, G. (2019). Primary sensory map formations reflect unique needs and molecular cues specific to each sensory system. F1000Research *8*.

Fritzsch, B., Erives, A., Eberl, D.F., and Yamoah, E.N. (2020). Genetics of mechanoreceptor evolution and development. In Reference Module in Neuroscience and Biobehavioral Psychology (Academic Press, Elsevier), vol. 2, pp. 277–310.

Fritzsch, B., and Northcutt, G. (1993). Cranial and spinal nerve organization in amphioxus and lampreys: evidence for an ancestral craniate pattern. Cells Tissues Organs *148*, 96–109.

Fritzsch, B., Sarai, P., Barbacid, M., and Silos-Santiago, I. (1997a). Mice with a targeted disruption of the neurotrophin receptor trkB lose their gustatory ganglion cells early but do develop taste buds. International Journal of Developmental Neuroscience *15*, 563–576.

Fritzsch, B., Silos-Santiago, I., Bianchi, L.M., and Farinas, I. (1997b). The role of neurotrophic factors in regulating the development of inner ear innervation. Trends in Neurosciences 20(4), 159–164.

Glover, J.C., Elliott, K.L., Erives, A., Chizhikov, V.V., and Fritzsch, B. (2018). Wilhelm His' lasting insights into hindbrain and cranial ganglia development and evolution. Developmental Biology *444*, S14–S24.

Glover, J.C., Renaud, J.S., and Rijli, F.M. (2006). Retinoic acid and hindbrain patterning. Journal of Neurobiology *66*, 705–725.

Goodman, J.M., and Bensmaia, S.J. (2020). The neural mechanisms of touch and proprioception at the somatosensory periphery. In The Senses, B. Fritzsch, ed. (Elsevier), pp. 2–27.

Gray, P. (2013). Transcription factors define the neuroanatomical organization of the medullary reticular formation. Frontiers in Neuroanatomy *7*.

He, L., Si, G., Huang, J., Samuel, A.D.T., and Perrimon, N. (2018). Mechanical regulation of stem-cell differentiation by the stretch-activated Piezo channel. Nature *555*, 103–106.

Hellard, D., Brosenitsch, T., Fritzsch, B., and Katz, D.M. (2004). Cranial sensory neuron development in the absence of brain-derived neurotrophic factor in BDNF/Bax double null mice. Developmental Biology *275*, 34–43.

Hernandez-Miranda, L.R., Müller, T., and Birchmeier, C. (2017). The dorsal spinal cord and hindbrain: From developmental mechanisms to functional circuits. Developmental Biology *432*, 34–42.

Holland, L. (2020). Invertebrate origins of vertebrate nervous systems. In Evolutionary Neuroscience (Elsevier), pp. 51–73.

Hu, Y., Wang, Z., Liu, T., and Zhang, W. (2019). Piezo-like gene regulates locomotion in Drosophila larvae. Cell Reports *26*, 1369–1377. e1364.

Huang, E.J., Liu, W., Fritzsch, B., Bianchi, L.M., Reichardt, L.F., and Xiang, M. (2001). Brn3a is a transcriptional regulator of soma size, target field innervation and axon pathfinding of inner ear sensory neurons. Development *128*, 2421–2432.

Huang, E.J., and Reichardt, L.F. (2003). Trk receptors: roles in neuronal signal transduction. Annual Review of Biochemistry *72*, 609–642.

Imayoshi, I., and Kageyama, R. (2014). bHLH factors in self-renewal, multipotency, and fate choice of neural progenitor cells. Neuron *82*, 9–23.

Iskusnykh, I.Y., Steshina, E.Y., and Chizhikov, V.V. (2016). Loss of Ptf1a leads to a widespread cell-fate misspecification in the brainstem, affecting the development of somatosensory and viscerosensory nuclei. Journal of Neuroscience *36*, 2691–2710.

Iwasato, T., and Erzurumlu, R.S. (2018). Development of tactile sensory circuits in the CNS. Current Opinion in Neurobiology *53*, 66–75.

Jahan, I., Kersigo, J., Elliott, K.L., and Fritzsch, B. (2021). Smoothened overexpression causes trochlear motoneurons to reroute and innervate ipsilateral eyes. Cell and Tissue Research, *384*: 59–72.

Kageyama, R., Shimojo, H., and Ohtsuka, T. (2019). Dynamic control of neural stem cells by bHLH factors. Neuroscience Research *138*, 12–18.

Kim, S.E., Coste, B., Chadha, A., Cook, B., and Patapoutian, A. (2012). The role of Drosophila piezo in mechanical nociception. Nature *483*, 209–212.

Kim, W.-Y., Fritzsch, B., Serls, A., Bakel, L.A., Huang, E.J., Reichardt, L.F., Barth, D.S., and Lee, J.E. (2001). NeuroD-null mice are deaf due to a severe loss of the inner ear sensory neurons during development. Development *128*, 417–426.

Kolb, B., Mychasiuk, R., Muhammad, A., Li, Y., Frost, D.O., and Gibb, R. (2012). Experience and the developing prefrontal cortex. Proceedings of the National Academy of Sciences of the United States of America *109*, 17186–17193.

Kramer, I., Sigrist, M., de Nooij, J.C., Taniuchi, I., Jessell, T.M., and Arber, S. (2006). A role for Runx transcription factor signaling in dorsal root ganglion sensory neuron diversification. Neuron *49*, 379–393.

Kratochwil, C.F., Maheshwari, U., and Rijli, F.M. (2017). The long journey of pontine nuclei neurons: from rhombic lip to cortico-ponto-cerebellar circuitry. Frontiers in Neural Circuits *11*, 33.

Kratochwil, C.F., and Rijli, F.M. (2020). The Cre/Lox system to assess the development of the mouse brain. In Brain Development (Springer), pp. 491–512.

Krubitzer, L., Kunzle, H., and Kaas, J. (1997). Organization of sensory cortex in a Madagascan insectivore, the tenrec (*Echinops telfairi*). Journal of Comparative Neurology *379*, 399–414.

Kung, C., Martinac, B., and Sukharev, S. (2010). Mechanosensitive channels in microbes. Annual Review of Microbiology *64*, 313–329.

Lai, H.C., Seal, R.P., and Johnson, J.E. (2016). Making sense out of spinal cord somatosensory development. Development *143*, 3434–3448.

Lassiter, R.N., Stark, M.R., Zhao, T., and Zhou, C.J. (2014). Signaling mechanisms controlling cranial placode neurogenesis and delamination. Developmental Biology *389*, 39–49.

Lee, H.-K., Lee, H.-S., and Moody, S.A. (2014). Neural transcription factors: from embryos to neural stem cells. Molecules and Cells *37*, 705.

Lee, S., Danielian, P.S., Fritzsch, B., and McMahon, A.P. (1997). Evidence that FGF8 signalling from the midbrain-hindbrain junction regulates growth and polarity in the developing midbrain. Development *124*, 959–969.

Leitch, D.B., and Gaede, A.H. (2020). Speciliazed somatosensory systems revealed. In The Senses, B. Fritzsch, ed. (Elsevier), pp. 445–461.

Lessmann, V., Gottmann, K., and Malcangio, M. (2003). Neurotrophin secretion: current facts and future prospects. Progress in Neurobiology *69*, 341–374.

Liao, C.-C., Qi, H.-X., Reed, J.L., and Kaas, J.H. (2020). The somatosensory system of primates. In The Senses, B. Fritzsch, ed. (Elsevier), pp. 180–197.

Lipovsek, M., Ledderose, J., Butts, T., Lafont, T., Kiecker, C., Wizenmann, A., and Graham, A. (2017). The emergence of mesencephalic trigeminal neurons. Neural Development *12*, 1–13.

López-Bendito, G., Flames, N., Ma, L., Fouquet, C., Di Meglio, T., Chedotal, A., Tessier-Lavigne, M., and Marín, O. (2007). Robo1 and Robo2 cooperate to control the guidance of major axonal tracts in the mammalian forebrain. Journal of Neuroscience *27*, 3395–3407.

Ma, Q., Chen, Z., del Barco Barrantes, I., De La Pompa, J.L., and Anderson, D.J. (1998). neurogenin1 is essential for the determination of neuronal precursors for proximal cranial sensory ganglia. Neuron *20*, 469–482.

Macova, I., Pysanenko, K., Chumak, T., Dvorakova, M., Bohuslavova, R., Syka, J., Fritzsch, B., and Pavlinkova, G. (2019). Neurod1 is essential for the primary tonotopic organization and related auditory information processing in the midbrain. Journal of Neuroscience *39*, 984–1004.

Mao, Y., Reiprich, S., Wegner, M., and Fritzsch, B. (2014). Targeted deletion of Sox10 by Wnt1-cre defects neuronal migration and projection in the mouse inner ear. PloS One *9*, e94580.

Maricich, S.M., Wellnitz, S.A., Nelson, A.M., Lesniak, D.R., Gerling, G.J., Lumpkin, E.A., and Zoghbi, H.Y. (2009). Merkel cells are essential for light-touch responses. Science *324*, 1580–1582.

Mastick, G.S., Fan, C.M., Tessier-Lavigne, M., Serbedzija, G.N., McMahon, A.P., and Easter Jr, S.S. (1996). Early deletion of neuromeres in Wnt-1-/-mutant mice: Evaluation by morphological and molecular markers. Journal of Comparative Neurology *374*, 246–258.

Messlinger, K., Balcziak, L.K., and Russo, A.F. (2020). Cross-talk signaling in the trigeminal ganglion: role of neuropeptides and other mediators. Journal of Neural Transmission, 1–14.

Mishima, Y., Lindgren, A.G., Chizhikov, V.V., Johnson, R.L., and Millen, K.J. (2009). Overlapping function of Lmx1a and Lmx1b in anterior hindbrain roof plate formation and cerebellar growth. Journal of Neuroscience *29*, 11377–11384.

Moody, S.A., and LaMantia, A.-S. (2015). Transcriptional regulation of cranial sensory placode development. In Current Topics in Developmental Biology (Elsevier), pp. 301–350.

Moreno-Juan, V., Filipchuk, A., Antón-Bolaños, N., Mezzera, C., Gezelius, H., Andrés, B., Rodríguez-Malmierca, L., Susín, R., Schaad, O., Iwasato, T., *et al.* (2017). Prenatal thalamic waves regulate cortical area size prior to sensory processing. Nature Communications *8*, 14172.

Moroni, M., Servin-Vences, M.R., Fleischer, R., Sanchez-Carranza, O., and Lewin, G.R. (2018). Voltage gating of mechanosensitive PIEZO channels. Nature Communications *9*, 1096.

Murthy, S.E., Loud, M.C., Daou, I., Marshall, K.L., Schwaller, F., Kühnemund, J., Francisco, A.G., Keenan, W.T., Dubin, A.E., and Lewin, G.R. (2018). The mechanosensitive ion channel Piezo2 mediates sensitivity to mechanical pain in mice. Science Translational Medicine *10*, eaat9897.

Nieuwenhuys, R., and Puelles, L. (2015). Towards a New Neuromorphology (Springer).

Noden, D.M., and Trainor, P.A. (2005). Relations and interactions between cranial mesoderm and neural crest populations. Journal of Anatomy *207*, 575–601.

Nothwang, H.G. (2016). Evolution of mammalian sound localization circuits: A developmental perspective. Progress in Neurobiology *141*, 1–24.

O'Connor, R., and Tessier-Lavigne, M. (1999). Identification of maxillary factor, a maxillary process–derived chemoattractant for developing trigeminal sensory axons. Neuron *24*, 165–178.

O'Neill, P., Mak, S.-S., Fritzsch, B., Ladher, R.K., and Baker, C.V. (2012). The amniote paratympanic organ develops from a previously undiscovered sensory placode. Nature Communications *3*, 1–11.

Oury, F., Murakami, Y., Renaud, J.-S., Pasqualetti, M., Charnay, P., Ren, S.-Y., and Rijli, F.M. (2006). Hoxa2-and rhombomere-dependent development of the mouse facial somatosensory map. Science *313*, 1408–1413.

Pallas, S.L. (2001). Intrinsic and extrinsic factors that shape neocortical specification. Trends in Neurosciences *24*, 417–423.

Pani, A.M., Mullarkey, E.E., Aronowicz, J., Assimacopoulos, S., Grove, E.A., and Lowe, C.J. (2012). Ancient deuterostome origins of vertebrate brain signalling centres. Nature *483*, 289–294.

Penfield, W., and Boldrey, E. (1937). Somatic motor and sensory representation in the cerebral cortex of man as studied by electrical stimulation. Brain *60*, 389–443.

Pierce, M.L., Weston, M.D., Fritzsch, B., Gabel, H.W., Ruvkun, G., and Soukup, G.A. (2008). MicroRNA-183 family conservation and ciliated neurosensory organ expression. Evolution & Development *10*, 106–113.

Pombal, M.A., and Megías, M. (2019). Development and functional organization of the cranial nerves in lampreys. The Anatomical Record *302*, 512–539.

Poole, K., Moroni, M., and Lewin, G.R. (2015). Sensory mechanotransduction at membrane-matrix interfaces. Pflügers Archiv-European Journal of Physiology *467*, 121–132.

Qi, H.-X., Liao, C.-C., Reed, J.L., and Kaas, J.H. (2020). Cortical and Subcortical Plasticity After Sensory Loss in the Somatosensory System of Primates. In The Senses, B. Fritzsch, ed. (Elsevier), pp. 399–418.

Reiprich, S., and Wegner, M. (2015). From CNS stem cells to neurons and glia: Sox for everyone. Cell and Tissue Research *359*, 111–124.

Riddiford, N., and Schlosser, G. (2016). Dissecting the pre-placodal transcriptome to reveal presumptive direct targets of Six1 and Eya1 in cranial placodes. Elife *5*, e17666.

Ridone, P., Pandzic, E., Vassalli, M., Cox, C.D., Macmillan, A.M., Gottlieb, P.A., and Martinac, B. (2019). Cholesterol-Dependent Piezo1 Clusters are Essential for Efficient Cellular Mechanotransduction. Biophysical Journal *116*, 377a.

Saotome, K., Murthy, S.E., Kefauver, J.M., Whitwam, T., Patapoutian, A., and Ward, A.B. (2018). Structure of the mechanically activated ion channel Piezo1. Nature *554*, 481.

Sarlak, G., and Vincent, B. (2016). The roles of the stem cell-controlling Sox2 transcription factor: from neuroectoderm development to Alzheimer's disease? Molecular Neurobiology *53*, 1679–1698.

Sharma, N., Flaherty, K., Lezgiyeva, K., Wagner, D.E., Klein, A.M., and Ginty, D.D. (2020). The emergence of transcriptional identity in somatosensory neurons. Nature *577*, 392–398.

Shuda, M., Guastafierro, A., Geng, X., Shuda, Y., Ostrowski, S.M., Lukianov, S., Jenkins, F.J., Honda, K., Maricich, S.M., and Moore, P.S. (2015). Merkel cell polyomavirus small T antigen induces cancer and embryonic Merkel cell proliferation in a transgenic mouse model. PloS One *10*, e0142329.

Steiner, L., Federspiel, A., Slavova, N., Wiest, R., Grunt, S., Steinlin, M., and Everts, R. (2020). Functional topography of the thalamo-cortical system during development and its relation to cognition. NeuroImage *223*, 117361.

Suryanarayana, S.M., Pérez-Fernández, J., Robertson, B., and Grillner, S. (2020). The evolutionary origin of visual and somatosensory representation in the vertebrate pallium. Nature Ecology & Evolution *4*, 639–651.

Syeda, R., Xu, J., Dubin, A.E., Coste, B., Mathur, J., Huynh, T., Matzen, J., Lao, J., Tully, D.C., and Engels, I.H. (2015). Chemical activation of the mechanotransduction channel Piezo1. Elife *4*, e07369.

Szczot, M., Liljencrantz, J., Ghitani, N., Barik, A., Lam, R., Thompson, J.H., Bharucha-Goebel, D., Saade, D., Necaise, A., and Donkervoort, S. (2018). PIEZO2 mediates injury-induced tactile pain in mice and humans. Science translational medicine *10*, eaat9892.

Ter-Avetisyan, G., Dumoulin, A., Herrel, A., Schmidt, H., Strump, J., Afzal, S., and Rathjen, F.G. (2018). Loss of axon bifurcation in mesencephalic trigeminal neurons impairs the maximal biting force in Npr2-deficient mice. Frontiers in Cellular Neuroscience *12*, 153.

Thiery, A., Buzzi, A.L., and Streit, A. (2020). Cell fate decisions during the development of the peripheral nervous system in the vertebrate head. Current Topics in Developmental Biology *139*, 127–167.

Toyoda, R., Assimacopoulos, S., Wilcoxon, J., Taylor, A., Feldman, P., Suzuki-Hirano, A., Shimogori, T., and Grove, E.A. (2010). FGF8 acts as a classic diffusible morphogen to pattern the neocortex. Development *137*, 3439–3448.

Vermeiren, S., Bellefroid, E.J., and Desiderio, S. (2020). Vertebrate sensory ganglia: common and divergent features of the transcriptional programs generating their functional specialization. Frontiers in Cell and Developmental Biology *8*, 587699.

Watson, C., Shimogori, T., and Puelles, L. (2017). Mouse Fgf8-Cre-LacZ lineage analysis defines the territory of the postnatal mammalian isthmus. Journal of Comparative Neurology *525*, 2782–2799.

Welker, E., Armstrong-James, M., Bronchti, G., Ourednik, W., Gheorghita-Baechler, F., Dubois, R., Guernsey, D., Van der Loos, H., and Neumann, P. (1996). Altered sensory processing in the somatosensory cortex of the mouse mutant barrelless. Science *271*, 1864–1867.

Wong, P., and Kaas, J.H. (2009). An architectonic study of the neocortex of the short-tailed opossum (*Monodelphis domestica*). Brain, Behavior and Evolution *73*, 206–228.

Woolsey, T.A., and Van der Loos, H. (1970). The structural organization of layer IV in the somatosensory region (SI) of mouse cerebral cortex: the description of a cortical field composed of discrete cytoarchitectonic units. Brain Research *17*, 205–242.

Yoshikawa, M., Masuda, T., Kobayashi, A., Senzaki, K., Ozaki, S., Aizawa, S., and Shiga, T. (2016). Runx1 contributes to the functional switching of bone morphogenetic protein 4 (BMP4) from neurite outgrowth promoting to suppressing in dorsal root ganglion. Molecular and Cellular Neuroscience *72*, 114–122.

Zhao, Q., Zhou, H., Chi, S., Wang, Y., Wang, J., Geng, J., Wu, K., Liu, W., Zhang, T., and Dong, M.-Q. (2018). Structure and mechanogating mechanism of the Piezo1 channel. Nature *554*, 487.

Ziermann, J.M., Diogo, R., and Noden, D.M. (2018). Neural crest and the patterning of vertebrate craniofacial muscles. Genesis *56*, e23097.

Zilles, K., and Palomero-Gallagher, N. (2020). The architecture of somatosensory cortex. In The Senses, B. Fritzsch, ed. (Elsevier), pp. 225–260.

5 Taste Buds Explained
From Taste Sensing to Taste Processing in the Forebrain

Stephen D. Roper, Robin F. Krimm, Bernd Fritzsch

CONTENTS

5.1 INTRODUCTION TO TASTE: AN OVERVIEW

Taste buds contain chemosensory transducing cells and are the peripheral end organs of gustation. These sensory organs transduce stimuli represented by water soluble chemical compounds and ions into signals that can be transmitted to the brain to generate taste perception (Lundy Jr and Norgren, 2015; Roper, 2020). Among vertebrates, taste buds are principally in the oral cavity, specifically on the tongue and palate (Kirino et al., 2013; Witt and Reutter, 2015). Taste buds are uniquely innervated by three specific epibranchial placodes (geniculate of the facial, petrosal of the glossopharyngeal, nodose of the vagus nerve) that all depend on *Neurog2* (*Neurog1* in chicken) for neuronal formation (Alsina, 2020; Fode et al., 1998; O'Neill et al., 2012). Developing taste buds depend on *Sox2* (Martin et al., 2016; Okubo et al., 2006; Zhang et al., 2021). *Sox2* interacts with *Shh* and *Gli1* to orchestrate taste bud development (Castillo-Azofeifa et al., 2017; 2018; Lu et al., 2018; Zhang et al., 2020). All taste bud neurons depend on support by neurotrophins, in particular *BDNF* and *TrkB* (*Ntrk2*) for viability and target innervation (Fritzsch et al., 1997; Nosrat et al., 1997; 2012; Rios-Pilier and Krimm, 2019).

In mammals, taste buds contain 30–150 taste cells. There are three distinct fields of taste buds on the tongue:

1. Fungiform papillae on the anterior tongue, innervated by the facial nerve
2. Foliate papillae on the lateral margins towards the posterior, innervated by overlapping branches of the facial and glossopharyngeal nerves
3. (Circum)vallate papillae on the posterior tongue, innervated by the glosso-pharyngeal nerve (Barlow and Klein, 2015; Witt and Reutter, 2015)

In addition, there are taste buds on the soft palate, innervated by the greater super-ficial petrosal nerve, and on the epiglottis (Figure 5.1), innervated by the vagus nerve (Ache and Young, 2020; Finger and Kinnamon, 2011; Krasteva-Christ et al., 2020). Cyclostomes and gnathostomes have somewhat different innervation patterns

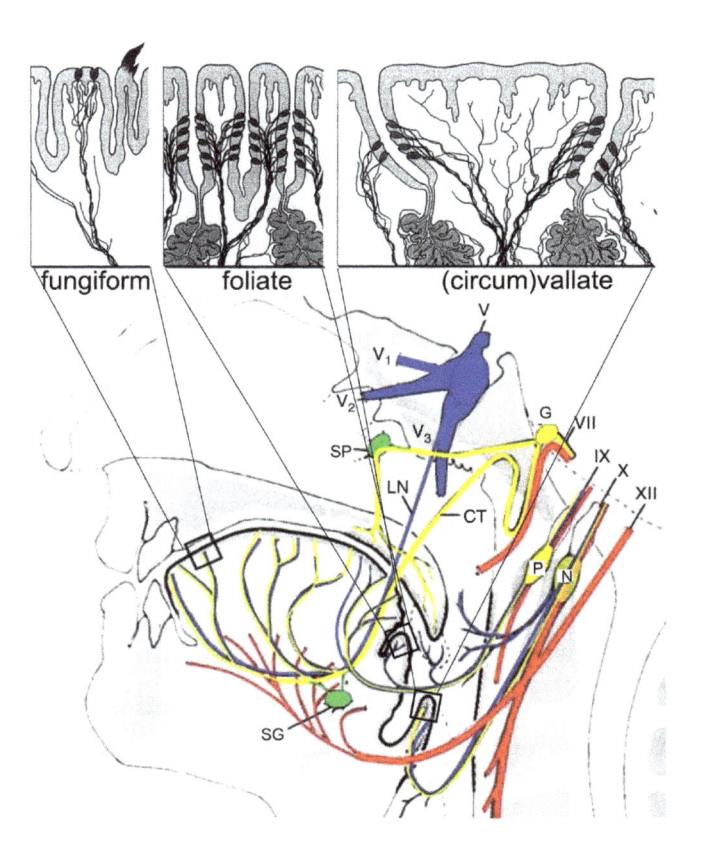

FIGURE 5.1   The organization of the tongue is provided by three distinct papillae types, the fungiform, foliate, and (circum)vallate papillae (top). The geniculate (G); petrosal (P); and nodose (N) ganglia innervate the anterior 2/3 of the tongue and the palate. These neurons bypass the sphenopalatine (SP) and the submandibular ganglion (SG). Petrosal nerve fibers reach the foliate papillae and the vallate papillae, while the nodose ganglia innervate taste buds on the epiglottis and pharynx. (Modified from Witt and Reutter, 2015.)

(Braun, 1998; Daghfous et al., 2020; Fain, 2019; Kirino et al., 2013), including the unique taste buds of frogs (Witt and Reutter, 2015).

Different taste buds were once perceived to provide distinct input based on location. There are at least five basic tastes for mammalian taste buds: sweet, salty, sour, bitter, and umami (Fain, 2019; Kinnamon and Finger, 2019; Roper, 2020; Witt, 2020). Taste buds on different regions of the mammalian tongue were once believed to differ in their sensitivity to these tastes. However, this concept of a 'tongue map' arose from a misinterpretation of early human psychophysical studies (Bartoshuk and Pangborn, 1993; Lindemann, 1999).

There are four different types of cells in mammalian taste buds (Yang et al., 2020). Chemosensory transduction mechanisms for the five basic tastes differ and involve different members of these four cell types. For example, specific G protein-coupled taste receptors (TasR1s, TasR2s) initiate an intracellular transduction cascade for sweet, bitter, and umami tastes in Type II taste bud cells. Salt taste (NaCl) is a result of activating different taste cells. There appears to be a subset of Type II cells that respond to the pleasant taste of low to moderate concentrations of NaCl, at least in rodents (Nomura et al., 2020). Other cells, possibly Type III taste bud cells, respond to the aversive taste of high NaCl concentrations. Type III taste bud cells also respond to sour (acid) stimuli (Huang et al., 2006; Huang et al., 2008). Transmission of taste responses is mediated by the synaptic release of ATP and activation of purinergic P2X2/3 receptors expressed on the afferent (postsynaptic) nerve fibers (Finger et al., 2005; Roper and Chaudhari, 2017, Rodriguez et al., 2021). Type IV taste bud cells are immature cells that will eventually differentiate. Type I taste bud cells have features similar to those of glial cells (Kinnamon and Finger, 2019; Roper and Chaudhari, 2017). Types II and III taste bud cells generate action potentials and release neurotransmitters. Type II cells, in particular, utilize a unique, non-vesicular mechanism for transmitter (ATP) release, namely secretion through large-pore, voltage-dependent *CALHM1/3* ion channels (Ma et al., 2018; Taruno et al., 2013). By contrast, Type III cells appear to utilize conventional vesicular transmitter release to secrete transmitters, including serotonin (5-HT), GABA, and norepinephrine (Kinnamon and Finger, 2019; Roper and Chaudhari, 2017).

Taste information projects from three cranial nerves (facial, VII; glossopharyngeal, IX; vagus, X) to produce terminal fields in the solitary tract that retain a rough orotopic organization (Fritzsch et al., 2019; Lundy Jr and Norgren, 2015; May and Hill, 2006). Taste afferents are capable of reaching the general region of their target in the absence of their specific post-synaptic targets in the solitary nucleus (Qian et al., 2001). The expression of the solitary nucleus specifying transcription factor, *Tlx3*, is directed by *BMP g*radients (Glover et al., 2018; Hernandez-Miranda et al., 2017).

Functionally, all taste buds transduce all taste stimuli, with varying thresholds (Kinnamon and Finger, 2019; Roper, 2020). It remains unclear what specific information the rough orotopic projection of afferents extracts and how the differential activity of each taste bud contributes to different concentrations of tastants can be used to sharpen the taste map (Lundy Jr and Norgren, 2015; Schier and Spector, 2019). The development of taste buds and sensory neurons depends on molecular guidance for taste afferents to form the orotopic organization provided to the solitary nuclei. It remains unclear how the orotopic projection will likely develop in the absence of taste buds (Barlow, 2015; Travers et al., 2018) or may be reorganized in the absence of *Tlx3* (Qian et al., 2001).

The orotopic organization is lost in higher-order projections, making the need of the orotopic primary map unclear (Lundy Jr and Norgren, 2015). Highly conserved second-order neurons project taste information (Lundy Jr and Norgren, 2015; Staszko and Boughter, 2020; Vendrell-Llopis and Yaksi, 2015) that is combined with tongue-related somatosensation and olfaction into an integrated experience related to food intake (Schier and Spector, 2019; Shepherd, 2006). The major terminal field in the insula region is known to interact with other inputs, in addition to taste (Fain, 2019; Staszko et al., 2020).

This review will describe the origin of the three sets of neurons in mammals that innervate the taste buds and the central solitary tract and nucleus. It will describe the functional and anatomical characteristics of different cell types in the periphery. Lastly, it will provide an overview of the central pathways from the nucleus of the solitary tract to the insula in the cortex, which are used for processing taste information.

5.2 TASTE SENSORY NEURONS AND TASTE BUDS: MOLECULAR BASE OF DISTINCT NEUROSENSORY FORMATION AND TASTE BUD-NEURON INTERACTIONS

In gnathostomes, three distinct epibranchial placodes give rise to the gustatory ganglia, which contain the taste bud-innervating neurons (Alsina, 2020; Krimm et al., 2015; Northcutt, 2004; Witt and Reutter, 2015). In rodents, taste buds are the most numerous on the tongue and are located in specialized epithelial structures called fungiform, foliate, and circumvallate papillae. The taste buds within the fungiform papillae are located in the front two-thirds of the tongue and derive from the mandibular division of the first branchial arch. Taste buds in the fungiform papillae are innervated by the chorda tympani, whose cell bodies are located in the geniculate ganglion. The geniculate ganglion also innervates taste buds located in the soft palate and in the nasoincisive papilla via the greater superficial petrosal nerve (Miller and Spangler, 1982). Taste buds within the circumvallate and foliate papillae are located in the caudal portion of the tongue, which derives from the third branchial arch, and are innervated by the glossopharyngeal nerves of the petrosal ganglia. The nodose ganglion innervates taste buds located in the epiglottis, larynx, and pharynx (Prescott et al., 2020; Stedman et al., 1983; Suzuki and Takeda, 1983). Similarly, cyclostomes, which lack a tongue and the associated facial taste system, present with taste buds in the pharynx that are innervated by the vagal ganglion (Barreiro-Iglesias et al., 2010). In contrast to specific gnathostomes that lose the facial taste system entirely (i.e., lampreys), the facial nerves of the taste system innervate the skin of certain bony fish (Finger and Kinnamon, 2011; Northcutt, 2004), resulting in the entire animal being covered in taste buds. In spite of these variations in taste bud location and innervation patterns, all of the ganglion neurons that innervate taste buds arise from epibranchial placodes. Epibranchial placodes are generated in the ectoderm and develop through a stepwise progression of neuronal generation (Figure 5.2). These neurons then delaminate from the epithelium and migrate to achieve a final position near the location where the three cranial nerves (facial, glossopharyngeal, and vagus) enter the brainstem.

Gustatory ganglion formation depends on the sequential expression of specific genes, resulting in epibranchial placode formation, delamination, migration, and

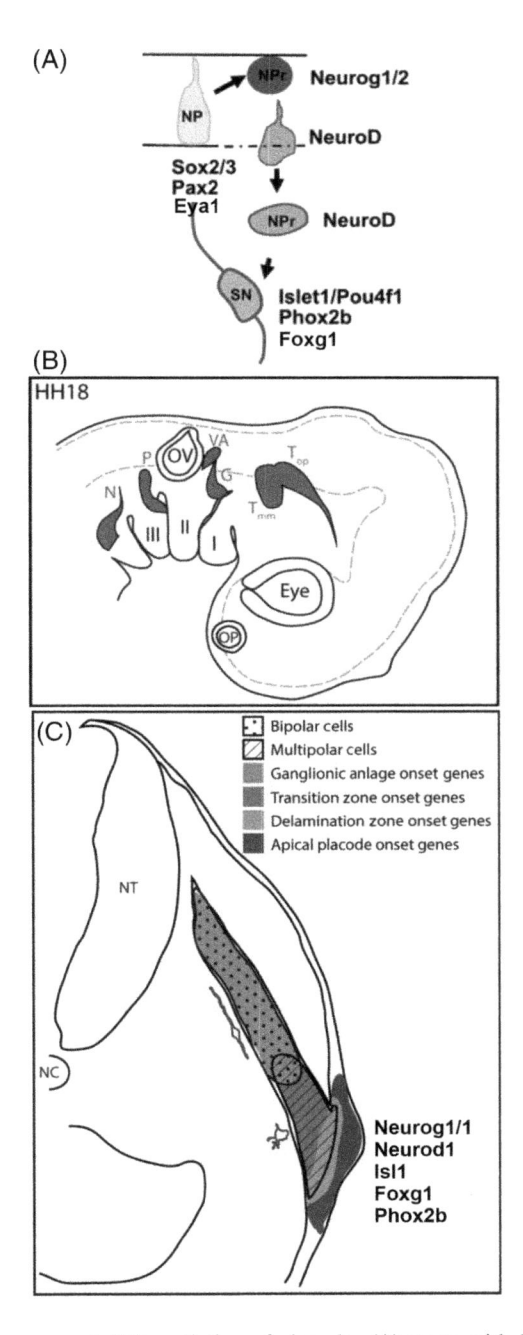

FIGURE 5.2 The neuronal differentiation of placodes (A) starts with *Eya/Six* followed by *Pax2* and *Sox2*, which initiate transformation of the epibranchial placodes (geniculate, G; petrosal, P; nodose, N) indicated at HH18 old chicken (B). As neuroblasts migrate from the ectoderm (C) to deeper locations, additional factors are expressed in sequence, *Neurog2* (in mice, *Neurog1* in chicken) followed by *Neurod1*, *Isl1*, *Foxg1*, *Pou4f1*, and *Phox2b*. (Modified from Alsina, 2020; Smith et al., 2015.)

cellular differentiation. Initial placode development depends on the expression of the transcription factor *Foxi3*, which is essential for the development of the ear and epibranchial placodes. In the absence of *Foxi3* expression, the ectoderm fails to thicken, and all placodes fail to form (Birol et al., 2016). *Six1/2/4* and *Eya1/2* are essential for the specific formation of epibranchial placodes (Moody and LaMantia, 2015; Zhang et al., 2021) downstream of *Foxi3*. *Sox2*, which is ubiquitously expressed in the epibranchial placodes, is required for neuron development (Dvorakova et al., 2020; Imayoshi and Kageyama, 2014; Kageyama et al., 2019). Epibranchial placodes are patterned into rostral and caudal domains by *Notch*-signaling (Wang et al., 2020) which is regulated by *Eya1* (Zhang et al., 2017). From the *Sox2*+ precursor pool, placode cells divide into a non-neural population and neuroblasts defined by the transcription factor *Neurog2* in mammals (Zhang et al., 2017). *Neurog2* is critical for neuronal development (Fan et al., 2019; Fode et al., 2000) and *Neurog1* plays a similar role in chickens (Freter et al., 2012; Smith et al., 2015). The activation of *Notch*-signaling discourages neuronal fate (Zhang et al., 2017) in favor of a non-neuronal cell fate. In both mammals and chickens, the transcription factor, *Neurod1,* is also expressed during the early placode development (Fode et al., 2000; Smith et al., 2015), and is likely required for neuron survival during differentiation (Kim et al., 2001). Downstream of *Neurod1, Isl1, Pou4f1,* and *Phox2a/b* interact to regulate neuronal migration and differentiation (Dykes et al., 2011; Fan et al., 2019; Huang et al., 2001; Smith et al., 2015). Although all neuroblasts express the pro-neural transcription factor *Isl1*, in developing geniculate ganglion, *Phox2b* is required specifically for a visceral sensory neuron (taste) fate (D'Autreaux et al., 2011). *Foxg1, Phox2a,* and *Phox2b* are expressed in the placodes during delamination (Smith et al., 2015), followed by the set of genes *Coe1, Drg11,* and *Dcx,* which are only activated after the migrating cells have left the placode (Freter et al., 2012; Smith et al., 2015). Once the final position is achieved (facial, glossopharyngeal, vagus ganglion) the neurons grow a single process that branches to innervate the targeted taste bud cells and sends the proximal innervation to reach distinct areas of the hindbrain (Chizhikov et al., 2021; O'Neill et al., 2012).

Following neuronal differentiation, developing gustatory neurons express *Ntrk2* (*TrkB*) which is required for the survival of all epibranchial neurons (Fritzsch et al., 1997). Interestingly, the ligands for the *TrkB* receptor, brain-derived neurotrophic factor (*BDNF*), and neurotrophin 4 (*NT-4*) are expressed in the epibranchial neuron placodes, as well as the taste buds (Huang and Krimm, 2010; Nosrat et al., 1997; 2012; von Bartheld and Fritzsch, 2006). For the axons of these neurons to reach the correct target, they must make multiple decisions at several different locations along the path to the taste buds. For example, these axons must exit the correct location in the ganglion. The axons of the chorda tympani and the greater superficial petrosal nerve each exit the geniculate ganglion from different locations (Yang et al., 2014). Similarly, specific branches of the glossopharyngeal nerve (petrosal or inferior branch) innervate the foliate taste buds, and the circumvallate taste buds. Once these axons reach the general region of their targets, *BDNF* and *NT-4* in the ganglia are down-regulated (Hellard et al., 2004; Huang and Krimm, 2010). At this point, *BDNF* (but not *NT-4*) from the taste, epithelial target supports taste neuron survival (Patel and Krimm, 2010). In addition, *BDNF* is also both necessary and sufficient for axons to locate and innervate the taste placodes in the tongue, where taste buds will form (Lopez and Krimm, 2006; Ma et al., 2009). Overall, the final distinctive patterns of gustatory innervation to the tongue are regulated by both by proximal events that control neuron differentiation and the location from which axons exit from the ganglion and distal trophic interactions between the developing taste epithelia and nerve processes (Dvorakova et al., 2020; Sudiwala and Knox, 2019)

Similar to the taste ganglia, taste bud development is orchestrated by a specific sequence of gene expression and trophic interactions with nerve fibers. In the tongue, taste buds are located in papillae, which develop from placodes that arise prior to innervation and taste bud formation (Barlow, 2015; Paulson et al., 1995). The initial signals that orchestrate taste bud development and establish their patterns on the tongue arise from the tongue epithelium (Barlow, 2015; Barlow and Northcutt, 1997). Interestingly, signals patterning the location of taste buds differ depending on the tongue region (i.e. fungiform papillae vs circumvallate papillae) likely because these tissues arise from different branchial arches. Specifically, fungiform papillae are patterned during development by sonic hedgehog (*Shh*) and Wnt-signaling (Hall et al., 2003; Iwatsuki et al., 2007; Liu et al., 2007; Mistretta et al., 2003), whereas the circumvallate papilla is regulated by fibroblast growth factor 10 (*FGF10*) and its receptors *Spry1-2* (Pauley et al., 2003; Petersen et al., 2011). The cells within the developing taste epithelial placode that express *Shh* differentiate into taste buds during development (Thirumangalathu et al., 2009). Prior to differentiation, these *Shh+* placodal cells become innervated, and this innervation is required to maintain *Sox2*-expression (Dvorakova et al., 2020; Ito and Nosrat, 2009). Both *Sox2* expression and innervation are required for continued taste bud development (Fan et al., 2019; Okubo et al., 2006). For example, the loss taste bud innervation following the knockout of either the neurotrophin, *BDNF*, or its receptor, *TrkB* results in a loss of taste buds (Fritzsch et al., 1997; Mistretta et al., 1999; Rios-Pilier and Krimm, 2019). The factors that are produced by the neurons to support continued taste bud development are unclear; however *Shh* and R-spondin are likely possibilities (Castillo-Azofeifa et al., 2017; Lin et al., 2021; Lu et al., 2018). Although the taste system matures, some developmental processes continue into adulthood, including taste bud cell differentiation. One unique feature of taste buds is that cells are replaced every 8–20 days, depending on the specific cell types (Beidler and Smallman, 1965; Okubo et al., 2009; Perea-Martinez et al., 2013). The stem cells that give rise to adult taste buds express *Lgr6* in the fungiform papillae, but express *Lgr5* within the circumvallate papillae (Ren et al., 2014; Takeda et al., 2013). *Lgr5/6* cells give rise to all taste bud cell types (Ren et al., 2014). The absence of *Neurog2* leads to reduced *Sox2* expression (Ohmoto et al., 2017; Okubo et al., 2006) and disrupts taste bud formation beyond a limited differentiation of small, single taste buds (Fan et al., 2019). Taste bud cells that transduce bitter express *Eya1* suggesting that it may be involved in the differentiation of this cell type (Ohmoto et al., 2021). Similarly, the differentiation of type II cells is dependent on the transcription factor, *Pou2f3*, while type III cells depend on *Ascl1* for their differentiation (Kito-Shingaki et al., 2014; Matsumoto et al., 2011; Ohmoto et al., 2021; Seta et al., 2011).

In summary, the development of both taste neurons and taste buds is controlled by a series of gene expression events that occur independently. However, once the axons of peripheral neurons reach the taste epithelium these two cell types become interdependent. Thus, the formation of the peripheral taste system represents the interaction between early and late genes that regulate cell fate and the trophic interactions that occur between taste buds and nerve fibers.

5.3 TASTE BUD CELLS

Taste buds are compact sensory organs distributed in the stratified squamous epithelium of the tongue, palate, and epiglottis (Figure 5.3). Thirty to 150 tall, columnar cells are tightly clustered within each taste bud (Roper, 2020). The apical tips of taste cells are studded with microvilli and extend into a taste pore that opens onto the mucosal surface (Yang et al., 2020).

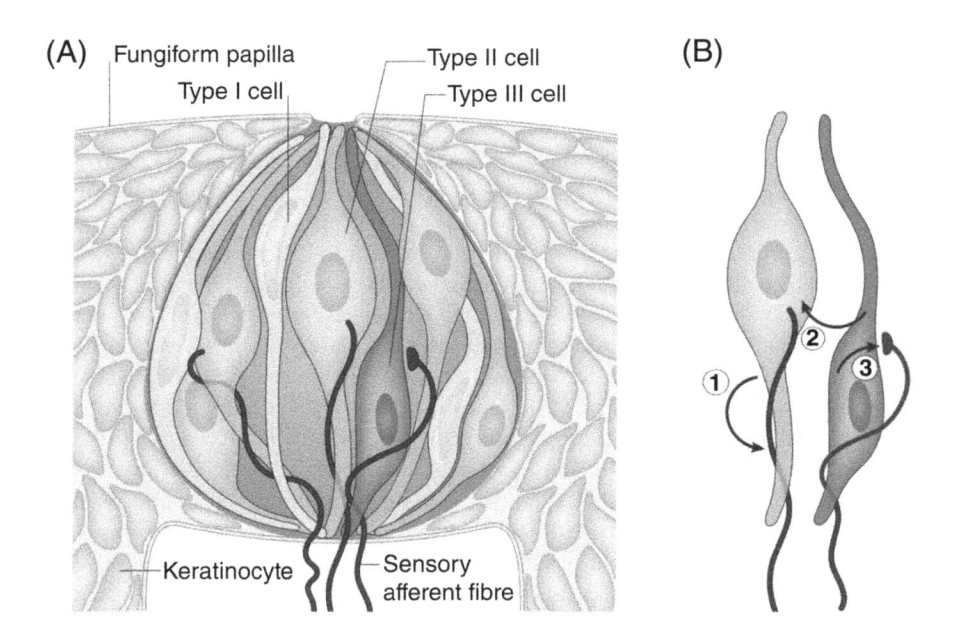

FIGURE 5.3 Taste buds have four types of cell types and transmit signals with several neurotransmitters. A, Type I taste bud cells are glial-like cells; Type II and Type III cells are chemosensory cells for sweet, bitter, salty, sour, and umami tastes; Type IV cells (not shown) are undifferentiated basal cells. B, In response to taste stimuli, the transmitter ATP released from Type II cells acts on afferent nerve fibers as well as provides autocrine positive feedback (1). Cell–cell communication between Type II and Type III cells is mediated by ATP, serotonin, and GABA (2). Interactions between type III cells and gustatory afferent fibres include serotonergic (feedforward) and glutamatergic (feedback) transmission (3). C, Type III taste cells transduce acid (sour) stimuli. D, Type II cells respond to sweet, bitter, or umami. AtypMito, atypical mitochondria; ER, endoplasmic reticulum; Gα, alpha subunit of G protein; Gβγ, beta-gamma subunits of G protein; HAc; organic acid; IP 3, inositol trisphosphate; IP 3R3, inositol trisphosphate receptor isoform 3; PLCβ2, phospholipase C isoform β2; TRPM4, transient receptor potential cation channel subfamily M member 4; TRPM5, transient receptor potential cation channel subfamily M member 5; VGCC, voltage-gated calcium channel; VG-Na +, voltage-gated sodium channel. *(Continued)*

As stated previously, there are four types of taste bud cells (Yang et al., 2020; Fain, 2019; Kinnamon and Finger, 2019; Roper and Chaudhari, 2017). Among the four cell types, the largest population (about 50% of cells) are Type I cells. Type I cells have glia-like functions. For instance, the ecto-ATPase on Type I cells prevents extracellular accumulation of neurotransmitter (ATP) released during taste bud stimulation. Type I cells also take up and distribute K+ (spatial buffering) released by action potentials in Types II and III taste cells during taste stimulation (Dvoryanchikov et al., 2009). Type I taste cells might represent a heterogeneous population (Kinnamon and Finger, 2019; Roper and Chaudhari, 2017; Yang et al., 2020). Types II and Type III cells are receptor cells for the 5 basic taste qualities (sweet, salty, sour, bitter, umami; Figure 5.3), detailed below. Type IV cells are immature cells that do not extend to the taste pore. These cells differentiate into other cell types over a few to 20 days (Ohmoto et al., 2021; Perea-Martinez et al., 2013). Type IV cells express *Shh* (Fan et al., 2019; Ren et al., 2017). The relative numbers of Type II, III and IV is about 19, 15, and 14% of taste bud cells, respectively (Yang et al., 2020)

Type II taste cells are chemosensors for sugars, amino acids and bitter compounds and perhaps NaCl (see below). Type II cells express G protein-coupled taste receptors (Tas1Rs and Tas2Rs) for these substances. There are 3 Tas1R receptors (Tas1R1, R2, R3) and, in mammals, 25 Tas2R receptors. For instance, sugars bind to the dimeric receptor Tas1R2+Tas1R3. Umami compounds (amino acids) activate the dimeric receptor Tas1R1+Tas1R3 as well as two taste-specific metabotropic glutamate receptors, mGluR1 and mGluR4. Bitter substances stimulate different members of the Tas2R receptor family. It is not known with certainty whether Tas2Rs

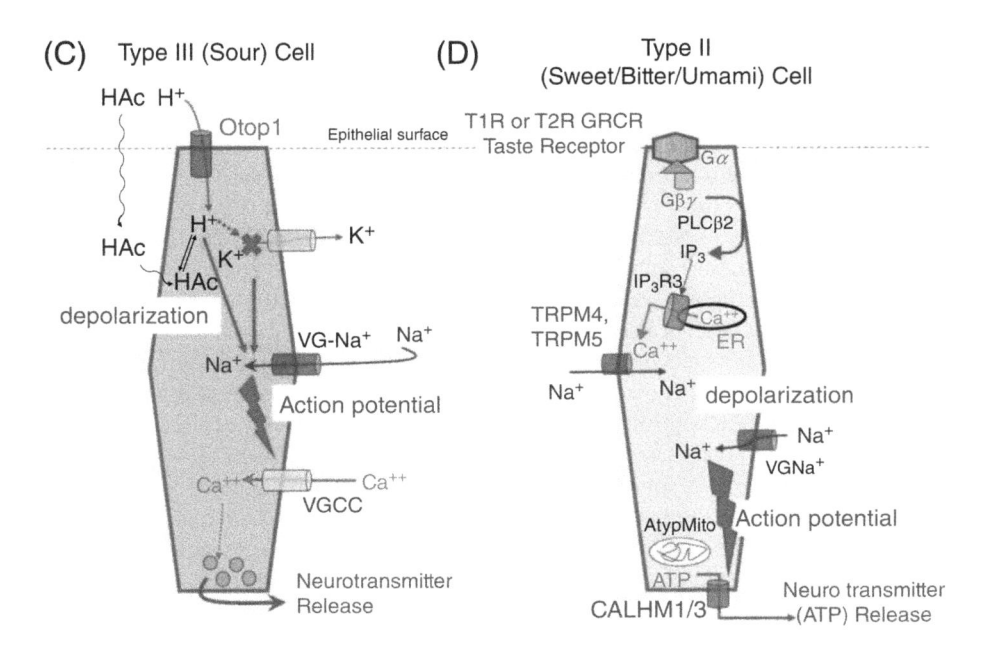

FIGURE 5.3 *(Continued)*

form dimers (Kinnamon and Finger, 2019; Ohmoto et al., 2021; Roper, 2020). There is little co-expression of Tas1Rs and Tas2Rs in individual Type II taste cells, consistent with the observation that separate Type II cells respond to sweet, umami, or bitter. Tas1Rs and Tas2Rs alike use a common transduction pathway that includes a phospholipase C (*Plcβ2*), IP3-mediated intracellular Ca release, and activation of TRPM4 and TRPM5, cation channels that depolarize the cell. Notably, Type II cells lack voltage-gated Ca^{2+} currents (Roper, 2020)

Sweet: Taste buds transduce sugars through Tas1R2+Tas1R3. Mice lacking Tas1r2+3, TaslR3 are taste blind to sugars or artificial sweeteners. Interestingly, certain sugars (glucose, sucrose, but not fructose) also activate sugar transporters on Type II cells and may stimulate transmitter secretion via mechanisms similar to those found in pancreatic Islets of Langerhans cells. That is, intracellular glucose metabolism following sugar uptake ultimately leads to blockage of K^+ channels and consequently, to membrane depolarization (Roper and Chaudhari, 2017). This sugar transduction pathway appears to be important for cephalic phase insulin release (Glendinning et al., 2017).

Umami: Taste buds respond to umami taste stimuli such as glutamate (e.g., monosodium glutamate) and aspartate by activating Tas1R1+Tas1R3 receptors as well as taste-specific variants of the metabotropic glutamate receptors mGluR1 and mGluR4 (Kinnamon and Finger, 2019; Roper and Chaudhari, 2017). Mice lacking Tas1r1+Tas1r3 have reduced, but not eliminated taste responses to umami.

Bitter: Taste buds respond to a large variety of bitter compounds which activate Tas2R receptors. There are approximately 25 genes that encode Tas2Rs in humans. A given Tas2R can bind and respond to several different bitter compounds, and a given Type II taste bud cell can express a number of different Tas2R receptors. This provides taste buds with a broad range of sensitivity for bitter tasting chemicals. It is not yet known whether Tas2R receptors form homo- or heterodimers. The *Eya1* gene, a transcription factor involved in the development of a large variety of neurons (Xu et al., 1997; Zhang et al., 2021; Zou et al., 2004), is expressed at high level in bitter-responding Type II taste bud cells and in putative immature cells. This suggests that *Eya1* is a candidate for bitter cell differentiation (Ohmoto et al., 2021).

Salty: NaCl taste is an enigma. Recent findings indicate that a subset of Type II cells in mouse taste buds express the Na^+ channel ENaC and are stimulated by low (appetitive) concentrations of NaCl to release neurotransmitter (ATP) (Ma et al., 2018). High (aversive) concentrations of NaCl appear to activate other taste bud cells via alternative, and as yet unknown, receptor mechanisms (Bigiani, 2020). Chloride, the anion component of salt taste, is also believed to play an important role in salt taste (Roebber et al., 2019). Salt taste mechanisms in humans may differ fundamentally from those in rodents and even less is known about NaCl transduction in our species (Bigiani, 2020).

Type III taste cells form conventional synapses with sensory afferent fibers and respond to ***sour*** (acid) tastes (Roper, 2020). Type III cells secrete serotonin and GABA in response to sour taste stimulation and these transmitters inhibit Type II taste cells within the taste bud (Roper, 2020; Roper and Chaudhari, 2017). It is unclear whether Type III cells also release ATP as a transmitter, though results from isolated taste cells suggest not (Huang et al., 2007). Type III cells transduce sour taste by a combination of (1) intracellular acidification from permeation of acid molecules through the cell membrane (Roper, 2020) and (2) proton (H^+) influx through apical

OTOP1 channels (Teng et al., 2019; Zhang et al., 2020). Intracellular acidification is particularly important for sensing organic acids such as citric acid (e.g., lemons) and acetic acid (e.g., vinegar), both of which are highly membrane permeable. H^+ influx through OTOP1 channels is important for mineral acids such as HCl and H_2SO_4, which dissociate in solution and generate free protons. Intracellular protons block inwardly rectifying K^+ channels, thereby causing the cell to depolarize. Additionally, proton influx through OTOP1 produces a depolarizing inward current.

Interestingly, Type III cells also receive excitatory input from Type II cells via purinergic synapses (Tomchik et al., 2007). Thus, Type III cells are directly stimulated by acids (above) as well as indirectly by tastes that activate Type II cells (Figure 5.3). Excitatory and inhibitory interactions between Types II and III cells provide feed-forward and feedback signals that may be important for the output of taste buds onto primary afferent terminals (Roper, 2020).

In summary, two taste cell types directly transduce chemosensory stimuli: Type II cells (sweet, bitter, umami) and Type III cells (sour). The cells responsible for salt (NaCl) sensing, especially in humans, remain unclear. Additionally, synaptic interactions between Types II and III taste cells suggest that there is some degree of information processing in taste buds before signals are transmitted to sensory afferent fibers.

5.4 FROM TASTE BUD RECEPTOR CELLS TO SENSORY AFFERENT NERVE FIBERS

At least six neurotransmitters have been identified in taste buds: ATP, 5-HT, GABA, ACh, glutamate, and noradrenaline (Roper and Chaudhari, 2017). The main transmitter, ATP, acts on $P2X_2$ and $P2X_3$ purinoceptors expressed on the (postsynaptic) sensory afferent terminals within the taste bud. Taste-evoked release of ATP from Type II cells, and 5-HT and GABA from Type III cells has been clearly demonstrated. How the other transmitters are involved in taste transmission remains somewhat unclear, though cell-cell synaptic interactions within the taste bud appear to be important (Roper, 2020).

In *Type III* taste cells, sour taste stimulation evokes Ca^{2+} influx through voltage-gated Ca^{2+} channels and initiates canonical vesicular transmitter release via SNARE mechanisms (Figure 5.4). By contrast, in *Type II* cells, bitter, sweet, umami, and NaCl taste stimulation elicits action potentials (without Ca^{2+} influx) that directly open CALHM1/3 channels through which ATP is secreted (Taruno et al., 2013; Ma et al., 2018). Thus, ATP is acknowledged as a primary transmitter for Type II cell tastes. ATP also transmits sour taste information (Finger and Kinnamon, 2011). However, the source of ATP for *sour* information remains unclear insofar as Type III taste bud cells that transduce sour taste are not believed to secrete ATP (Huang et al., 2007).

A major issue in taste research concerns how gustatory information is coded *en route* to the central nervous system. It is highly controversial whether individual taste cells and gustatory afferent nerve fibers respond to single tastes (sweet, salty, sour, etc.) or respond more broadly across several taste qualities (Wooding and Ramirez, 2020). Some researchers have posited there are dedicated taste cells and neurons that respond to a single taste, i.e. 'labeled line' coding (Barretto et al., 2015; Yarmolinsky et al., 2009). In contrast, others point out that some taste bud cells, many gustatory sensory neurons, and a majority of gustatory cortical neurons respond to multiple

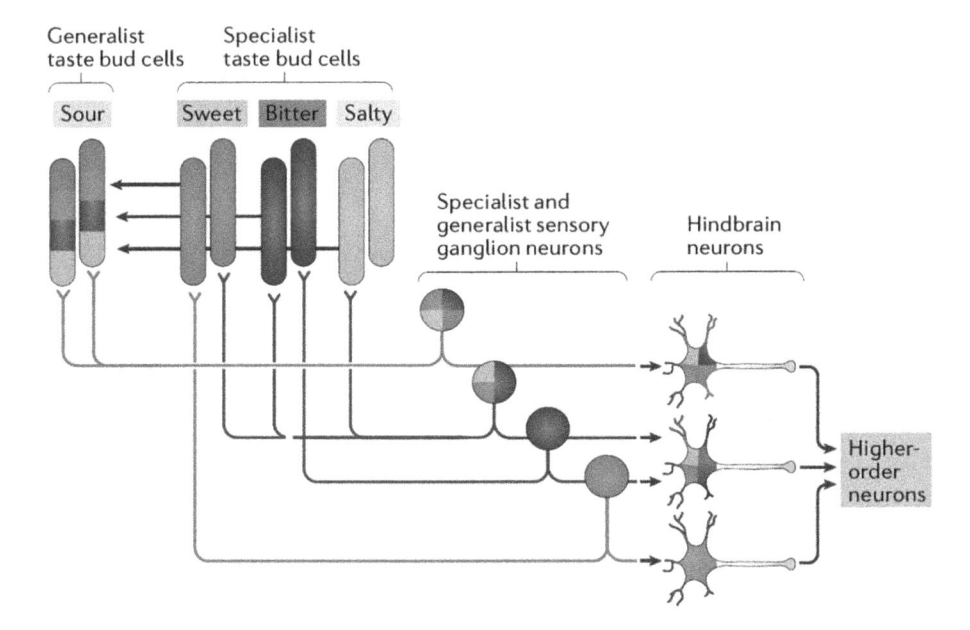

FIGURE 5.4 Individual Type II taste bud cells are mostly tuned to one taste quality (for example, bitter, sweet or salty): that is, they are 'specialists'. Type III cells sense sour tastes and also respond secondarily to other taste stimuli via cell-to-cell (paracrine) communication within the taste bud (represented by the arrows between the taste bud cells). Thus, Type III cells can be termed 'generalists'. Some afferent ganglion neurons receive input from taste cells that respond to a single taste quality and hence would be specialist neurons. Other afferent ganglion neurons receive input from many taste cells or from Type III cells and thus are multiply sensitive 'generalist' neurons. In the hindbrain, sensory ganglion cells converge on neurons in the nucleus of the solitary tract. (Taken from Roper and Chaudhari, 2017.)

taste qualities, making a simple labeled line taste coding hypothesis untenable (Ohla et al., 2019). Those findings are more compatible with combinatorial or population taste coding (Roper, 2020).

In addition to chemical stimuli, taste buds also sense touch and temperature (Mistretta and Bradley, 2021). Interestingly, single cell RNAseq analysis of sensory neurons in the geniculate ganglion identified a subset of neurons that innervate taste buds and that express the mechanosensitive channel Piezo1 (Dvoryanchikov et al., 2017); these gustatory sensory neurons may play a role in tactile responsiveness of taste buds. A full description of the functional specificity of sensory ganglion neurons that innervate taste buds and how this information is encoded to discern and discriminate taste stimuli remain open questions.

5.5   SOLITARY TRACT INFORMATION REVEALS AN OROTOPIC ORGANIZATION

The first central relay for taste information is in the nucleus of the solitary tract (nST) in the hindbrain. The hindbrain is patterned via a series of transcription factors in a similar manner as the spinal cord (Hernandez-Miranda et al., 2017). A set of genes are

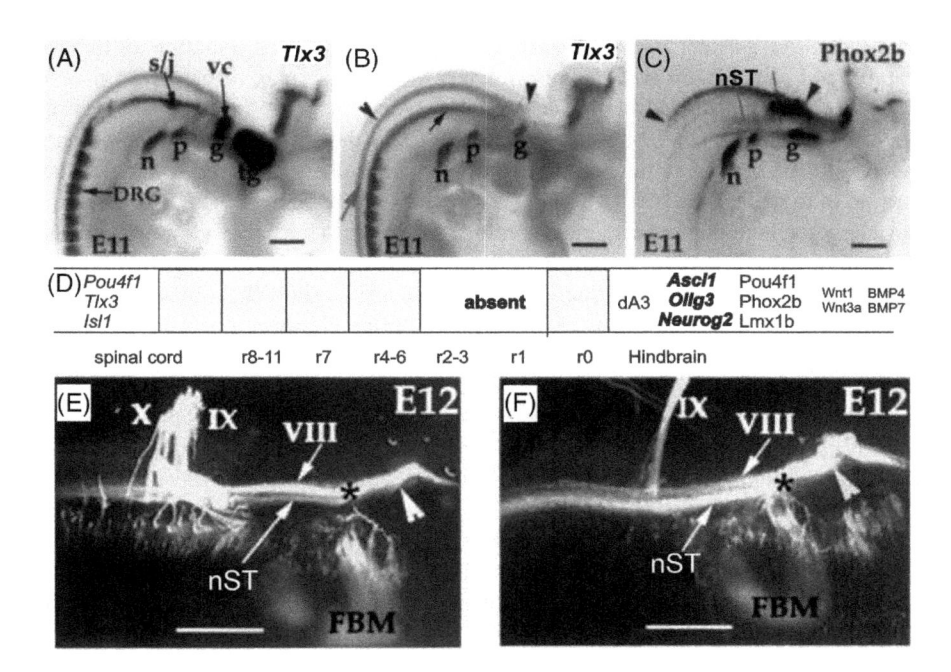

FIGURE 5.5 *Tlx3* and *Phox2b* are expressed in the nucleus solitary tract (nST, [A], [C]) which is more obvious in the *Neurog1* null mouse due to the loss of the vestibular-cochlear (vc) and superior/jugular ganglia (s/j) (B). Areas of overlap between in *Tlx3* and *Phox2b* expression are shown (arrowheads) (b,c). Rostral to caudal representation of Tlx3 (D) expression at the level of dA3. *Ascl1, Olig3,* and *Neurog2* are bHLH genes for early differentiation, followed by additional downstream gene for differentiation. Note that *Tlx3* is absent from r1-3 (A), (B), (D), but is expressed in r0. DiI-labeling of central projections of the facial (FBN), vestibular (VIII), and IX, X nerve illustrates a near normal solitary tract (E), (F) that is enlarged in the *Tlx3* null mice (F) despite the loss of the nTS neurons. Bar indicates 200 μm (A–C) and 500 μm (E), (F). (Modified from Hernandez-Miranda et al., 2017; Qian et al., 2001.)

expressed in a distinct distribution that correlates with nST development (Hernandez-Miranda et al., 2017; Qian et al., 2001). On the dorsal to ventral axis a combination of bHLH genes *Ascl1, Neurog2,* and *Olig3* defines dA3, in the brainstem, which extends through the spinal cord (rostral to caudal) at a location which includes the developing nTS (Hernandez-Miranda et al., 2017). *In situ* hybridization showed co-expression of *Rnx (Tlx3)* and *Phox2b* in the neurons of the nST. Note that the expression of *Tlx3* in r4-11 is obvious (Figure 5.5A) in the absence of *Neurog1 Neurog2* (Figure 5.5B), because non-taste ganglia are lost. *Phox2b* expression within the nST extends rostral to reach r1 (Figure 5.5C). Projection neurons of nST are entirely absent in both *Tlx3 (Rnx)*, and *Phox2b* knockouts indicating that these neurons are dependent on these factors for their development (Qian et al., 2001). These genes appear to function independently of each other. However, *Rnx* expression is dependent on the dorsal influence of BMP signaling (Hornbruch et al., 2005) which is counteracted by *Shh*.

Axons from the geniculate, nodose and petrosal ganglia project to the brainstem and reach the solitary tract (nST) (Chizhikov et al., 2021; Qian et al., 2001) after *Neurog2+* ganglion neurons first differentiate (Fode et al., 1998; Smith et al., 2015). The factors that guide these axons along the tract are not known; however, members of the neuropilin and semaphorin families of guidance factors may play a role (Corson et al., 2013). In *Tlx3* null mice, the nucleus of the solitary tract is reduced in size whereas the solitary tract (nST) is larger (Figure 5.5D, E). In the absence of postsynaptic nST neurons, the axons of primary taste neurons project to vestibular regions which have moved in to replace the nST (Maklad and Fritzsch, 2003; Qian et al., 2001). This finding suggests that peripheral neurons continue to project to the correct brainstem region even in the absence of the post-synaptic neurons (Figure 5.5D, E). Detailed descriptions of central projections of three different taste nerves show that the terminal fields are initially entirely overlapping (May and Hill, 2006). However, postnatal refinement of these terminal fields results in each nerve occupying a discrete, yet overlapping, territory, suggesting a weak oral topography- and/or modality-based selectivity (Hill and May, 2007; Lundy Jr and Norgren, 2015; May and Hill, 2006). This topography continues along the rostral to caudal axis of the nST, since more caudal regions of the nST receive input from non-oral regions of the alimentary canal. Orotopy is much more pronounced in lampreys and lower gnathostomes than in mammals (Daghfous et al., 2020; Finger, 1978; Fritzsch and Northcutt, 1993), suggesting that it is an evolutionarily conserved feature of the system. In mammals, the question of functionality uniform input, referred as orotopic organization (Finger, 2008), is in need of further investigation (Lundy Jr and Norgren, 2015). For example c-Fos activation was not conclusive (Corson et al., 2012; Harrer and Travers, 1996; Travers et al., 2018).

In summary, taste information is provided by three cranial nerves (VII, IX, X) to the hindbrain. The hindbrain develops early through specific gene expression. Innervating nerves form a dorso-ventral and rostro-caudal overlapping afferent distribution in the solitary tract that retains a rough orotopic organization.

5.6 SECOND-ORDER PROJECTIONS TO VPM AND THE GUSTATORY (INSULA) PROJECTIONS INTEGRATE WITH HIGHER CORTICAL REGIONS

The nucleus of the solitary tract (nST) has been analyzed in adult rodents (Corson et al., 2012; Lundy Jr and Norgren, 2015; May and Hill, 2006; Staszko and Boughter, 2020). The taste area of the nST extends from the rostral tip of r4 and goes beyond r8-11 areas (Watson et al., 2012) extending beyond the calamus scriptorius and reaches beyond the choroid plexus of the rhombencephalon, defined by r1-r8 in mice (Glover et al., 2018). In rodents, some postsynaptic neurons of mostly the nST project to the parabrachial nuclei (PBN) in the dorsal pons to innervate ipsilateral PBN neurons, which then project to thalamus (Herbert et al., 1990; Huang et al., 2021; Lundy Jr and Norgren, 2015; Saper and Loewy, 1980). In contrast, in primates many neurons of the nST project directly to the thalamic gustatory relay (Beckstead et al., 1980). The PBN surrounds the superior cerebellar peduncle as it enters the brainstem (Huang et al., 2021). The PBN can be divided into a medial PBN and lateral PBN. These have in turn been subdivided into a dozen subnuclei that extend along the ventrolateral margin of the lateral parabrachial complex (Huang et al., 2021). Injecting the PBN with tracers labels the nST selectively, but the PBN also receives other inputs that will generate complex interactions within the nucleus (Palmiter, 2018).

PBN axons project ipsilaterally to the dorsal thalamus, specifically the medial tip of the ventral posteromedial nucleus (VPN), referred to as the somatosensory relay. The thalamic gustatory relay is adjacent and medial to the VPM, referred to as ventral posterior thalamic nucleus parvocellularis (VPPC), which is defined as a taste relay (Lundy Jr and Norgren, 2015). This unique ipsilateral connection breaks down into a complex interaction from the VPPC that receives input from the ipsi- and, to a lesser extent, the contralateral PBN (Figure 5.6). Projections reach the hypothalamus, zona incerta and internal capsule to reach the ventral forebrain. The primary PBN projections are to the medial septum, olfactory tubercle and receive projections from the gustatory cortex (Lundy Jr and Norgren, 2015; Saper and Loewy, 1980). PBN neurons reach the cortex, in addition to the taste area, that expands directly to the PBN.

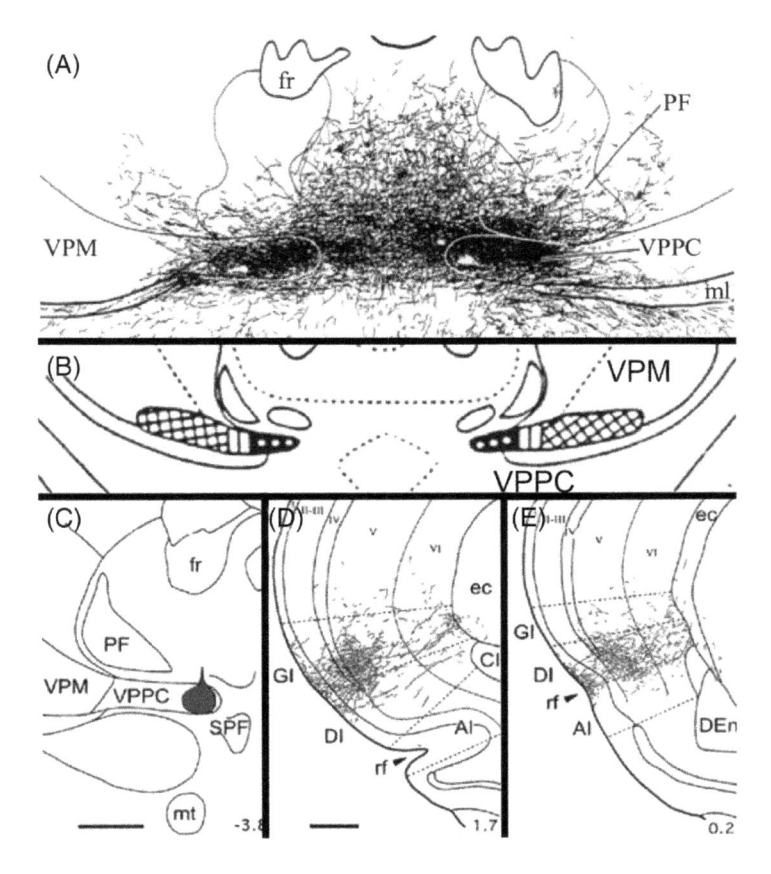

FIGURE 5.6 Projections from the parabrachial nuclei (PBN) reach the ipsilateral ventral posterior thalamic nucleus parvocellularis (VPPC), which is adjacent to the ventral postero-medial nucleus (VPM) and part of the somatosensory thalamus (A–E). Projections from VPPC (C) reach the cortex (D, E). AI, agranular insula cortex; CI, claustrum; DEn, dorsal endopiriform insular cortex; DI, dysgranular insular cortex; ec, external capsule; fr, fasciculus retroflexus; GI, granular insula cortex; mt, mammillothalamic olfactory bulb; PF, parafascicular nucleus; rf, rhinal fissure. (Modified from Lundy Jr and Norgren, 2015.)

Gustatory responses were elicited from the anterior tongue from which was evoked from chorda tympani stimulation. Posterior tongue and pharyngeal responses were located caudal to the anterior tongue (Hanamori et al., 1998). Reciprocal connections exist between the cortex, PBN and nST and indicate descending projections from forebrain areas that converge onto brainstem gustatory neurons, the PBN and the nST.

The orotopic organization in the nST is not preserved in higher-order nuclei, making the need of any orotopic primary map unclear (Lundy Jr and Norgren, 2015). Highly conserved second-order neurons project taste information to the cortex (Lundy Jr and Norgren, 2015; Staszko and Boughter, 2020; Vendrell-Llopis and Yaksi, 2015). Interactions with other sensory input combines with taste, somatosensation and olfaction into an integrated experience related to food intake (Schier and Spector, 2019; Shepherd, 2006). The major terminal field in the insula region is known to interact with other inputs, in addition to taste (Fain, 2019; Staszko et al., 2020).

In contrast to the clear-cut neuroanatomical pathways for taste, there is yet to emerge a uniform interpretation of taste coding in the insula cortex. Imaging studies have conflicting reports (Chen et al., 2011; 2021) and many studies report a broad distribution of taste neurons throughout the insula cortex with no spatial organization (Staszko et al., 2020), including fMRI studies in human subjects (Avery et al., 2020; Canna et al., 2019). Instead of a spatial map ordered by taste quality, there may be a hedonic or 'viscerotopic' map.

In summary, gustatory PBN projects via the VPPC to the taste cortex without mixed relay with thalamus and has reciprocal connections with gustatory and visceral afferents directly from the PBN. It remains unclear how the plastic and spatially disperse taste representation in cortex will ultimately be reconciled, based on controversial evidence.

5.7   SUMMARY AND CONCLUSION

A great deal is now known about gustatory responses in taste buds with respect to the basic tastes: sweet, salty (preferred), bitter, and umami tastes stimulate Type II taste bud cells; sour (acid) tastes directly stimulate Type III cells. Type II cells secrete ATP via unconventional synapses with primary gustatory afferent fibers and also secrete ATP to excite neighboring Type III cells. Type III cells secrete transmitters via vesicular release that presumably act on gustatory afferent fibers as well as on surrounding Type II taste bud cells (intragemmal cell-cell communication). Taste buds in different regions and on different lingual papillae (fungiform, foliate, vallate) are innervated by nerve fibers from the geniculate, petrosal, and nodose ganglia. The central terminals of these neurons project to the nucleus of the solitary tract (nST). Within the nST, these central projections from the geniculate, petrosal, and nodose ganglia terminate in partially overlapping/partially segregated projections. It remains difficult to reconcile the orotopic organization of the ganglion neuron terminal fields with their connectivity in the nST. In rodents, second-order projections from the nST send ipsilateral fibers to the parabrachial nucleus (PBN), which projects to the ipsilateral thalamus. Ultimately, taste information is projected to the rhinal fissure to reach the insular cortex. In humans, it isn't clear whether/how the PBN participates in the gustatory pathway. Second-order projections from the gustatory cortex interact with somatosensory and olfactory inputs to generate an interactive circuitry for food intake. Details of taste information processing and coding are widely debated and have not yet reached a uniform interpretation for many regions of the taste pathway, including the insula cortex.

REFERENCES

Ache, B., and Young, J. (2020). Phylogeny of chemical sensitivity. In The Senses, B. Fritzsch, ed. (Elsevier), pp. 4–23.

Alsina, B. (2020). Mechanisms of cell specification and differentiation in vertebrate cranial sensory systems. Current Opinion in Cell Biology *67*, 79–85.

Avery, J.A., Liu, A.G., Ingeholm, J.E., Riddell, C.D., Gotts, S.J., and Martin, A. (2020). Taste quality representation in the human brain. Journal of Neuroscience *40*, 1042–1052.

Barlow, L.A. (2015). Progress and renewal in gustation: new insights into taste bud development. Development *142*, 3620–3629.

Barlow, L.A., and Klein, O.D. (2015). Developing and regenerating a sense of taste. In Current Topics in Developmental Biology (Elsevier), pp. 401–419.

Barlow, L.A., and Northcutt, R.G. (1997). Taste buds develop autonomously from endoderm without induction by cephalic neural crest or paraxial mesoderm. Development *124*, 949–957.

Barreiro-Iglesias, A., Anadon, R., and Rodicio, M.C. (2010). The gustatory system of lampreys. Brain, Behavior and Evolution *75*, 241–250.

Barretto, R.P., Gillis-Smith, S., Chandrashekar, J., Yarmolinsky, D.A., Schnitzer, M.J., Ryba, N.J., and Zuker, C.S. (2015). The neural representation of taste quality at the periphery. Nature *517*, 373–376.

Bartoshuk, L.M., and Pangborn, R.M. (1993). The biological basis of food perception and acceptance. Food Quality and Preference *4*, 21–32.

Beckstead, R.M., Morse, J.R., and Norgren, R. (1980). The nucleus of the solitary tract in the monkey: projections to the thalamus and brain stem nuclei. Journal of Comparative Neurology *190*, 259–282.

Beidler, L.M., and Smallman, R.L. (1965). Renewal of cells within taste buds. Journal of Cell Biology *27*, 263–272.

Bigiani, M. (2020). Salt taste. In The Senses, B. Fritzsch, ed. (Elsevier), pp. 247–263.

Birol, O., Ohyama, T., Edlund, R.K., Drakou, K., Georgiades, P., and Groves, A.K. (2016). The mouse Foxi3 transcription factor is necessary for the development of posterior placodes. Developmental Biology *409*, 139–151.

Braun, C.B. (1998). Schreiner organs: A new craniate chemosensory modality in hagfishes. Journal of Comparative Neurology *392*, 135–163.

Canna, A., Prinster, A., Cantone, E., Ponticorvo, S., Russo, A.G., Di Salle, F., and Esposito, F. (2019). Intensity-related distribution of sweet and bitter taste fMRI responses in the insular cortex. Human brain mapping *40*, 3631–3646.

Castillo-Azofeifa, D., Losacco, J.T., Salcedo, E., Golden, E.J., Finger, T.E., and Barlow, L.A. (2017). Sonic hedgehog from both nerves and epithelium is a key trophic factor for taste bud maintenance. Development *144*, 3054–3065.

Castillo-Azofeifa, D., Seidel, K., Gross, L., Golden, E.J., Jacquez, B., Klein, O.D., and Barlow, L.A. (2018). SOX2 regulation by hedgehog signaling controls adult lingual epithelium homeostasis. Development *145*, dev164889.

Chen, X., Gabitto, M., Peng, Y., Ryba, N.J., and Zuker, C.S. (2011). A gustotopic map of taste qualities in the mammalian brain. Science *333*, 1262–1266.

Chen, K., Kogan, J.F., and Fontanini, A. (2021). Spatially distributed representation of taste quality in the gustatory insular cortex of behaving mice. Current Biology *31*, 247–256. e244.

Chizhikov, V.V., Iskusnykh, I.Y., Fattakhov, N., and Fritzsch, B. (2021). Lmx1a and Lmx1b are redundantly required for the development of multiple components of the mammalian auditory system. Neuroscience *452*, 247–264.

Corson, J., Aldridge, A., Wilmoth, K., and Erisir, A. (2012). A survey of oral cavity afferents to the rat nucleus tractus solitarii. Journal of Comparative Neurology *520*, 495–527.

Corson, S.L., Kim, M., Mistretta, C.M., and Bradley, R.M. (2013). Gustatory solitary tract development: a role for neuropilins. Neuroscience *252*, 35–44.

D'Autreaux, F., Coppola, E., Hirsch, M.R., Birchmeier, C., and Brunet, J.F. (2011). Homeoprotein Phox2b commands a somatic-to-visceral switch in cranial sensory pathways. Proceedings of the National Academy of Sciences of the United States of America *108*, 20018–20023.

Daghfous, G., Auclair, F., Blumenthal, F., Suntres, T., Lamarre-Bourret, J., Mansouri, M., Zielinski, B., and Dubuc, R. (2020). Sensory cutaneous papillae in the sea lamprey (Petromyzon marinus L.): I. Neuroanatomy and physiology. Journal of Comparative Neurology *528*, 664–686.

Dvorakova, M., Macova, I., Bohuslavova, R., Anderova, M., Fritzsch, B., and Pavlinkova, G. (2020). Early ear neuronal development, but not olfactory or lens development, can proceed without SOX2. Developmental Biology *457*, 43–56.

Dvoryanchikov, G., Hernandez, D., Roebber, J.K., Hill, D.L., Roper, S.D., and Chaudhari, N. (2017). Transcriptomes and neurotransmitter profiles of classes of gustatory and somatosensory neurons in the geniculate ganglion. Nature Communications *8*, 1–16.

Dvoryanchikov, G., Sinclair, M.S., Perea-Martinez, I., Wang, T., and Chaudhari, N. (2009). Inward rectifier channel, ROMK, is localized to the apical tips of glial-like cells in mouse taste buds. Journal of Comparative Neurology *517*, 1–14.

Dykes, I.M., Tempest, L., Lee, S.I., and Turner, E.E. (2011). Brn3a and Islet1 act epistatically to regulate the gene expression program of sensory differentiation. Journal of Neuroscience *31*, 9789–9799.

Fain, G.L. (2019). Sensory Transduction (Oxford University Press).

Fan, D., Chettouh, Z., Consalez, G.G., and Brunet, J.-F. (2019). Taste bud formation depends on taste nerves. Elife *8*, e49226.

Finger, T.E. (1978). Gustatory pathways in the bullhead catfish. II. Facial lobe connections. J Comp Neurol *180*, 691–705.

Finger, T.E. (2008). Sorting food from stones: the vagal taste system in Goldfish, *Carassius auratus*. Journal of Comparative Physiology A *194*, 135–143.

Finger, T.E., and Kinnamon, S.C. (2011). Taste isn't just for taste buds anymore. F1000 Biology Reports *3*.

Finger, T.E., Danilova, V., Barrows, J., Bartel, D.L., Vigers, A.J., Stone, L., Hellekant, G., and Kinnamon, S.C. (2005). ATP signaling is crucial for communication from taste buds to gustatory nerves. Science *310*, 1495–1499.

Fode, C., Gradwohl, G., Morin, X., Dierich, A., LeMeur, M., Goridis, C., and Guillemot, F. (1998). The bHLH protein NEUROGENIN 2 is a determination factor for epibranchial placode–derived sensory neurons. Neuron *20*, 483–494.

Fode, C., Ma, Q., Casarosa, S., Ang, S.-L., Anderson, D.J., and Guillemot, F. (2000). A role for neural determination genes in specifying the dorsoventral identity of telencephalic neurons. Genes & Development *14*, 67–80.

Freter, S., Muta, Y., O'Neill, P., Vassilev, V.S., Kuraku, S., and Ladher, R.K. (2012). Pax2 modulates proliferation during specification of the otic and epibranchial placodes. Developmental Dynamics *241*, 1716–1728.

Fritzsch, B., Elliott, K.L., and Pavlinkova, G. (2019). Primary sensory map formations reflect unique needs and molecular cues specific to each sensory system. F1000Research *8*, 345.

Fritzsch, B., and Northcutt, R.G. (1993). Cranial and spinal nerve organization in amphioxus and lampreys: evidence for an ancestral craniate pattern. Acta Anat (Basel) *148*, 96–109.

Fritzsch, B., Sarai, P., Barbacid, M., and Silos-Santiago, I. (1997). Mice with a targeted disruption of the neurotrophin receptor trkB lose their gustatory ganglion cells early but do develop taste buds. International Journal of Developmental Neuroscience *15*, 563–576.

Glendinning, J.I., Tang, J., Morales Allende, A.P., Bryant, B.P., Youngentob, L., and Youngentob, S.L. (2017). Fetal alcohol exposure reduces responsiveness of taste nerves and trigeminal chemosensory neurons to ethanol and its flavor components. Journal of Neurophysiology *118*, 1198–1209.

Glover, J.C., Elliott, K.L., Erives, A., Chizhikov, V.V., and Fritzsch, B. (2018). Wilhelm His' lasting insights into hindbrain and cranial ganglia development and evolution. Developmental Biology *444*, S14–S24.

Hall, J.M., Bell, M.L., and Finger, T.E. (2003). Disruption of sonic hedgehog signaling alters growth and patterning of lingual taste papillae. Developmental Biology *255*, 263–277.

Hanamori, T., Kunitake, T., Kato, K., and Kannan, H. (1998). Responses of neurons in the insular cortex to gustatory, visceral, and nociceptive stimuli in rats. Journal of Neurophysiology *79*, 2535–2545.

Harrer, M.I., and Travers, S.P. (1996). Topographic organization of Fos-like immunoreactivity in the rostral nucleus of the solitary tract evoked by gustatory stimulation with sucrose and quinine. Brain Research *711*, 125–137.

Hellard, D., Brosenitsch, T., Fritzsch, B., and Katz, D.M. (2004). Cranial sensory neuron development in the absence of brain-derived neurotrophic factor in BDNF/Bax double null mice. Developmental Biology *275*, 34–43.

Herbert, H., Moga, M.M., and Saper, C.B. (1990). Connections of the parabrachial nucleus with the nucleus of the solitary tract and the medullary reticular formation in the rat. Journal of Comparative Neurology *293*, 540–580.

Hernandez-Miranda, L.R., Muller, T., and Birchmeier, C. (2017). The dorsal spinal cord and hindbrain: From developmental mechanisms to functional circuits. Developmental Biology *432*, 34–42.

Hill, D.L., and May, O.L. (2007). Development and plasticity of the gustatory portion of nucleus of the solitary tract. In The Role of the Nucleus of the Solitary Tract in Gustatory Processing (CRC Press/Taylor & Francis).

Hornbruch, A., Ma, G., Ballermann, M.A., Tumova, K., Liu, D., and Cairine Logan, C. (2005). A BMP-mediated transcriptional cascade involving Cash1 and Tlx-3 specifies first-order relay sensory neurons in the developing hindbrain. Mechanisms of Development *122*, 900–913.

Huang, A.L., Chen, X., Hoon, M.A., Chandrashekar, J., Guo, W., Trankner, D., Ryba, N.J., and Zuker, C.S. (2006). The cells and logic for mammalian sour taste detection. Nature *442*, 934–938.

Huang, D., Grady, F.S., Peltekian, L., and Geerling, J.C. (2021). Efferent projections of Vglut2, Foxp2, and Pdyn parabrachial neurons in mice. Journal of Comparative Neurology *529*, 657–693.

Huang, T., and Krimm, R.F. (2010). Developmental expression of Bdnf, Ntf4/5, and TrkB in the mouse peripheral taste system. Developmental Dynamics *239*, 2637–2646.

Huang, E.J., Liu, W., Fritzsch, B., Bianchi, L.M., Reichardt, L.F., and Xiang, M. (2001). Brn3a is a transcriptional regulator of soma size, target field innervation and axon pathfinding of inner ear sensory neurons. Development *128*, 2421–2432.

Huang, Y.A., Maruyama, Y., Stimac, R., and Roper, S.D. (2008). Presynaptic (Type III) cells in mouse taste buds sense sour (acid) taste. Journal of Physiology *586*, 2903–2912.

Huang, Y.-J., Maruyama, Y., Dvoryanchikov, G., Pereira, E., Chaudhari, N., and Roper, S.D. (2007). The role of pannexin 1 hemichannels in ATP release and cell–cell communication in mouse taste buds. Proceedings of the National Academy of Sciences *104*, 6436–6441.

Imayoshi, I., and Kageyama, R. (2014). bHLH factors in self-renewal, multipotency, and fate choice of neural progenitor cells. Neuron *82*, 9–23.

Ito, A., and Nosrat, C.A. (2009). Gustatory papillae and taste bud development and maintenance in the absence of TrkB ligands BDNF and NT-4. Cell Tissue Research *337*, 349–359.

Iwatsuki, K., Liu, H.X., Gronder, A., Singer, M.A., Lane, T.F., Grosschedl, R., Mistretta, C.M., and Margolskee, R.F. (2007). Wnt signaling interacts with Shh to regulate taste papilla development. Proceedings of the National Academy of Sciences of the United States of America *104*, 2253–2258.

Kageyama, R., Shimojo, H., and Ohtsuka, T. (2019). Dynamic control of neural stem cells by bHLH factors. Neuroscience Research *138*, 12–18.

Kim, W.Y., Fritzsch, B., Serls, A., Bakel, L.A., Huang, E.J., Reichardt, L.F., Barth, D.S., and Lee, J.E. (2001). NeuroD-null mice are deaf due to a severe loss of the inner ear sensory neurons during development. Development *128*, 417–426.

Kinnamon, S.C., and Finger, T.E. (2019). Recent advances in taste transduction and signaling. F1000Research 8, doi: 10.12688/f1000research.21099.1

Kirino, M., Parnes, J., Hansen, A., Kiyohara, S., and Finger, T.E. (2013). Evolutionary origins of taste buds: phylogenetic analysis of purinergic neurotransmission in epithelial chemosensors. Open Biology *3*, 130015.

Kito-Shingaki, A., Seta, Y., Toyono, T., Kataoka, S., Kakinoki, Y., Yanagawa, Y., and Toyoshima, K. (2014). Expression of GAD67 and Dlx5 in the taste buds of mice genetically lacking Mash1. Chemical Senses *39*, 403–414.

Krasteva-Christ, G., Lin, W., and Tizzano, M. (2020). Extraoral taste receptors. In The Senses, B. Fritzsch, ed. (Elsevier), pp. 353–381.

Krimm, R.F., Thirumangalathu, S., and Barlow, L.A. (2015). Development of the taste system. Handbook of Olfaction and Gustation, 727–748.

Lindemann, B. (1999). Receptor seeks ligand: on the way to cloning the molecular receptors for sweet and bitter taste. Nature Medicine *5*, 381–382.

Lin, X., Lu, C., Ohmoto, M., Choma, K., Margolskee, R.F., Matsumoto, I., and Jiang, P. (2021). R-spondin substitutes for neuronal input for taste cell regeneration in adult mice. Proceedings of the National Academy of Sciences of the United States of America *118*.

Liu, F., Thirumangalathu, S., Gallant, N.M., Yang, S.H., Stoick-Cooper, C.L., Reddy, S.T., Andl, T., Taketo, M.M., Dlugosz, A.A., Moon, R.T., *et al.* (2007). Wnt-beta-catenin signaling initiates taste papilla development. Nature Genetics *39*, 106–112.

Lopez, G.F., and Krimm, R.F. (2006). Epithelial overexpression of BDNF and NT4 produces distinct gustatory axon morphologies that disrupt initial targeting. Developmental Biology *292*, 457–468.

Lu, W.-J., Mann, R.K., Nguyen, A., Bi, T., Silverstein, M., Tang, J.Y., Chen, X., and Beachy, P.A. (2018). Neuronal delivery of Hedgehog directs spatial patterning of taste organ regeneration. Proceedings of the National Academy of Sciences *115*, E200–E209.

Lundy Jr, R.F., and Norgren, R. (2015). Gustatory system. In The Rat Nervous System (Elsevier), pp. 733–760.

Ma, L., Lopez, G.F., and Krimm, R.F. (2009). Epithelial-derived brain-derived neurotrophic factor is required for gustatory neuron targeting during a critical developmental period. Journal of Neuroscience *29*, 3354–3364.

Ma, Z., Taruno, A., Ohmoto, M., Jyotaki, M., Lim, J.C., Miyazaki, H., Niisato, N., Marunaka, Y., Lee, R.J., and Hoff, H. (2018). CALHM3 is essential for rapid ion channel-mediated purinergic neurotransmission of GPCR-mediated tastes. Neuron *98*, 547–561. e510.

Maklad, A., and Fritzsch, B. (2003). Partial segregation of posterior crista and saccular fibers to the nodulus and uvula of the cerebellum in mice, and its development. Brain Res Dev Brain Res *140*, 223–236.

Martin, K.J., Rasch, L.J., Cooper, R.L., Metscher, B.D., Johanson, Z., and Fraser, G.J. (2016). Sox2+ progenitors in sharks link taste development with the evolution of regenerative teeth from denticles. Proceedings of the National Academy of Sciences *113*, 14769–14774.

Matsumoto, I., Ohmoto, M., Narukawa, M., Yoshihara, Y., and Abe, K. (2011). Skn-1a (Pou2f3) specifies taste receptor cell lineage. Nature Neuroscience *14*, 685–687.

May, O.L., and Hill, D.L. (2006). Gustatory terminal field organization and developmental plasticity in the nucleus of the solitary tract revealed through triple-fluorescence labeling. Journal of Comparative Neurology *497*, 658–669.

Miller, I.J., and Spangler, K.M. (1982). Taste bud distribution and innervation on the palate of the rat. Chemical Senses *7*, 99–108.

Mistretta, C.M., and Bradley, R.M. (2021). The Fungiform Papilla Is a Complex, Multimodal, Oral Sensory Organ. Current Opinion in Physiology *20*, 165–173.

Mistretta, C.M., Goosens, K.A., Farinas, I., and Reichardt, L.F. (1999). Alterations in size, number, and morphology of gustatory papillae and taste buds in BDNF null mutant mice demonstrate neural dependence of developing taste organs. Journal of Comparative Neurology *409*, 13–24.

Mistretta, C.M., Liu, H.X., Gaffield, W., and MacCallum, D.K. (2003). Cyclopamine and jervine in embryonic rat tongue cultures demonstrate a role for Shh signaling in taste papilla development and patterning: fungiform papillae double in number and form in novel locations in dorsal lingual epithelium. Developmental Biology *254*, 1–18.

Moody, S.A., and LaMantia, A.S. (2015). Transcriptional regulation of cranial sensory placode development. Current Topics in Developmental Biology *111*, 301–350.

Nomura, K., Nakanishi, M., Ishidate, F., Iwata, K., and Taruno, A. (2020). All-electrical Ca2+-independent signal transduction mediates attractive sodium taste in taste buds. Neuron *106*, 816–829. e816.

Northcutt, R.G. (2004). Taste buds: development and evolution. Brain, Behavior and Evolution *64*, 198–206.

Nosrat, C.A., Blomlof, J., ElShamy, W.M., Ernfors, P., and Olson, L. (1997). Lingual deficits in BDNF and NT3 mutant mice leading to gustatory and somatosensory disturbances, respectively. Development *124*, 1333–1342.

Nosrat, I.V., Margolskee, R.F., and Nosrat, C.A. (2012). Targeted taste cell-specific overexpression of brain-derived neurotrophic factor in adult taste buds elevates phosphorylated TrkB protein levels in taste cells, increases taste bud size, and promotes gustatory innervation. Journal of Biological Chemistry *287*, 16791–16800.

O'Neill, P., Mak, S.-S., Fritzsch, B., Ladher, R.K., and Baker, C.V. (2012). The amniote paratympanic organ develops from a previously undiscovered sensory placode. Nature Communications *3*, 1–11.

Ohla, K., Yoshida, R., Roper, S.D., Di Lorenzo, P.M., Victor, J.D., Boughter, J.D., Fletcher, M., Katz, D.B., and Chaudhari, N. (2019). Recognizing taste: coding patterns along the neural axis in mammals. Chemical Senses *44*, 237–247.

Ohmoto, M., Kitamoto, S., and Hirota, J. (2021). Expression of Eya1 in mouse taste buds. Cell and Tissue Research 383(3), 979–986.

Ohmoto, M., Ren, W., Nishiguchi, Y., Hirota, J., Jiang, P., and Matsumoto, I. (2017). Genetic lineage tracing in taste tissues using Sox2-CreERT2 strain. Chemical Senses *42*, 547–552.

Okubo, T., Clark, C., and Hogan, B.L. (2009). Cell lineage mapping of taste bud cells and keratinocytes in the mouse tongue and soft palate. Stem Cells *27*, 442–450.

Okubo, T., Pevny, L.H., and Hogan, B.L. (2006). Sox2 is required for development of taste bud sensory cells. Genes & Development *20*, 2654–2659.

Palmiter, R.D. (2018). The parabrachial nucleus: CGRP neurons function as a general alarm. Trends in Neurosciences *41*, 280–293.

Patel, A.V., and Krimm, R.F. (2010). BDNF is required for the survival of differentiated geniculate ganglion neurons. Developmental Biology *340*, 419–429.

Pauley, S., Wright, T.J., Pirvola, U., Ornitz, D., Beisel, K., and Fritzsch, B. (2003). Expression and function of FGF10 in mammalian inner ear development. Developmental dynamics: an official publication of the American Association of Anatomists *227*, 203–215.

Paulson, R.B., Alley, K.E., Salata, L.J., and Whitmyer, C.C. (1995). A scanning electron-microscopic study of tongue development in the frog Rana pipiens. Archives of Oral Biology *40*, 311–319.

Perea-Martinez, I., Nagai, T., and Chaudhari, N. (2013). Functional cell types in taste buds have distinct longevities. PLoS One *8*, e53399.

Petersen, C.I., Jheon, A.H., Mostowfi, P., Charles, C., Ching, S., Thirumangalathu, S., Barlow, L.A., and Klein, O.D. (2011). FGF signaling regulates the number of posterior taste papillae by controlling progenitor field size. PLoS Genet *7*, e1002098.

Prescott, S.L., Umans, B.D., Williams, E.K., Brust, R.D., and Liberles, S.D. (2020). An airway protection program revealed by sweeping genetic control of vagal afferents. Cell *181*, 574–589 e514.

Qian, Y., Fritzsch, B., Shirasawa, S., Chen, C.-L., Choi, Y., and Ma, Q. (2001). Formation of brainstem (nor) adrenergic centers and first-order relay visceral sensory neurons is dependent on homeodomain protein Rnx/Tlx3. Genes & Development *15*, 2533–2545.

Ren, W., Aihara, E., Lei, W., Gheewala, N., Uchiyama, H., Margolskee, R.F., Iwatsuki, K., and Jiang, P. (2017). Transcriptome analyses of taste organoids reveal multiple pathways involved in taste cell generation. Scientific Reports *7*, 1–13.

Ren, W., Lewandowski, B.C., Watson, J., Aihara, E., Iwatsuki, K., Bachmanov, A.A., Margolskee, R.F., and Jiang, P. (2014). Single Lgr5-or Lgr6-expressing taste stem/progenitor cells generate taste bud cells ex vivo. Proceedings of the National Academy of Sciences *111*, 16401––16406.

Rios-Pilier, J., and Krimm, R.F. (2019). TrkB expression and dependence divides gustatory neurons into three subpopulations. Neural Development *14*, 1–13.

Rodriguez, Y., Roebber, J., Dvoryanchikov, F., Makhoul, V., Roper, S.D., and Chaudhari, N. (2021). Tripartite synapses" in taste buds: a role for Type I glial-like taste cells. Journal of Neuroscience, in press.

Roebber, J.K., Roper, S.D., and Chaudhari, N. (2019). The role of the anion in salt (NaCl) detection by mouse taste buds. Journal of Neuroscience *39*, 6224–6232.

Roper, S. (2020). Microphysiology of taste buds. In The Senses, B. Fritzsch, ed. (Elsevier), pp. 187–210.

Roper, S.D., and Chaudhari, N. (2017). Taste buds: cells, signals and synapses. Nature Reviews Neuroscience *18*, 485–497.

Saper, C., and Loewy, A. (1980). Efferent connections of the parabrachial nucleus in the rat. Brain Research *197*, 291–317.

Schier, L.A., and Spector, A.C. (2019). The functional and neurobiological properties of bad taste. Physiological Reviews *99*, 605–663.

Seta, Y., Oda, M., Kataoka, S., Toyono, T., and Toyoshima, K. (2011). Mash1 is required for the differentiation of AADC-positive type III cells in mouse taste buds. Developmental Dynamics *240*, 775–784.

Shepherd, G.M. (2006). Smell images and the flavour system in the human brain. Nature *444*, 316–321.

Smith, A.C., Fleenor, S.J., and Begbie, J. (2015). Changes in gene expression and cell shape characterise stages of epibranchial placode-derived neuron maturation in the chick. Journal of Anatomy *227*, 89–102.

Smith, D.V., and Margolskee, R.F. (2001). Making sense of taste. Scientific American *284*, 32–39.

Staszko, S.M., Boughter Jr, J.D., and Fletcher, M.L. (2020). Taste coding strategies in insular cortex. Experimental Biology and Medicine *245*, 448–455.

Staszko, S., and Boughter, J. (2020). Taste pathways, representation and processing in the brain. In The Senses, B. Fritzsch, ed. (Elsevier), pp. 280–297.

Stedman, H.M., Mistretta, C.M., and Bradley, R.M. (1983). A quantitative study of cat epiglottal taste buds during development. Journal of Anatomy *136*, 821–827.

Sudiwala, S., and Knox, S.M. (2019). The emerging role of cranial nerves in shaping cranio-facial development. Genesis *57*, e23282.

Suzuki, Y., and Takeda, M. (1983). Ultrastructure and monoamine precursor uptake of taste buds in the pharynx, nasopalatine ducts, epiglottis and larynx of the mouse. Kaibogaku Zasshi *58*, 593–605.

Takeda, N., Jain, R., Li, D., Li, L., Lu, M.M., and Epstein, J.A. (2013). Lgr5 identifies progenitor cells capable of taste bud regeneration after injury. PLoS One *8*, e66314.

Taruno, A., Vingtdeux, V., Ohmoto, M., Ma, Z., Dvoryanchikov, G., Li, A., Adrien, L., Zhao, H., Leung, S., Abernethy, M., Koppel, J., Davies, P., Civan, M.M., Chaudhari, N., Matsumoto, I., Hellekant, G., Tordoff, M.G., Marambaud, P., and Foskett, J.K. (2013). CALHM1 ion channel mediates purinergic neurotransmission of sweet, bitter and umami tastes. Nature *495*, 223–226.

Teng, B., Wilson, C.E., Tu, Y.-H., Joshi, N.R., Kinnamon, S.C., and Liman, E.R. (2019). Cellular and neural responses to sour stimuli require the proton channel Otop1. Current Biology *29*, 3647–3656. e3645.

Thirumangalathu, S., Harlow, D.E., Driskell, A.L., Krimm, R.F., and Barlow, L.A. (2009). Fate mapping of mammalian embryonic taste bud progenitors. Development *136*, 1519–1528.

Tomchik, S.M., Berg, S., Kim, J.W., Chaudhari, N., and Roper, S.D. (2007). Breadth of tuning and taste coding in mammalian taste buds. Journal of Neuroscience *27*, 10840–10848.

Travers, S., Breza, J., Harley, J., Zhu, J., and Travers, J. (2018). Neurons with diverse phenotypes project from the caudal to the rostral nucleus of the solitary tract. Journal of Comparative Neurology *526*, 2319–2338.

Vendrell-Llopis, N., and Yaksi, E. (2015). Evolutionary conserved brainstem circuits encode category, concentration and mixtures of taste. Scientific Reports *5*, 17825.

von Bartheld, C.S., and Fritzsch, B. (2006). Comparative analysis of neurotrophin receptors and ligands in vertebrate neurons: tools for evolutionary stability or changes in neural circuits? Brain Behav Evol *68*, 157–172.

Wang, L., Xie, J., Zhang, H., Tsang, L.H., Tsang, S.L., Braune, E.B., Lendahl, U., and Sham, M.H. (2020). Notch signalling regulates epibranchial placode patterning and segregation. Development *147*.

Watson, C., Paxinos, G., and Puelles, L. (2012). The Mouse Nervous System (Academic Press).

Witt, M. (2020). Anatomy and development of the human gustatory and olfactory systems. In The Senses, B. Fritzsch, ed. (Elsevier), pp. 85–118.

Witt, M., and Reutter, K. (2015). Anatomy of tongue and taste buds. Handbook of Olfaction and Gustation, 637–664.

Wooding, s., and Ramirez, V. (2020). Taste genetics. In The Senses, B. Fritzsch, ed. (Elsevier), pp. 264–279.

Xu, P.-X., Woo, I., Her, H., Beier, D.R., and Maas, R.L. (1997). Mouse Eya homologues of the Drosophila eyes absent gene require Pax6 for expression in lens and nasal placode. Development *124*, 219–231.

Yang, R., Dzowo, Y.K., Wilson, C.E., Russell, R.L., Kidd, G.J., Salcedo, E., Lasher, R.S., Kinnamon, J.C., and Finger, T.E. (2020). Three-dimensional reconstructions of mouse circumvallate taste buds using serial blockface scanning electron microscopy: I. Cell types and the apical region of the taste bud. Journal of Comparative Neurology *528*, 756–771.

Yang, T., Jia, Z., Bryant-Pike, W., Chandrasekhar, A., Murray, J.C., Fritzsch, B., and Bassuk, A.G. (2014). Analysis of PRICKLE 1 in human cleft palate and mouse development demonstrates rare and common variants involved in human malformations. Molecular Genetics & Genomic Medicine *2*, 138–151.

Yarmolinsky, D.A., Zuker, C.S., and Ryba, N.J. (2009). Common sense about taste: from mammals to insects. Cell *139*, 234–244.

Zhang, Y., Lu, W.-J., Bulkley, D.P., Liang, J., Ralko, A., Han, S., Roberts, K.J., Li, A., Cho, W., and Cheng, Y. (2020). Hedgehog pathway activation through nanobody-mediated conformational blockade of the Patched sterol conduit. Proceedings of the National Academy of Sciences *117*, 28838–28846.

Zhang, H., Wang, L., Wong, E.Y.M., Tsang, S.L., Xu, P.X., Lendahl, U., and Sham, M.H. (2017). An Eya1-Notch axis specifies bipotential epibranchial differentiation in mammalian craniofacial morphogenesis. Elife *6*, e30126.

Zhang, T., Xu, J., and Xu, P.X. (2021). Eya2 expression during mouse embryonic development revealed by Eya2 lacZ knockin reporter and homozygous mice show mild hearing loss. Developmental Dynamics. *250*(10): 1450–1462.

Zou, D., Silvius, D., Fritzsch, B., and Xu, P.-X. (2004). Eya1 and Six1 are essential for early steps of sensory neurogenesis in mammalian cranial placodes. Development *131*, 5561–5572.

6 Assembly and Functional Organization of the Vestibular System

Karen L. Elliott, Hans Straka

CONTENTS

DOI: 10.1201/9781003092810-6

6.1   INTRODUCTION

The evolutionary emergence of motion detection systems accompanied the acquisition of non-sessile lifestyles of animals and promoted the advancement of propulsive strategies to maneuver in 3D space (Straka and Gordy, 2020). The basic design and functional principle of vertebrate motion sensors (Platt and Straka, 2020) is very similar to those of invertebrates (Schmitz, 2020) and likely derives from common origins and homologies in genetic regulatory networks during organ assembly. This suggests that recruitment and modification of pre-existent configuration elements and morphological differentiation dynamics created a primordial motion sensor that in vertebrates radiated into multi-compartmental inner ear cavities for the detection, spatio-temporal decomposition, and neuronal encoding of head/body motion (Fritzsch and Elliott, 2017; Fritzsch and Straka, 2014). Following an evolutionary acquisition of maximal complexity in number and phenotypic diversity of ducts and pouches in jawed fish, the basic blueprint of vestibular end organs remained essentially unchanged for the past 500 million years (Fritzsch and Straka, 2014; Platt and Straka, 2020).

The invention of a motion-sensitive vestibular organ facilitated the acquisition of locomotor proficiency by providing sensory information about motion in space that is indispensable for a non-sessile lifestyle. In fact, sensory-motor transformation of vestibular signals ensured the stability of the body position in space and of the eyes with respect to the environment (Straka and Gordy, 2020). The latter vestibulo-motor behavior became particularly important with increasing eco-physiological importance of the visual system since perceived visual motion can either derive from moving objects, from self- or passively induced motion, or from various combinations of both. By initiating compensatory eye movements through contractions of extraocular muscles (Walls, 1962), vestibular signals became necessary for resolving this ambiguity and for constantly maintaining visual acuity.

The primary requirement for vestibular signals to drive gaze- and posture-stabilization shaped the organizational principles for transduction and encoding of head/body motion by sensory receptor cells and corresponding central nervous elements (Chagnaud et al., 2017). The implementation of short-latency connections between the sensory periphery and skeletal or eye muscles produced a representation within the central nervous system (CNS) that differs from the typical sensory topographies of the visual or somatosensory systems (Chagnaud et al., 2017). In fact, vestibular pathways form phylogenetically conserved, target-specific premotor/motor maps along the hindbrain segmental scaffold (Glover, 2003; Straka, 2010). Much of this motor output-centered representation of signals from individual vestibular end organs derives from ontogenetic specification processes. The developmental assembly is controlled by conserved genetic regulatory networks, which transform nascent otic placodes into inner ear end organs with resultant influences on the formation of neuronal connections based on rhombomere identity (Fritzsch et al., 2002; Glover, 2003, Schlosser, 2010).

The unique role of the vestibular system requires particular transduction and processing strategies to decode head/body movements. Key to this task is the vectorial decomposition of head/body motion components (Glasauer and Knorr, 2020) and separate encoding within modality-specific and frequency-tuned channels (Straka et al., 2009). At variance with most unimodal sensory systems, central vestibular circuits integrate motion-related visual, proprioceptive, and motor efference copy signals (Straka et al., 2014) to produce a coherent self-motion percept. This review provides a concise summary of hallmark features and events that cause the 'vestibular system' to optimally stabilize gaze and posture, and to provide the reference frame for the establishment of a bilateral body symmetry (Lambert et al., 2009), as well as for updating circuits responsible for navigation and orientation in space (Cullen and Taube, 2017). By highlighting key ontogenetic regulatory events and molecular specification processes of sensory organs and associated neuronal pathway formation, the apparent complexity of vestibular circuits and signaling principles become conceptually accessible.

6.2 VESTIBULAR REFERENCE FRAME AND PRINCIPLES OF MOTION DETECTION

All extant vertebrates, except for hagfish (one semicircular canal duct with two sensory patches and a single otolith organ) and lamprey (two distinct vertical ducts and associated sensory epithelia and a single otolith organ) share a common blueprint with three orthogonally arranged ducts, and two or three otolith organs (Figure 6.1A) apart from a variety of species-specific auxiliary mechanoreceptive pouches mostly with acoustic sensitivity (Fritzsch and Straka, 2014; Platt and Straka, 2020). The ensemble of distinct vestibular end organs evolved from a simple gravistatic epithelium in pre-vertebrate ancestors by duplication and diversification events of genetic regulatory elements, which produced morphologies that became highly sensitive for detecting and encoding head/body movements (Straka and Gordy, 2020). For example, based on this implemented mechanical principle, semicircular canals are sensitive to angular acceleration, whereas otolith organs form accelerometers that measure linear acceleration (translation) and the orientation and motion of the head (tilt) within the gravitational field (Glasauer and Knorr, 2020). The three semicircular canals on both sides of the head, each with a unique directional sensitivity, are oriented perpendicular to each other and form an orthogonal reference frame (Figure 6.1A, B) that is remarkably conserved across vertebrates (Graf and Simpson, 1981; Simpson and Graf, 1985). The mirror-symmetric bilateral arrangement of the ducts within the head causes semicircular canals to constitute functional pairs across the two sides (Figure 6.1B). This generates a push-pull system that is optimally suited to detect and decompose 3D angular head/body movements while at the same time minimizing the redundancy regarding directional sensitivity (Glasauer and Knorr, 2020). The structural orthogonality of the semicircular canals matches the pulling directions of the eye muscles, which form major vestibulo-motor targets (Figure 6.1D; Horn and Straka, 2021). In addition, the head-centered semicircular canal configuration is aligned

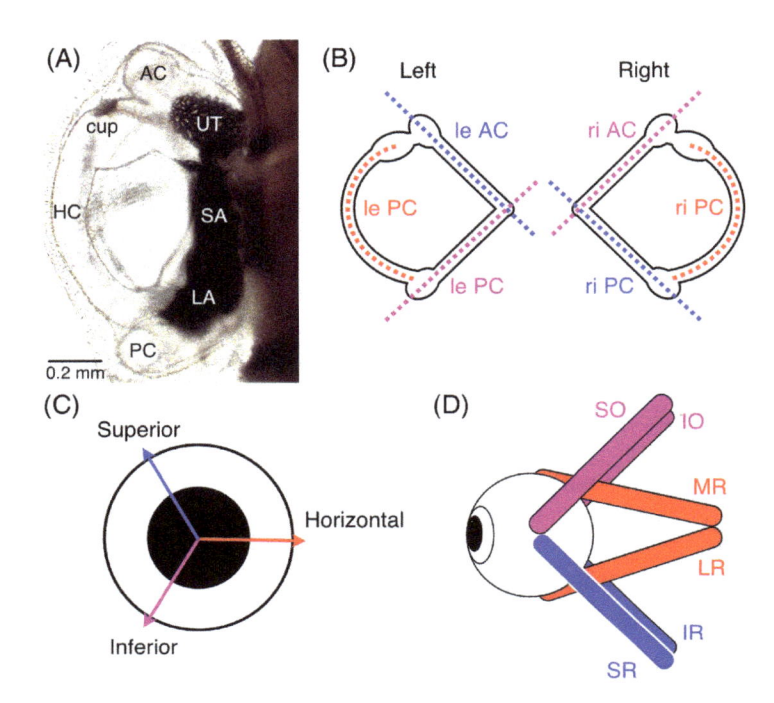

FIGURE 6.1   **Orthogonality of sensory-motor reference frames.** (A) Photomicrograph of the left inner ear of a stage 49 *Xenopus laevis* tadpole from dorsal depicting the spatial arrangement of the three semicircular canals and the three otolith organs. (**B–D**) Schematics depicting in (**B**) the spatially aligned semicircular canals on the left (le) and right (ri) side, forming bilateral coplanar pairs with push-pull functionality, in (**C**) the principal axes for visual motion encoding and in (**D**), the pulling directions of the six extraocular muscles of the left eye; note that antagonistic pairs of eye muscles are approximately aligned with the planes of coplanar semicircular canals and principal axes for visual motion encoding. AC, PC, HC, anterior vertical, posterior vertical, horizontal semicircular canal; cup, cupula; IO, SO, inferior, superior oblique eye muscle; IR, SR, inferior, superior rectus eye muscle; LR, MR, lateral, medial rectus eye muscle; LA, lagena; SA, saccule; UT, utricle. ([a]) modified from Lambert et al., 2008.)

with the reference frame of the accessory optic system (Figure 6.1C). This latter motion-sensitive sensory system triggers optokinetic responses, which complement vestibulo-ocular reflexes during gaze stabilization by providing the relevant feedback signals about residual retinal image slip (Graf and Simpson, 1981; Simpson and Graf, 1985). Even organisms with a foveated visual system, where such accessory optic pathways are quickly supplanted by smooth pursuit computations, rely on optokinetic reflex contributions to assist stabilization of image slip during motion (Horn and Straka, 2021).

The organizational principle that governs the shape and arrangement of the different otolith organs inside the head is less obvious than the orthogonality of the six semicircular canals. This is in part related to the sensitivities of otolith organs to sensory modalities other than detection of head/body motion-related linear acceleration components (Platt and Straka, 2020). The utricle has a consistently confirmed role as a vestibular end organ, sensitive to linear translation and head tilt with a few confirmed reports of non-vestibular function, though mostly in fishes (Platt and Straka, 2020). In contrast, the saccule and lagena, while variably contributing to detection and encoding of linear head/body acceleration, also serve as highly sensitive end organs for substrate-vibration and airborne sound with species- and lifestyle-specific functional adaptations (Lewis et al., 1985; Platt and Straka, 2020). Generally, the horizontally oriented utricle is located dorsally in the inner ear and is innervated along with the anterior vertical and horizontal semicircular canals by the *ramus anterior* of the VIIIth nerve in all species (de Burlet, 1929). In contrast, the saccule and lagena (except for therian mammals, which lack the latter end organ) are largely positioned vertically and more ventral in the inner ear and are typically innervated along with the posterior vertical semicircular canal and papilla organs by the *ramus posterior* of the VIIIth nerve (de Burlet, 1929). Despite the preferential spatial orientation, the epithelial arrangement of hair cell polarization vectors in each otolith organ covers a directional sensitivity across 360° in the respective plane (Platt and Straka, 2020). Accordingly, spatially-specific responses of otolithic afferent fibers, in contrast to semicircular canal afferents, strictly depend on peripheral epithelial origins, obscuring a distinct otolithic reference frame for motion detection.

6.3 ONTOGENY OF THE INNER EAR AND FORMATION OF SEMICIRCULAR CANALS AND OTOLITH ORGANS

All inner ear structures and cellular elements are of placodal origin, in contrast to other sensory systems, like vision or somatosensation (Fritzsch et al., 2019; Schlosser, 2006). The otic placode forms out of a panplacodal ectodermal primordium, which broadly expresses *Eya1*, *Six1*, and *Six4* (Schlosser, 2010). Additional transcription factors become differentially confined to subregions of this panplacodal region (Schlosser and Ahrens, 2004). This process causes a subdivision into an anterior and a posterior compartment with the latter giving rise to the otic placode (Schlosser and Ahrens, 2004). *Foxi3* is expressed in the panplacodal region but constitutes an essential regulatory element for the induction of the posterior placodal area, including the otic placode (Birol et al., 2016; Khatri et al., 2014). In addition, signals from the surrounding mesoderm and the hindbrain, such as FGF signaling molecules, are indispensable for otic placode induction. Mice lacking both *Fgf3* and *Fgf10* fail to form otic vesicles (Alvarez et al., 2003; Wright and Mansour, 2003). These latter two key regulatory elements induce *Pax2*, *Pax8*, and *Sox3* expression in the otic placode, which as an ensemble are required to produce all inner ear elements (Alvarez et al., 2003; Hans et al., 2004; Mackereth et al., 2005; Schlosser, 2010; Solomon et al., 2003, 2004; Sun et al., 2007; Wright and Mansour, 2003) (Figure 6.2A). Interestingly, loss of either *Pax2* or *Pax8* alone is unable to completely disrupt inner ear development (Burton et al., 2004; Christ

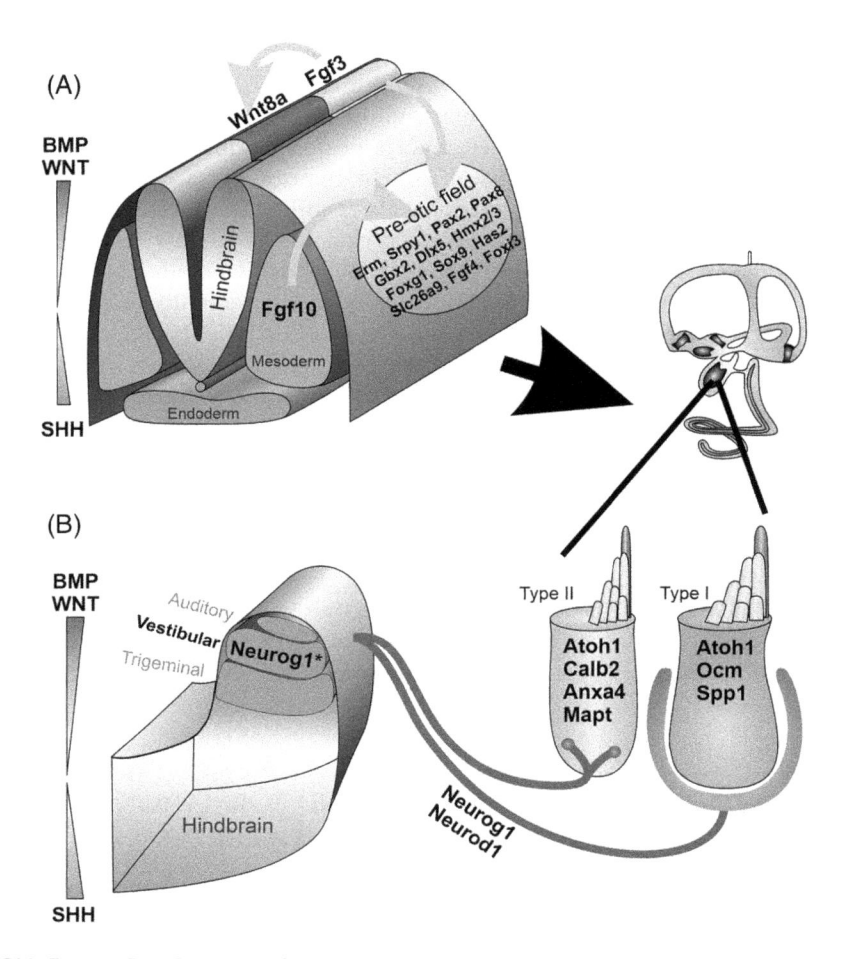

FIGURE 6.2   Development of the vestibular system. (**A**) Expression and release of *Fgf3* from the hindbrain and *Fgf10* from the mesoderm induce the formation of the otocyst, causing upregulation of a cascade of genes in the developing otocyst, such as *Pax2* and *Pax8* that together are essential for otic development; *Wnt*, *BMP*, and *Shh* further define dorsal and ventral portions of the developing ear. (**B**). Both types of vestibular hair cells in amniotes express and critically depend on *Atoh1*; type I hair cells uniquely express *Ocm* and *Spp1*, whereas type II hair cells express *Calb2*, *Anxa4*, and *Mapt*; vestibular neurons express and depend on *Neurog1* and *Neurod1* for their development and differentiation; cells in the central vestibular nucleus transiently express *Neurog1* (*). (Modified from Birol et al., 2016; Elliott et al., 2018; Urness et al., 2010.)

et al., 2004). However, in *Pax2/Pax8* double null mice, the developing inner ear becomes arrested at the otocyst stage, suggesting a functional redundancy of these genes (Bouchard et al., 2010). While *Pax* genes are indispensable, many additional genes are required for an undisturbed inner ear formation (Torres and Giráldez, 1998).

Following invagination of the otic placode and formation of a vesicle, the oto-cyst undergoes anterior-posterior as well as dorso-ventral patterning. The ante-rior portion gives rise to neurosensory structures, whereas the posterior portion produces non-sensory elements and solely one sensory epithelium, the future posterior semicircular canal crista (Wu and Kelley, 2012). Dorso-ventral pattern-ing depends upon signals from the hindbrain, such as *Wnt*, *BMP*, and *Shh* (Bok et al., 2007; Riccomagno et al., 2005). *Wnt1* and *Wnt3a*, which are expressed in dorsal regions of the hindbrain, are important for the formation of semicircular canals and associated cristae (Riccomagno et al., 2005). In addition, *Shh*, which specifies the ventral portion, may also have a role in the formation of the semicir-cular canals and the saccule, though the effect of *Shh* on the ducts is likely indi-rect (Brown and Epstein, 2011; Riccomagno et al., 2002; Wu and Kelley, 2012). The three semicircular canals derive from two perpendicularly oriented pouches. The vertical pouch forms the anterior and posterior vertical semicircular canals, whereas the horizontal pouch gives rise to the horizontal semicircular canal (Wu and Kelley, 2012). As these pouches grow during development, the central portions of the ducts become fused. The cells located in the center finally are resorbed, thereby producing a joint central duct (*common crus*) associated with both vertical semicircular canals (Martin and Swanson, 1993). In addition to *Wnt1* and *Wnt3a*, other genes such as *Dlx5*, *Hmx2/3*, and *Bmp4* are also involved in semicircular canal formation (Ketchum et al., 2020; Riccomagno et al., 2005). Interestingly, the horizontal semicircular canal requires *Otx1* expression, con-firmed by its absence in *Otx1* null mice (Morsli et al., 1998). While lampreys possess an *Otx* gene, it is not expressed in the inner ear in compliance with the lack of a horizontal semicircular canal in these animals (Tomsa and Langeland, 1999). Given the general presence of *Otx* in lamprey it is likely that the horizon-tal semicircular canal did not evolve with the *Otx* gene itself, but only with the evolutionary implementation of a downstream regulatory program (Elliott and Gordy, 2020; Higuchi et al., 2019).

The utricle develops in close proximity to the anterior and horizontal semicircular canal cristae while the saccule forms more ventrally, adjacent to those structures that, during vertebrate evolution, acquired acoustic function. The bHLH gene, *Neurog1*, is important for the development of sensory epithelia in general and for the saccular epithelium in particular (Ma et al., 2000). In contrast, the zinc finger transcription factor, *Gata3*, is necessary for the formation of most sensory epithelia, with the exception of the saccule (Duncan et al., 2011). Normally, the utricle and saccule are separated by the utriculo-saccular foramen. However, loss of *Otx1*, *N-myc*, or *Lmx1a* results in a fusion of both maculae (Fritzsch et al., 2001; Kopecky et al., 2011; Morsli et al., 1998; Nichols et al., 2008, 2020). A third otolith organ, the lagena, present in most vertebrates except for therian mammals develops in its own recess, adjacent to the saccule (Fritzsch et al., 2013; Lewis et al., 1985). All three otolith organs (utri-cle, saccule, lagena) are covered with calcium carbonate crystals termed otoconia, that are suspended above a gelatinous matrix (Ketchum et al., 2020), though with species- and end organ-specific compositions, such as differently sized grains and correspondingly variable molecular specification, which determines the principal sensitivity of the end organ to particular modalities (head/body motion, acoustic).

6.4 VESTIBULAR HAIR CELLS

While sensory neurons are the first to differentiate in the ear, the vestibular system is best understood in the context of functional assembly. The hair cell, the peripheral receptor cell, forms the sensory interface responsible for the mechano-electrical transduction of head/body orientation and motion into neuronal signals (Lysakowski, 2020). All vertebrate vestibular hair cells are intermingled with supporting cells and share a common blueprint that includes the lack of an axon (at variance with insect mechanoreceptor cells), the presence of a morphologically polarized cilial bundle protruding from the apical surface and extending into the epithelial coverage (semicircular canal cupula, otolithic membrane), and a ribbon synapse at the basal surface that is contacted by afferent fiber(s) (Fritzsch and Straka, 2014; Lysakowski, 2020).

6.4.1 Development of Hair Cells and Cilial Bundle Polarity

The ontogeny of hair cells has been particularly well characterized in mice and thus serves as a model to describe major genetic regulatory and molecular events responsible for establishing this key cellular element of the vestibular sensory periphery. Early proneurosensory domains within the otic vesicle are specified by *Eya1*, *Six1*, and *Sox2* expression and deletion of these genes affects the development of hair cells as well as sensory neurons (Elliott et al., 2021; Kiernan et al., 2005; Xu et al., 2021). Hair cells within the maculae and cristae begin to exit the cell cycle around embryonic stage E10.5-14.5 in mice and continue exiting the cell cycle through early postnatal days (Fritzsch et al., 2019; Ruben, 1967). Cell cycle exit begins in the center region and progresses toward the periphery of the epithelia (Slowik and Bermingham-McDonogh, 2016). Unlike in the cochlea, where cell cycle exit and differentiation of auditory hair cells are uncoupled, vestibular hair cells also differentiate in a central to peripheral gradient, suggesting that this process may occur shortly after cycle exit (McInturff et al., 2018; Slowik and Bermingham-McDonogh, 2016). One of the earliest markers of differentiated hair cells is the bHLH gene, *Atoh1*, which is expressed as early as E11.5 in mice (Raft et al., 2007). Regardless of the specific type, all inner ear hair cells require *Atoh1* for differentiation (Bermingham et al., 1999; Pan et al., 2011). *Atoh1* initiates an expression cascade of several other genes in hair cells, such as *Pou4f3* (Xiang et al., 2003) and *Gfi1* (Hertzano et al., 2004; Matern et al., 2020; Shroyer et al., 2005). However, some hair cell types can differentiate in the absence of *Pou4f3* or *Gfi1* expression (Hertzano et al., 2004; Matern et al., 2020). Specific genes responsible for the two different vestibular hair cell subtypes in mammals have not been fully elucidated; however, recent work demonstrated unique expression profiles, respectively. Accordingly, type II hair cells in mice uniquely express *Calb2*, *Anxa4*, and *Mapt*, whereas type I hair cells express *Ocm* and *Spp1* (McInturff et al., 2018) (Figure 6.2B). While both types differentiate around the same time, questions remain as to the details of the specific mechanisms (Burns and Stone, 2017).

The orientation of hair cells within the utricle of most vertebrates is highly conserved, at variance with the distribution of saccular hair cell directional sensitivities. While the utricle has only two regions of hair cells with opposing polarities, the saccule has anywhere between two and six different regions, depending on the species (Ladich and Schulz-Mirbach, 2016; Lewis et al., 1985). The epithelial region of hair cells with opposing functional polarity is termed striola. This area is centered in each macula and corresponds to the line of polarity reversal (Eatock and Songer, 2011). In the utricular epithelium, hair cell polarities are oriented toward the striola, whereas in the mammalian saccule, hair cell polarities are facing away from each other (Figure 6.3A). The polarity of individual hair cells is regulated by planar cell polarity (PCP) genes, such that all hair cells in a given region have the same preferential directionality (Deans, 2013; May-Simera and Kelley, 2012). Several PCP genes have been identified that play a role in hair cell polarity, such as *Vangl2*, *Prickle2*, *Fzd3*, *Fzd6*, and *Celsr1* (Figure 6.3B) (Curtin et al., 2003; Duncan et al., 2017; Montcouquiol et al., 2003; Wang et al., 2006; Yin et al., 2012), with different genes being responsible for the preferential directional specificity in different epithelial regions. For example, loss of *Vangl2* or *Celsr1* only affects hair cell polarity at the striola (Duncan et al., 2017; Yin et al., 2012). Interestingly, while the orientation of hair cells changes at the line of polarity reversal, PCP genes retain their localized cellular expression pattern, suggesting that while PCP proteins are important for coordinating adjacent cell orientation, they are not interpreted in the same way by a particular hair cell (Deans, 2013). Thus, the functional influence of PCP proteins depends on the side of the line of polarity reversal. A candidate gene for a differential interpretation of PCP proteins is *Emx2*. Mouse mutants for *Emx2* lack a line of polarity reversal and instead, all hair cells are arranged with the same cilial bundle directionality (Holley et al., 2010). Thus, collectively the shape and orientation of the macula and the expression of PCP genes allow for hair cells to be oriented in such a way that a perception of 360° of movement is possible (Deans, 2013).

6.4.2 Hair Cell Function

Within all vestibular end organs, hair cells exhibit epithelial region-specific morphophysiological variations, that collectively determine motion sensitivity, response adaptation, and as emerging trait, the effective sensory modality (Eatock and Songer, 2011). Variations in somatic shape and hair cell-afferent synapse structure exist between anamniote and amniote vertebrates, with the latter possessing large calyx-type synapses on type I hair cells while the ubiquitous type II hair cells present in all vertebrates are contacted by bouton-type synapses (Burns and Stone, 2017; Desai et al., 2005a, 2005b; Wersäll, 1956). The broad range of physiological properties distributed between type I and II hair cells of amniotes are, however, matched by a comparably broad functional and dynamic diversity for the single type II hair cells of anamniotes (Eatock and Songer, 2011). This largely excludes calyx-type synapses as the sole origin for the high dynamic signal structure of the associated afferent fibers (Paulin and Hoffman, 2019).

Vestibular hair cells are functionally polarized based on the directionality of the cilial bundle and the eccentric position of the kinocilium at the apical surface. During development, the kinocilium moves from a central position surrounded by microvilli toward one side while the microvilli elongate to become stereocilia (Figure 6.3C). The stereocilia closest to the kinocilium are the tallest and those farther away are

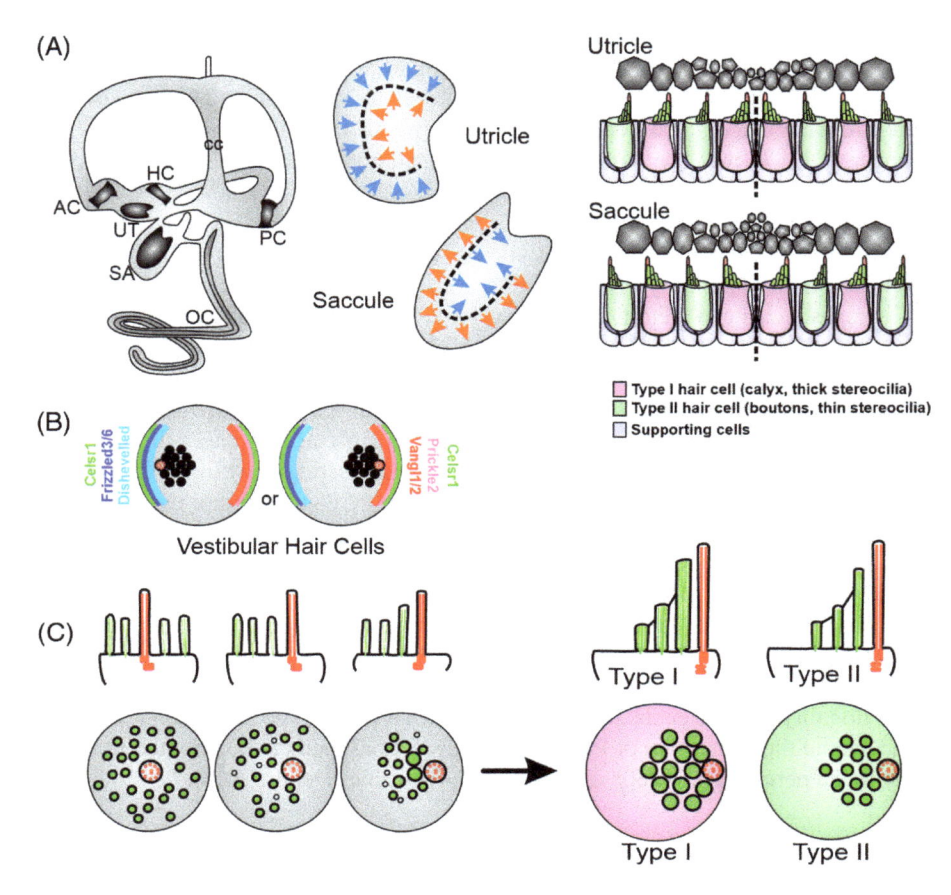

FIGURE 6.3   Polarity of vestibular hair cells. (A) Schematic of a mammalian inner ear (left panel) depicting the location and arrangement of the five vestibular sensory epithelia, the utricle (UT), saccule (SA), anterior semicircular canal crista (AC), horizontal semicircular canal crista (HC), and posterior semicircular canal crista (PC), and the one auditory epithelium, the organ of Corti (OC); common cruz (cc); vestibular hair cells are polarized; note the opposite arrangement across a line of polarity reversal (dotted line in middle and right panels) with hair cell stereocilia pointing toward this line in the utricle and away from it in the saccule; afferents innervating hair cells in the medial utricle and lateral saccule project to the vestibular nuclei (orange arrows), whereas afferents innervating hair cells in the lateral utricle and medial saccule project to the cerebellum (blue arrows). **(B)** Differential expression of Wnt/planar cell polarity (PCP) genes in vertebrate hair cells; in vestibular hair cells, *Prickle2* and *Vangle1/2* are located on one side of the cell and *Frizzled3/6* and *Dishevelled* on the other; *Celsr1* is expressed on both sides of the cells; although the distribution of PCP genes remains constant, stereocilia orient in opposing polarities at the line of polarity reversal. **(C)** Development of hair cell polarity from a central position of the kinocilium (red) to one with an asymmetric location; as development progresses, microvilli develop into stereocilia (green); differential development generates type I and type II hair cells with stereocilia of different diameters. (Modified from Elliott et al., 2018.)

shorter, producing a staircase-like arrangement (Elliott et al., 2018; May-Simera and Kelley, 2012). Mechanical deflection of the stereocilia, interconnected by tip links, towards the kinocilium increases the spontaneous K^+ influx from the K^+-rich endo-lymphatic fluid into the hair cell, depolarizes the membrane potential, and causes an increase in transmitter release (Corey and Hudspeth, 1983). Deflection of the ste-reocilia away from the kinocilium decreases the K^+-influx and thus hyperpolarizes the hair cell, causing a reduction of the spontaneous transmitter release (Corey and Hudspeth, 1983). While this mechano-electrical transduction principle applies to all vertebrate hair cells, the polarization dynamics varies between hair cell types in cor-respondence with cilial bundle morphology and ion channel endowment (e.g., Baird, 1992; Eatock and Songer, 2011).

In correspondence with the length, shape, and stiffness of the kinocilium, hair cells differentially express ion conductances that cause the electrical responses upon cilial deflection to be more phasic or more tonic (Baird and Lewis, 1986; Lewis and Li, 1975). Hair cell subtypes with different morphophysiological features, such as more phasic (type E in Figure 6.4B$_1$) or more tonic membrane properties (type B in Figure 6.4B$_1$) occupy preferential positions within the sensory epithelium of the semicircular canals and otolith organs (Baird and Lewis, 1986). Accordingly, hair cells along the striola of the utricle, for example, (Figure 6.4B$_1$) are characterized by a low input resistance and resonating membrane currents that collectively render these cells well suited for the encoding of head/body acceleration (Beraneck and Straka, 2011). In a complementary fashion, extrastriolar hair cells have a higher input resistance and non-adapting potas-sium currents that are best suited for proportionally encoding head position (Beraneck and Straka, 2011). Similar physiological differences of corresponding hair cells were encountered in the semicircular canal cristae at central (phasic properties) and periph-eral (tonic properties) locations (Figure 6.4B$_1$). Such differential properties assign to hair cells low- and high-pass filter-like properties that allow a decomposition of body motion into neuronal motion components with different signal dynamics (Goldberg, 2000). This differential physiological organization represents the origin of frequency-tuned, parallel pathways that form a major principle of vestibular signal transmission and transformation of head/body motion signals into motor commands (Beraneck and Straka, 2011; Straka et al., 2009).

Vestibular hair cells in all vertebrates use glutamate as transmitter. Glutamate is spontaneously released at rest, causing the postsynaptic vestibular afferent fibers to produce a resting discharge (Figure 6.4B$_2$; Goldberg, 2000; Eatock et al., 2008) that differs in regularity and rate between fiber types (Paulin and Hoffman, 2019; Figure 6.4C). The glutamatergic excitation of vestibular afferents is mediated by AMPA as well as NMDA receptors providing a tonic drive at rest and an increase in firing rate during hair cell depolarization (Annoni et al., 1984; Cochran and Correia, 1995). While the pharmacological profile applies to all hair cells (Figure 6.4B$_2$, upper scheme), phasic hair cells in the center of the semicircular canal crista (Figure 6.4B$_2$, lower scheme) co-release glutamate and GABA (Holstein et al., 2004A, 2004B). In this scheme, release of GABA causes a GABA$_B$ receptor-activated, and second messenger-mediated, delayed inhibition that truncates the faster ionotropic AMPA/NMDA receptor-evoked excitation (Figure 6.4B$_2$, lower scheme). This mechanism is exclusive to phasic hair cells rendering their transient responses even more dynamic

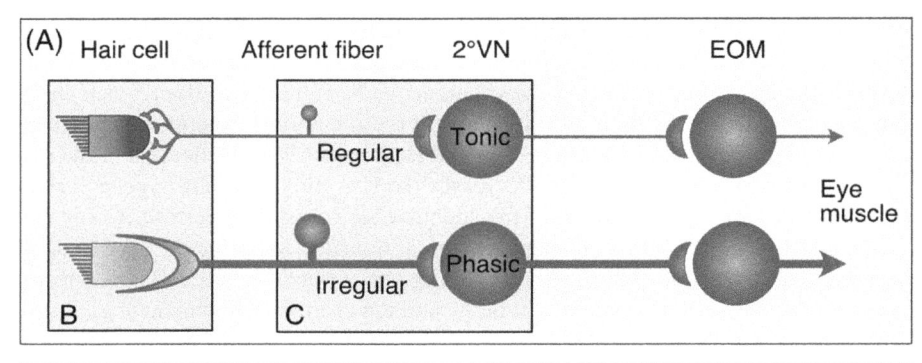

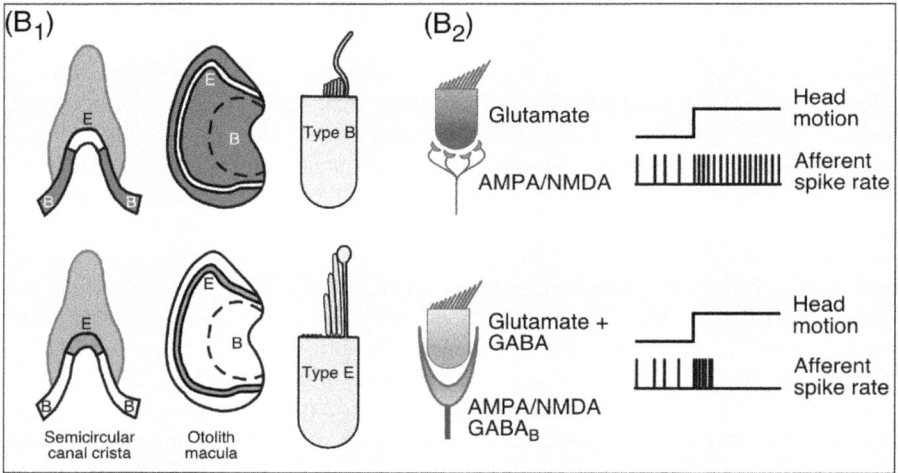

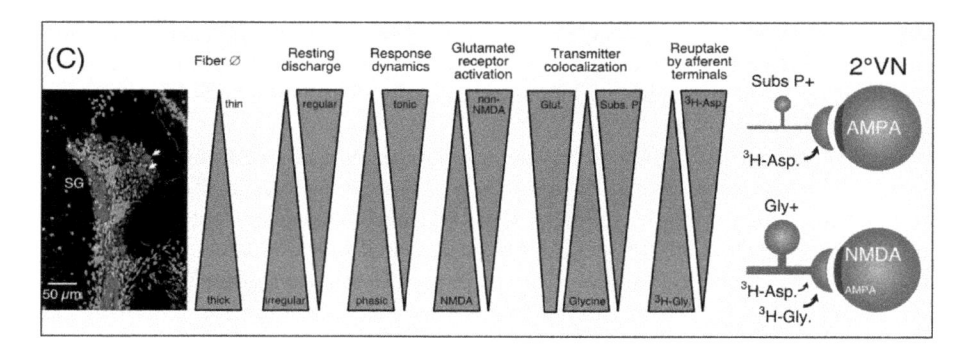

FIGURE 6.4   Frequency-tuned neuronal elements form parallel channels for vestibulo-ocular reflexes. (A). Schematic depicting the three-neuronal VOR pathway between inner ear and eye muscles as dynamically different channels with tonic (upper) and phasic (lower) properties at the extreme ends. **(B).** Schematic illustrating epithelial distributions of dynamically different hair cells **(B₁)** and tentative modes of synaptic transmission onto afferent fibers **(B₂)**; tonic (Type B) hair cells predominate in peripheral areas of the semicircular canal cristae and extrastriolar regions of the otolith maculae; phasic (Type E)

FIGURE 6.4   *(Continued)*   hair cells predominate in central crista and striolar regions. Tonic hair cells release glutamate, activate glutamatergic AMPA/NMDA receptors and produce a tonic increase in firing rate upon step-like head motion in the afferent fiber (upper); phasic hair cells co-release glutamate/GABA, activate glutamatergic AMPA/NMDA and GABA$_B$ receptors and produce a transient burst discharge upon step-like head motion in the afferent fiber (lower). (**C**). Cross-section through Scarpa's ganglion (SG) of a *Xenopus laevis* tadpole (left) depicting Kv1.1 immunopositivity predominantly of large cells (arrows), interrelated properties of vestibular afferents (middle) and different modes of synaptic transmission (right) between thin *versus* thick afferent fibers and second-order vestibular neurons (2°VN). ³H-Asp, tritiated aspartate; ³H-Gly, tritiated glycine; AMPA, α-amino-3-hydroxy-5-methyl-4-isoxazolepropionic acid; EOM, extraocular motoneuron. NMDA, N-methyl-D-aspartate. (Histological section in [**C**] modified from I Gusti Bagus et al., 2019.)

by introducing a mathematical derivative as part of the neuronal encoding of angular motion (Holstein et al., 2004a). In addition to the conventional quantal transmission, calyx-type synapses between amniote type I hair cells and corresponding afferent fibers utilize the functional consequence of cation accumulation in the calyx cleft that depolarize and electrically couple both synaptic elements (Contini et al., 2020). Thus, the frequency tuning of hair cell-afferent fibers derives from co-adapted intrinsic membrane and emerging synaptic properties responsible for a dynamic signal separation that starts at the sensory periphery and recurrently continues along the hierarchical pathway elements (Straka et al., 2009).

The efficacy of the hair cell-afferent synaptic transmission is controlled by the central nervous system through an efferent pathway that originates in the hindbrain and terminates in the inner ear epithelia on vestibular hair cells and afferent fibers (Holt, 2020; Mathews et al., 2020). These efferent neurons supposedly provide a context-specific modulation of the hair cell synaptic transmission (Holt, 2020) and afferent encoding of head/body motion during e.g., arousal, predation, or plasticity processes (Mathews et al., 2020). While clear evidence for the contextual implications is more suggestive than real, vestibular efferent fibers in *Xenopus* have been shown to relay spinal motor efference copies during locomotion, which serve to attenuate the sensory responsiveness of vestibular afferents for head motion during locomotor activity. Such a relay of predictive motor signals through inner ear efferent neurons likely presents with the goal to adjust stimulus statistics and/or prevent overstimulation (Chagnaud et al., 2015).

6.5   VESTIBULAR SENSORY NEURONS

In continuation with the peripheral cellular elements of the vestibular system, afferent sensory neurons constitute the next step required for sensory transmission of self-motion stimuli. These afferent fibers form synaptic contacts with hair cells and transmit the mechano-electrically transduced information about head/body orientation and motion in space as frequency-modulated spike trains (Goldberg, 2000). Following the merging of axons from all end organs into the VIIIth cranial nerve, afferent fibers predominantly terminate on second-order vestibular neurons (2°VN) in the dorsal hindbrain, with a relatively minor number of afferents which as mossy fibers innervate the cerebellum (Hoffman and Paulin, 2020). Afferent fibers establish separate information pathways that reinforce the vectorial and dynamic decomposition performed by end organs and hair cells, respectively, and distribute the signals in a distinct spatiotemporal pattern among CNS target neurons (Straka and Gordy, 2020).

6.5.1 Development of Vestibular Sensory Neurons

Although they are the second cell in the relay of vestibular information, vestibular neurons are the first to develop from the otocyst. After delaminating from the anteroventral portion of the otocyst, both vestibular and auditory sensory neurons initially form the cochleo-vestibular ganglion (Carney and Silver, 1983; Delacroix and Malgrange, 2015; Fariñas et al., 2001; Noden and Van De Water, 1986). One of the earliest genes expressed in these developing neurons is *Neurog1*. This highly conserved gene is necessary for inner ear neuronal specification and formation (Ma et al., 1998, 2000). In the absence of *Neurog1*, no inner ear neurons ever form (Ma et al., 2000). *Neurog1* induces the expression of a downstream gene, *Neurod1*, which is important for neuronal differentiation (Ma et al., 1998) (Figure 6.2B). Not only does a loss of *Neurod1* result in the absence of many inner ear neurons, but also in the development of ectopic 'hair cell-like' cells within vestibular ganglia (Jahan et al., 2010a, 2010b; Macova et al., 2019). This suggests that *Neurod1* normally downregulates *Atoh1* expression to promote a neuronal fate. Genes upregulated downstream of *Neurod1* that further contribute to neuronal differentiation include *Pou4f1* (Huang et al., 2001), *Islet1* (Radde-Gallwitz et al., 2004), *bHLHb5* (Brunelli et al., 2003), and *Tlx3* (Qian et al., 2001).

Molecular separation of vestibular *versus* auditory identity from the common developmental ganglia is apparent, and in part due to a dependence on *Sox2* expression in auditory, but not vestibular, neurons (Dvorakova et al., 2020). In addition, *Neurog1* expression has been demonstrated in vestibular neurons prior to an expression in auditory neurons, suggesting that these two populations have different cell cycle exit times (Koundakjian et al., 2007). Thus, the distinction between vestibular and auditory neurons occurs in time and place within the shared ganglion. Further differences include the dependence of auditory but not of vestibular neurons on the expression of *Lmx1a/b* (Chizhikov et al., 2021) and the requirement of all auditory, but only few vestibular neurons on the expression of *Gata3* (Duncan et al., 2011; Duncan and Fritzsch, 2013; Karis et al., 2001; Lawoko-Kerali et al., 2004).

During development, vestibular and auditory neurons physically segregate from each other to form separate vestibular and auditory ganglia (Delacroix and Malgrange, 2015); however, in the absence of *Neurod1*, auditory and vestibular neurons fail to fully separate (Filova et al., 2020; Jahan et al., 2010b; Macova et al., 2019). Unlike in the auditory system, where spiral ganglion neurons maintain the tonotopic organization of the associated hair cells, end organ-specific vestibular neurons within the vestibular ganglion are much less clustered (Appler and Goodrich, 2011; Fritzsch, 2003; Fritzsch et al., 2019). The vestibular ganglion and its afferents are organized into two divisions: superior/anterior and inferior/posterior, each innervating a distinct and generally conserved set of vestibular end organs (Vidal et al., 2015). The anterior and horizontal semicircular canal cristae and the utricle are innervated by vestibular afferents projecting along the *ramus anterior* of the vestibular nerve, whereas the posterior semicircular canal crista is innervated by afferents traversing along the *ramus posterior.* The saccule is innervated by portions of both branches with few species-specific exceptions (de Burlet, 1929; Fritzsch et al., 2002; Gacek and Rasmussen, 1961; Lewis et al., 1985; Maklad and Fritzsch, 1999). The lagena, present in elasmobranchs, bony fish, *Latimeria*, amphibians, sauropsids, and non-therian mammals, is innervated by an additional offshoot of the posterior branch of the VIIIth nerve (Fritzsch et al., 2013; Schultz et al., 2017).

6.5.2 Function of Vestibular Afferents

6.5.2.1 Hair Cell-Afferent Synaptic Connectivity

Afferent fibers in amniote vertebrates contact hair cells through calyx-type synapses onto type I hair cells, through bouton-type synapses onto type II hair cells, or through both types simultaneously as dimorphic fibers (Goldberg, 2000). Despite the presence of the latter group, afferents appear to distinguish into two broad classes with either an irregular or a regular discharge at rest (Goldberg, 2000; Paulin and Hoffman, 2019). This separation based on resting rate regularity correlates with a number of morphophysiological parameters (Goldberg, 2000; Figure 6.4C) such as a differential presence of ion conductances in the parent ganglion cells (Eatock and Songer, 2011). Accordingly, irregular firing afferents are of large caliber and are endowed with prominent voltage-dependent potassium conductances that render these neurons particularly suitable for transmitting information about fast and highly phasic motion components (Eatock and Songer, 2011; Goldberg, 2000). In contrast, regularly firing afferents are rather thin and characterized by ion conductances that allow sustenance of ongoing spike discharge as required for e.g., encoding of tonic positional deviations of the head/body (Eatock and Songer, 2011; Goldberg, 2000). Despite the lack of type I hair cells and calyces in anamniotes, afferent fibers of these vertebrates, which only form bouton-type synaptic endings onto type II hair cells, can also be distinguished into irregularly and regularly firing fibers with corresponding differences in fiber diameters, response dynamics and motion sensitivities (Dlugaiczyk et al., 2019). This organizational principle, based on firing rate regularity decouples the obviously ubiquitous interrelated morphophysiological features from the presence of calyx-type synapses.

6.5.2.2 Peripheral Innervation Pattern

Irrespective of vertebrate species, and thus of the presence or absence of calyx-type synapses, irregular and regular firing afferent fibers terminate on hair cells with corresponding response dynamics, respectively (Baird and Lewis, 1986; Goldberg, 2000; Lewis and Li, 1975). According to this scheme, irregular firing afferents innervate hair cells with phasic properties along the striola of the otolith organs and at the central region of the semicircular canal cristae (Figure 6.4B$_1$). In contrast, regular firing afferent fibers terminate predominantly on hair cells with tonic properties located at extrastriolar regions of the otolith organs and at peripheral areas of the semicircular canal cristae (Figure 6.4B$_1$). This region-specific connectivity between hair cells and afferent fibers with corresponding signal processing capacities generates distinct signaling pathways that allow separate encoding and transmission of dynamically different head/body motion components to central target neurons (Straka et al., 2009). These parallel information streams are generated by co-adapted morphophysiological properties that as an entity are responsible for the signal encoding capacities (Figure 6.4C). The different properties form a continuum rather than distinct subclasses (Paulin and Hoffman, 2019) and extend the frequency-tuned pathways into the brain, where the mode and pharmacological profile of the afferent synaptic transmission onto 2°VN is part of the differential dynamic tuning (Figure 6.4C).

6.5.2.3 Functional Properties of Vestibular Afferent Fibers

The difference in ion channel composition of the two vestibular afferent subtypes has considerable functional implications, and have been best evaluated in primates (Jamali et al., 2013; Sadeghi et al., 2007; Schneider et al., 2015). Accordingly, irregularly firing semicircular canal afferents encode high-frequency stimuli with high gains, while regularly firing afferents precisely transmit information about the timing of the motion stimulus (Cullen, 2012). This causes regularly firing afferents in primates to express a lower sensitivity but such a cost is offset by the ability to transmit twice as much information as irregular firing afferents (Sadeghi et al., 2007). This situation is different for otolithic afferents where adjustments in respective sensitivities and discharge variabilities cause neuronal thresholds to become independent of stimulus frequency and resting discharge regularity (Jamali et al., 2013). Importantly, however, both vestibular afferent subtypes constitute a functional complementarity that ensures transmission of naturally occurring head/body motion profiles with species-specific variations in overall bandwidth and range of peak performance.

6.5.2.4 Pharmacology of Vestibular Afferent Signal Transmission

Vestibular afferents terminate on the soma and dendrites of central vestibular neurons and elicit monosynaptic excitatory postsynaptic potentials (EPSPs) by the release of glutamate and activation of postsynaptic AMPA and NMDA receptors in all vertebrates (Straka and Dieringer, 2004). The extent to which the two glutamate receptor subtypes are recruited depends on the afferent fiber type (Straka et al., 2009; Figure 6.4C). Irregularly firing thick fibers monosynaptically excite 2°VN mainly *via* NMDA receptors with little contribution from AMPA receptors, a condition that is supported by the co-release of glutamate and glycine, with the latter amino acid acting as co-agonist for the activation of NMDA receptors (Straka and Dieringer, 2004). The combination of NMDA receptor activation, dendritic calcium spikes and voltage-dependent potassium conductance in the target neurons (Pfanzelt et al., 2008) promotes the activation of highly transient responses upon sensory stimulation in agreement with the expected central organization of frequency-tuned pathways (see below). In contrast, regularly firing, thin afferents predominantly recruit AMPA receptors following the exclusive release of glutamate (Straka and Dieringer, 2004), although with a gradual transition between the two pharmacological variants

(Figure 6.4C). Thus, the synaptic organization of afferent synaptic inputs onto individual 2°VN complies with a dissociation of ascending vestibular pathways as frequency-tuned information channels that preferentially mediate and transmit signals according to dynamic content.

6.5.2.5 Central Projections of Afferent Fibers

Vestibular afferent fibers terminate predominantly in the dorsal hindbrain with species-specific supplementary projections to the cerebellum and reticular formation (Straka and Dieringer, 2004). For the otolith end organs, whether afferents project to vestibular nuclei or to the cerebellum depends on the specific polarity of the associated hair cells on a particular side of the line of polarity reversal (Figure 6.3A). Afferents that innervate the medial half of the utricle will project nearly entirely to the vestibular nuclei whereas those innervating the lateral area will project mainly to the cerebellum (Maklad et al., 2010). For the mammalian saccule, which has hair cells aligned in opposition to those of the utricle across the line of polarity reversal, a similar, oppositely targeted central innervation scheme is observed. Accordingly, afferents innervating the medial half of the saccular epithelium project primarily to the cerebellum while those afferents that innervate the lateral area project to the vestibular nuclei (Maklad et al., 2010). A few genes have been identified that are involved in guiding vestibular afferents to their central targets. For instance, loss of *Neurod1* results in a highly disorganized and mixed projection of vestibular afferent fibers together with auditory fibers into the cochlear nucleus (Filova et al., 2020; Jahan et al., 2010b; Macova et al., 2019). In addition, reduction or loss of the *Wnt* receptor, *Fzd3*, results in erroneous targeting of some vestibular afferents to the lateral line nucleus in frogs and to the cochlear nucleus in mice (Duncan et al., 2019). Furthermore, in the absence of *Fzd3*, there is a higher degree of overlap in projections of specific end organs (Duncan et al., 2019). However, if *Fzd3* absence has consequences on the sensory-motor transduction ability is currently unknown, particularly given that end organ-specific overlap occurs normally in development and is believed to be functionally mitigated by synaptic pruning and/or silencing mechanisms (Elliott and Gordy, 2020; Straka et al., 2014). However, whether the downstream *Wnt/PCP* signaling molecule, *Prickle1*, also affects central guidance of vestibular afferents as it does with spiral ganglion neurons (Yang et al., 2017) has not yet been confirmed.

Collectively, the projections from the different end organs outline the overall extent of the central vestibular nuclei and its classical subdivisions (Horn, 2020). However, these projections lack a topographic representation of fibers from individual semicircular canal and otolith end organs (e.g., Birinyi et al., 2001; Figure 6.5A). This obvious absence of a distinct sensory map is at variance with the central representation of most other sensory systems including the closely related lateral line system of aquatic anamniotes (Chagnaud et al., 2017). Even though some vestibular end organs have topographical hot spots of afferent projections within the dorsal hindbrain, the extensive overlap of fibers from different end organs (Figure 6.5A) suggests that the spatial specificity of vestibular signals, achieved by vectorial decomposition of head/body motion through the plane-specific arrangement of the end organs is discontinued centrally.

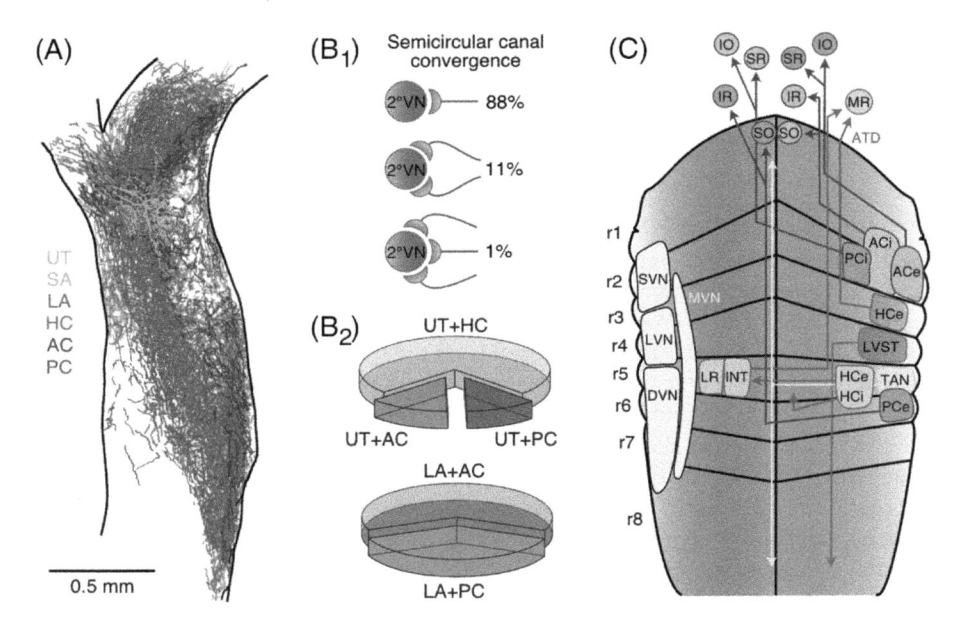

FIGURE 6.5 Spatial representation of semicircular canal and otolith signals in central vestibular circuits. (A) Overlay of afferent projections from the three semicircular canals and the three otolith organs on a horizontal section through the frog vestibular nuclei; note the large overlap of afferent terminations from the different end organs. (B) Monosynaptic semicircular canal and otolith afferent convergence in frog second-order vestibular neurons (2°VN), indicating maintained separation of semicircular canal signals in most 2°VN (**B₁**); utricular (UT; horizontal otolith organ) signals converge preferentially with horizontal semicircular canal (HC) afferent inputs and lagenar (LA; vertical otolith organ in amphibians) signals converge exclusively with anterior (AC) and posterior vertical semicircular canal (PC) inputs (**B₂**). (C) Schematic depicting the segmental arrangement of vestibulo-ocular and vestibulo-spinal projections along the rhombomeric (r) scaffold of a frog hindbrain. ACe, ACi, HCe, HCi, PCe, PCi, excitatory (e) and inhibitory (i) semicircular canal-related vestibular neurons; ATD, ascending tract of Deiters; DVN, LVN, MVN, SVN, descending, lateral, medial, superior vestibular nucleus; INT, abducens internuclear neurons; IO, IR, LR, MR, SO, SR, motoneuronal populations innervating the inferior oblique, inferior rectus, lateral rectus, medial rectus, superior oblique, superior rectus eye muscles; LVST, lateral vestibulospinal tract; SA, saccule; TAN, tangential vestibular nucleus. (Summary in [A] obtained from data reported by Birinyi et al., 2001.)

However, this is rather not the case because 2°VN, despite the availability of afferent synaptic inputs from most, if not all, end organs throughout the vestibular nuclei are highly selective in accepting only those that match the spatial signaling profile of their respective target neurons (see below). In addition to the considerable overlap of afferent fibers from different vestibular end organs, afferents with different axon diameters also terminate within the same area. The formation of synaptic contacts of thin and thick fibers with the same target neurons suggests that signals with different dynamics are mixed in individual 2°VN. This is however not the case, because the synapses from fibers with different calibers segregate along the soma-dendritic extent of the neurons and thus allow respective modifications of the signaling dynamics based on soma-dendritic variations of the membrane properties (Sato and Sasaki, 1993). Accordingly, this arrangement permits cell compartment-specific time constants to differentially integrate synaptic inputs and thus to monosynaptically combine head motion-related components with different dynamics. This specific selectivity described here is believed to be generated by an activity-based refinement of early synaptic connections (Elliott and Gordy, 2020). Though direct evidence for such mechanisms remained historically elusive, central tuning of vestibular neurons has been previously alluded to (Straka et al., 1997, 2002) and recently demonstrated in zebrafish (Liu et al., 2020). Remarkably, the degree of similarity in spatio-temporal convergence was shown to correlate with the complexity of tuning of specific central targets (Liu et al., 2020).

6.6 FUNCTIONAL ORGANIZATION OF CENTRAL VESTIBULAR NUCLEI

Central vestibular neurons form the key elements for multimodal integration of head/body motion-related sensory feedback and predictive feedforward signals. Understanding the morphophysiological organization of vestibular operations requires insight into ontogenetic, topographic, and functional concepts of central vestibular signal processing. The emerging properties of the vestibular nuclei allow computations that are at the origin of the vectorial reconstruction of 3D motion and the processing principle for signals across a wide range of motion dynamics. Following spatio-temporal transformation, these nuclei form the major hub for distributing neuronal representations of motion along intermediate pathways to higher-order centers.

6.6.1 VESTIBULAR NUCLEUS DEVELOPMENT

The vestibular nucleus develops along the rostro-caudal extent of the dorsal hindbrain, spanning a region that extends throughout most of the rhombomeric scaffold (Cambronero and Puelles, 2000; Glover, 2020; Hernandez-Miranda et al., 2017; Marín and Puelles, 1995; Pasqualetti et al., 2007). Very early in development, anterior-posterior patterning of the hindbrain is accomplished by *Hox* genes (Hernandez-Miranda et al., 2017). While multiple *Hox* genes are supposedly involved in vestibular nucleus patterning, so far only the role of *Hoxb1* has been investigated. *Hoxb1* is expressed only in rhombomere (r) 4 where the lateral vestibular nucleus originates (Chen et al., 2012). Loss of *Hoxb1* does not affect the phenotype of r4 vestibular neurons, but results in severe disruption in their vestibulo-spinal projections (Chen et al., 2012; Di Bonito et al., 2015). Additional genes are expressed in a dorso-ventral gradient in the hindbrain that corresponds with the longitudinal columns of hindbrain nuclei (Glover et al., 2018). In the developing vestibular nuclei, there is a transient expression of *Neurog1* (Fritzsch et al., 2006; Glover et al., 2018; Ray and Dymecki, 2009) (Figure 6.2B); however, not all vestibular nucleus neurons develop from the *Neurog1*-expressing region (Glover et al., 2018). Additional genes expressed in vestibular nucleus neurons and likely related to morphophysiological specification processes include, *Lbx1*, *Phox2b*, *Pou3f1*, *Lhx1*, *Ascl1*, *Ptf1a*, and *Neurog2* (Hernandez-Miranda et al., 2017; Lunde et al., 2019).

6.6.2 Principles Governing the Processing of Spatial Motion Vectors

The overlapping termination of vestibular nerve afferents from different end organs within the vestibular nuclei (Figure 6.5A) suggests substantial convergence of afferent signals in individual central vestibular neurons. However, this is rather not the case as demonstrated in various vertebrate species (Kasahara and Uchino, 1974; Straka et al., 1997, 2002; Uchino et al., 1981; Wilson and Felpel, 1972; Zhang et al., 2001). Afferent inputs from individual semicircular canal and otolith organs are combined in 2°VN with an evolutionary conserved pattern (Straka et al., 2014). Within this scheme, most 2°VN receive monosynaptic inputs from only one ipsilateral semicircular canal (Figure 6.5B$_1$), while a small proportion is activated by afferents from two or all three semicircular canals (Straka et al., 1997). This indicates that the vectorial decomposition of angular acceleration components, achieved by the orthogonal arrangement of the semicircular canals, is maintained at the central vestibular level. This representation scheme is complemented by a similar separation of monosynaptic afferent inputs from the different otolith organs (Straka et al., 2002). Accordingly, afferent fibers from the utricle and saccule (in mammals) or from the utricle and lagena (in e.g., amphibians) activate monosynaptic responses in 2°VN in an almost complementary manner (Straka et al., 2002; Uchino, 2001). Thus, the dissociation of linear acceleration components, accomplished by the differential arrangement of the otolith organs in the inner ear, is also retained in central vestibular neurons. In contrast, semicircular canal signals converge with those from otolith organs in an end organ-specific manner (Figure 6.5B$_2$; Straka et al., 2002). Accordingly, afferent inputs from the horizontal semicircular canal are preferentially integrated with those from the utricle, while afferent inputs from the anterior and the posterior vertical semicircular canals converge predominantly with afferent signals from the vertical linear acceleration-sensitive otolith organ (Figure 6.5B$_2$). The emerging pattern appears to be evolutionary conserved, but does present with some species-specific variations in the extent of signal convergence from distinct semicircular canal and otolith organs (Straka et al., 2002; Uchino, 2001). Nonetheless, as a general principle, similar proportions of 2°VN mediate signals exclusively from a semicircular canal or an otolith organ (~25% each), while ~50% of the central vestibular neurons transmit a combination of signals from the two types of end organs (Dickman and Angelaki, 2002; Straka et al., 2002).

In order to obtain a single motion percept and to trigger coherent bilateral motor responses, signals from both inner ears are combined through brainstem commissural pathways, interconnecting the vestibular nuclei on the two sides (Straka, 2020). This computation, in addition to, and beyond, the traditional processing with bilaterally organized brainstem circuits allows signals from semicircular canal and otolith organs on one side to be reciprocally integrated with spatially matching signals from the other side through midline-crossing connections (Kasahara and Uchino, 1974; Malinvaud et al., 2010). Key to this processing is again the mirror-symmetric arrangement of vestibular end organs (Figure 6.1B; Graf and Simpson, 1981). Typically, semicircular canal commissural connections are plane-specific and consist of an inhibition from the contralateral coplanar semicircular canal. Commissural connections that relay signals from otolith organs, in particular from the utricle, transmit inhibitory signals from epithelial sectors with opposite hair cell polarity (Uchino et al., 2001; Uchino and Kushiro, 2011). In contrast, saccular 2°VN generally lack a commissural inhibition but instead exhibit ipsilateral cross-striolar inhibition from the same saccular epithelium (Uchino, 2004). Commissural pathways are either formed by midline-crossing GABAergic neurons or by excitatory commissural neurons that activate local glycinergic and/or GABAergic interneurons (Malinvaud et al., 2010). Functionally, these connections ensure a bilaterally balanced vestibular activity and enhance the motion sensitivity and output gain of central vestibular neurons.

6.6.3 PRINCIPLES GOVERNING THE PROCESSING OF MOTION DYNAMICS

Following peripheral decomposition of motion dynamics and separate encoding of temporally different aspects by hair cells, neuronal correlates of motion components are transmitted as modulated spike trains along frequency-tuned afferent pathways to neurons in the vestibular nuclei (Straka et al., 2009). This causes the population of 2°VN to be constantly confronted with semicircular canal and otolith afferent inputs with dynamically different signatures (Goldberg, 2000). These inputs with different temporal properties require processing by neurons with adequate intrinsic membrane properties that match the dynamic structure of the afferent signals. Accordingly, central vestibular neurons of all vertebrates form differently tuned functional subgroups rather than a single prototypic neuronal substrate (Straka et al., 2005). Although specific details regarding the endowment of vestibular neurons with particular ion conductances differ between species, a common theme is the separation into a more tonic and a more phasic neuronal subtype (Straka et al., 2005). Thus, descriptions of type A/B neurons in the medial vestibular nucleus of rodents/guinea pigs (Serafin et al., 1991), of tonic/phasic neurons in frogs (Straka et al., 2004), of tonic/kinetic neurons in cat (Shimazu and Precht, 1965) and elongate/principal cells in the chicken tangential vestibular nucleus (Peusner and Giaume, 1997) appear to represent only species-specific variations of a conserved physiological duality (Straka et al., 2005).

Obvious key features that distinguish between dynamically different vestibular cell types are spike shape and magnitude and form of the after-hyperpolarization indicating differences in the expression of voltage-dependent K^+-conductances such as Kv-channels (Straka et al., 2005). The differences in ion channel composition cause mammalian type A neurons to express low-frequency capabilities (tonic cells) and type B neurons to exhibit high-frequency dynamics (kinetic cells). This dynamic differentiation appears to be a common property since central vestibular neurons of other vertebrates such as frogs also subdivide into two distinct populations with different intrinsic membrane properties, discharge dynamics, and synaptic response characteristics (Beraneck and Straka, 2011; Rössert and Straka, 2011). Accordingly, tonic vestibular neurons form low-pass filters with membrane properties that promote amplification of synaptic inputs, while phasic vestibular neurons form bandpass filters that allow a frequency-dependent shunting of inputs. Thus, differential, yet complementary endowments with particular conductances render tonic neurons most suitable for signal integration and phasic neurons for synaptic differentiation and event detection (Beraneck and Straka, 2011).

The functional consequences of such membrane properties are reinforced by a differential insertion of the two cell types into microcircuits formed by local interneurons (Figure 6.6A, B; Straka and Dieringer, 2004). Integration of the monosynaptic afferent excitation with a delayed GABAergic and/or glycinergic inhibition from the same peripheral origin in phasic but not in tonic vestibular neurons (Figure 6.6A) truncates the spike discharge and thus renders inputs even more phasic (Biesdorf et al., 2008), comparable in outcome to hair cells that co-release glutamate and GABA (see above). In contrast, embedding of tonic vestibular neurons in local circuits of excitatory interneurons (Figure 6.6B) causes further amplification of vestibular inputs and thus a reinforcement of signal integration (Straka and Dieringer, 2004). The synergy of cellular and circuit properties therefore creates sets of neuronal elements with particular filter characteristics that form flexible, frequency-tuned, components optimized for the transformation of the wide dynamic range of head/body motion-related vestibular signals. The separate processing of different dynamic aspects subsequently extends onto the motor elements of vestibular reflexes as demonstrated for the extraocular motor system (Horn and Straka, 2021).

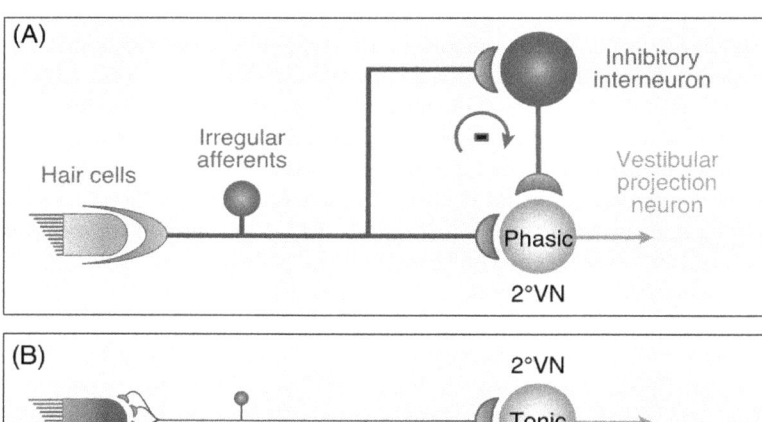

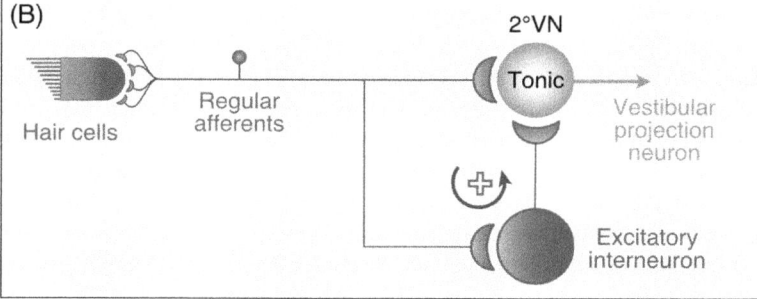

FIGURE 6.6   Dynamic tuning of responses in second-order vestibular neurons. (**A**) and (**B**) Schematics depicting tentative feed-forward inhibitory (−) and excitatory (+) central vestibular circuitries, reinforcing the intrinsic membrane properties of phasic (**A**) and tonic (**B**) second-order vestibular neurons (2°VN), respectively; note the differential activation of the local inhibitory and excitatory interneuronal side-loops by thick, irregular (**A**) and thin regular firing afferents (**B**), respectively.

6.6.4 Principles Governing the Topographical Organization of Vestibular Pathways

The considerable overlap of afferent projections from different end organs (Figure 6.5A) in the vestibular nuclei essentially precludes a classical sensory map as organizational principle (Chagnaud et al., 2017). The selectivity and spatially specific convergence of semicircular canal and otolith afferent inputs in individual 2°VN, however, generates subgroups of neurons which encode particular head/body motion vectors (Straka et al., 2002). These neurons are distributed in all vertebrates throughout the vestibular nuclei, which in the traditional view form anatomically and cytoarchitectonically more or less distinguishable subnuclei (Figure 6.5C; Horn, 2020). This classical nomenclature has been refined by the alignment of the latter with the rhombomeric scaffold (Cambronero and Puelles, 2000; Glover, 2020; Straka et al., 2014). Generally, the superior vestibular nucleus (SVN) derives from r1/r2, the lateral vestibular nucleus (LVN) from r3/r4, the descending vestibular nucleus (DVN) from r5-r8 and the major part of the medial vestibular nucleus (MVN) from r3 to r7, representing the conserved blueprint of early vertebrates (Figure 6.5C; Straka and Baker, 2013). Within the segmental scaffold, particular vestibular phenotypes originate from distinct segments and generate identifiable populations based on rhombomeric origin of the soma, axonal trajectory (ipsilateral, contralateral, ascending, descending), and targets such as extraocular and spinal motoneurons, or the cerebellum (Diaz and Puelles, 2019; Glover, 2000, 2003, Straka et al., 2014).

According to this scheme, VOR projections to vertical and oblique extraocular motoneurons in the oculomotor and trochlear nuclei originate from r1–r3 and r5, respectively, giving rise to uncrossed inhibitory and crossed excitatory projections (Figure 6.5C; Straka et al., 2014). VOR projections to horizontal extraocular motoneurons in the abducens nucleus derive from r5 and/or r6 and are complemented by abducens internuclear pathways to ensure conjugate horizontal movements of both eyes (Figure 6.5C; Straka et al., 2014). These VOR circuits are supplemented by projections from r3 (ascending tract of Deiters) and r5 (tangential vestibular nucleus). A corresponding segment-specific organization is also implemented for spinal projections through two main vestibulo-spinal tracts (Glover, 2020). The origin of the lateral vestibulospinal tract largely resides in r4, while r5 contains a nucleus with separate, bidirectional ascending and descending projections with homology to the tangential vestibular nucleus of fish (Figure 6.5C; Straka and Baker, 2013). This projection is supplemented by a further, evolutionary conserved, crossed and uncrossed medial vestibulo-spinal tract (MVST) originating in r5 and r6 (Glover, 2020).

Commissural vestibular pathways required for signal amplification (see above) were also found to originate from distinct rhombomeric positions (Straka, 2020). Accordingly, commissural neurons form two populations arising from r1 to r3 and from r5 to r8, while the r4 compartment, which corresponds to the LVN, is largely devoid of commissural neurons (Malinvaud et al., 2010; Straka et al., 2001). Projections to the cerebellum, as a major contributor to adaptive plasticity of vestibular reflexes (Gittis and du Lac, 2006), also derive from specific locations within the rhombomeric scaffold (Horn, 2020). This vestibular population includes cell groups in r1–r3 and r6–r8 with either ipsi- or contralateral axonal projections.

In addition, projections from the r7 compartment, containing the nucleus *prepositus hypoglossus* mediate signals related to the velocity storage of visuo-vestibular motion signals [(Figure 6.7); Horn, 2020].

The segmental origins of vestibular projection pathways are rather conserved, although with homeotic variations between different vertebrate groups (Glover, 2020). The emergence from distinct rhombomeres suggests that the phenotypic identity of subgroups derives from the combinatorial expression of genetically regulated transcription factors that determine the formation of functional subdivisions along

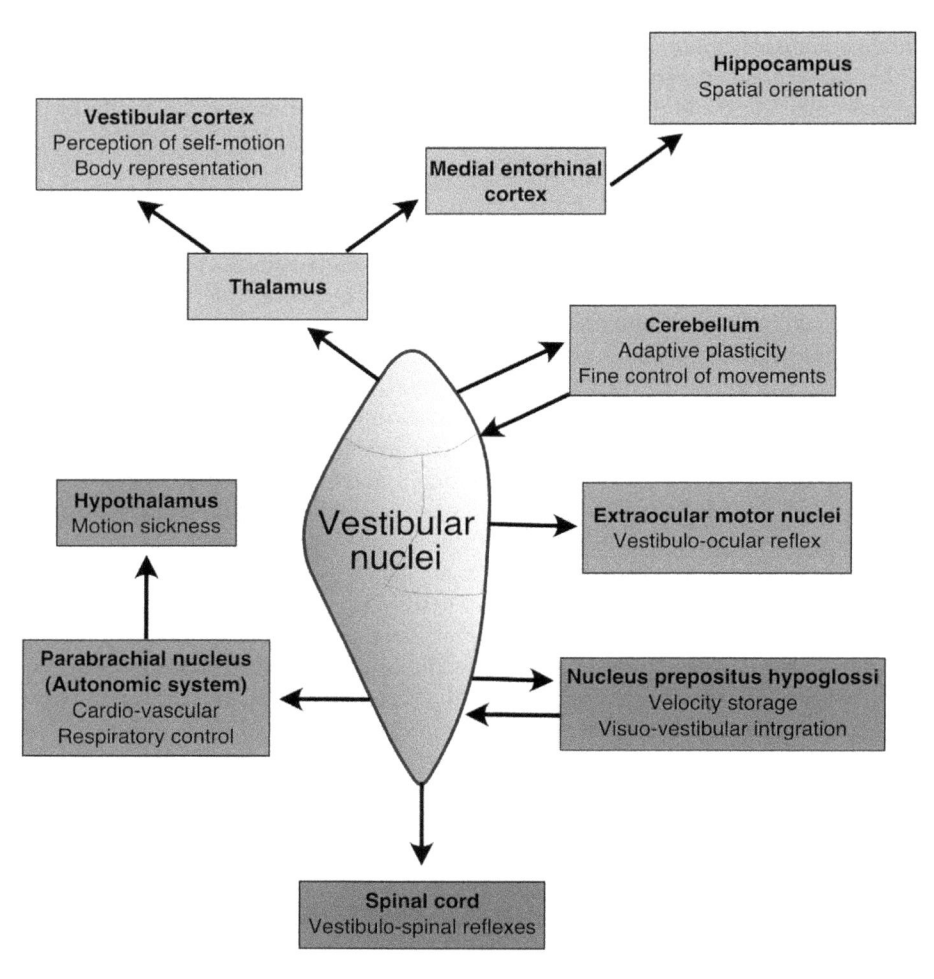

FIGURE 6.7   Major targets of central vestibular projections. Vestibular nuclei provide neuronal commands for gaze and posture stabilization (extraocular motor nuclei, spinal cord), for eye/head motion velocity storage (nucleus *prepositus hypoglossi*) and adaptive plasticity (cerebellum) through reciprocal connections, for vegetative reactions (parabrachial nucleus, hypothalamus) and for navigation, orientation and spatial representation of motion in space (thalamus, vestibular and entorhinal cortex, hippocampus).

the rostro-caudal axis of the vestibular nucleus (Glover, 2020). While so far specified only for vestibulo-spinal neurons (Lunde et al., 2019), this concept likely applies to other vestibular projections as well as for dorso-ventral and medio-lateral compartmental positions within the hindbrain. However, independent of specification rules, the central vestibular topography is not formed as a sensory representation of afferent inputs (sensory map) but emerges from the rhombomeric alignment of vestibulo-motor projection neurons that form a segment-specific premotor/motor map based on the identity of the target neurons and the respective constraints for mediating a particular motion vector (Chagnaud et al., 2017).

6.6.5 PRINCIPLES GOVERNING THE PHARMACOLOGICAL PROFILE OF VESTIBULAR NEURONS

6.6.5.1 Transmitter Profile

Excitatory vestibular neurons in different vertebrates use glutamate as neurotransmitter (Reichenberger et al., 1997; Walberg et al., 1990). The population of glutamate-immunopositive vestibular neurons extends across all sizes and shapes without regional predominance. This is in compliance with the ubiquitous role of this amino acid as excitatory transmitter of small local interneurons as well as large projection neurons (Reichenberger et al., 1997). With respect to the rhombomeric scaffold or the classical vestibular nuclear classification scheme, glutamatergic vestibular neurons exhibit no particular preferential segmental nor nuclear abundance. In contrast, GABA as well as glycine as dominating inhibitory transmitters of the CNS express regionally specific distribution patterns in the vestibular nuclei independent of vertebrate species (Reichenberger et al., 1997; Walberg et al., 1990). GABAergic vestibular neurons are particularly abundant in r1–r3 (Soupiadou et al., 2018), largely corresponding to the SVN. In contrast, glycinergic vestibular neurons predominate in r4–7, a hindbrain region that coincides with parts of the LVN, MVN and DVN (Reichenberger et al., 1997; Soupiadou et al., 2018; Walberg et al., 1990). This complementary distribution of inhibitory transmitters of central vestibular neurons also explains the differential pharmacological profile of inhibitory vestibulo-ocular projections for vertical and horizontal eye movements (Soupiadou et al., 2018). Accordingly, as a general principle, GABAergic neurons in r1-3 (SVN) project to the ipsilateral vertical and oblique extraocular motoneurons in the oculomotor and trochlear nuclei, whereas glycinergic neurons in r5 and r6 (MVN) project to the ipsilateral abducens nucleus (Soupiadou et al., 2018).

6.6.5.2 Ion Channel Profile

The different computational dynamics of central vestibular neurons derives from specific conductance profiles along with auxiliary influences by transmitter receptors and the emerging consequences of the network in which the respective neurons are embedded (Straka et al., 2005). A particular hallmark feature of phasic 2°VN is the presence of a prominent voltage-dependent potassium conductance, such as produced by Kv1.1. channels and predominantly present in large vestibular neurons (Beraneck and Straka, 2011). Although all vestibular regions and subnuclei contain cells of different diameters, large cells are particularly abundant in the LVN and the magnocellular part of the MVN (Beraneck and Straka, 2011). These vestibular areas give rise to vestibulo-spinal tracts, known to preferentially transmit phasic signals to spinal rather than to extraocular motor targets (Highstein et al., 1987; Straka et al., 2004). This suggests that ion channel profiles form an indirect classifier of the central vestibular architecture through a preferential presence in neurons involved in specific computational tasks (Kodama et al., 2012, 2020). Similar denominators of vestibular subdivisions are transcription factors that likely produce microdomains of vestibular neurons with specific spatio-temporal signatures or output targets (Lunde et al., 2019).

6.6.6 Projections to Higher-Order Centers

Following spatially specific fusion of bilateral inner ear inputs and reinforcement of temporal signatures, central vestibular neurons distribute these signals to several CNS regions involved in a wide range of computational tasks (Figure 6.7; Horn, 2020). Connections between the vestibular nuclei and extraocular motor nuclei in the mid- and hindbrain or spinal motor targets elicit short-latency reflexes for gaze and posture stabilization (Figure 6.7). Reciprocal connections with the nucleus *prepositus hypoglossi* and various areas of the cerebellum ensure velocity storage and fusion of visuo-vestibular signals and motor efference copies, thus extracting the sensory consequences of self-motion as requirement for the computation of responses from concurrent reafferent and exafferent head/body motion signals (Cullen, 2012, 2019). In addition, efferent cerebellar projections onto central vestibular neurons serve as gain control that allows context-dependent adjustments of vestibular reflex parameters.

The role of vestibular signals in motor reactions is supplemented by an implication in a number of other neuronal computations for which head/body motion signals are indispensable (Cullen, 2016). Vestibulo-autonomic pathways initiate alterations of blood pressure, heart rate, respiration, and gastrointestinal motility during postural shifts, including orthostatic challenges (Figure 6.7; Holstein, 2020). In this context, the parabrachial nucleus constitutes a dominant region of the autonomic nervous system that integrates vestibular, visceral, and emotional signals. By linkage with the hypothalamus and limbic system through reciprocal connections, this nucleus is also involved in the generation of motion sickness (Holstein, 2020).

The influence of vestibular signaling on perception is matched by an equally large influence on cognitive aspects as well as on orientation, navigation, and spatial map formation (Brandt and Dieterich, 2019; Cullen and Taube, 2017). Head direction cells in several limbic areas form a reference frame as function of directional heading in the environment. The formation of these signals critically depends on egocentric vestibular signals with online update by motor efference copies and information about the visual environment. The involved pathways for these computations include the tegmental nucleus, mammillary nucleus, anterodorsal thalamus, postsubiculum and entorhinal cortex, with a variety of feedback loops at different hierarchical levels (Figure 6.7; Cullen and Taube, 2017; Brandt and Dieterich, 2019). Path integration properties at high spatio-temporal resolution by the vestibular system is therefore key to orientation and navigation and illustrates the ubiquitous role of head motion signals beyond the apparently simple generation of motor commands.

6.7 MULTIMODAL FUSION OF HEAD/BODY MOTION SIGNALS

Three-dimensional head/body motion generates multisensory signals that during self-initiated movements are supplemented by neuronal copies of the motor commands. Accordingly, vestibular signals, which represent direct motor outcomes of these movements, are supplemented by visual motion signals and in case of actively induced movements in addition by proprioceptive feedback and motor efference copies (Straka et al., 2016). The combination of synergistic modalities ensures spatio-temporally appropriate neural representations of head/body motion by jointly exploiting individual sensitivities, dynamic working ranges and task-specific differential availability of each feedback and feedforward signal component.

6.7.1 VISUAL MOTION SIGNALS

Active and passive head/body motion within the visual environment evokes concurrent retinal image motion that is counteracted by gaze-stabilizing eye movements through a spatio-temporally adequate VOR (Figure 6.7; Straka and Dieringer, 2004). The VOR, however, operates in an open-loop mode and thus requires appropriate feedback to optimize image stability. These signals are provided by concurrent visuo-motor responses such as the optokinetic reflex (OKR), which is triggered by large-field visual image motion signals and activates eye movements that aim at minimizing the residual image slip that remains uncompensated by the open-loop VOR (Collewijn, 1969; Horn and Straka, 2021). This visuo-vestibular interaction depends on the functional complementarity of the two sensory systems in the frequency and time domain (Straka and Dieringer, 2004). While vestibular signals contribute to gaze stability preferentially at high frequency, high acceleration head movements, visual motion signals are particularly dominant at low frequencies and slow image motion velocities (Straka and Dieringer, 2004). Thus, both reflex components act synergistically with the overall goal to stabilize retinal images. The fusion of vestibular and visual motion signals occurs through multisynaptic connections largely within the cerebellum (Blazquez et al., 2020). The cerebellar circuitry with mossy and climbing fiber inputs from the inferior olive detects residual image slip during execution of the VOR and initiates compensation of the mismatch by providing the required visuo-motor commands (Blazquez et al., 2020). While beneficial for optimal gaze stability, visuo-vestibular signals are forwarded to cortical areas (Figure 6.7) where these signals serve as motion estimators required for orientation, navigation, and motion perception (Cullen and Taube, 2017; Laurens and Angelaki, 2018).

6.7.2 PROPRIOCEPTIVE SIGNALS

Active head movements are initiated by neck muscles that produce proprioceptive feedback signals related to the timing and pattern of the respective muscle contractions (Cullen and Zobeiri, 2021). Neck proprioceptive signals are relayed by afferent fibers through the cervical spinal cord with collaterals that terminate in the vestibular nuclei and the cerebellum (Cullen, 2016). While simultaneously providing information about strength and duration of the muscle contractions and updating the somatosensory cortex, proprioceptive signals are combined with vestibular signals to provide information on the origin and dynamics of the head motion. In fact, neck proprioceptive signals serve to dissociate active from passive head movements (Cullen, 2012). This computation is achieved by gating reafferent signals related to the active motion component. In fact, the discharge profile of spinal cord-projecting vestibular-only neurons but not of gaze-related position-vestibular-pause neurons differs between actively induced and passive head movements (Cullen, 2011, 2012). Vestibular-only neurons respond robustly during passive head movements, while the firing rate modulation is markedly attenuated during active head movements due to cancelation of the latter neuronal motion correlate. The gating is beneficial given that activation of vestibulo-spinal reflexes during active head movements is often counterproductive and potentially in opposition to intended head movements (Cullen, 2011, 2012). The context-dependent interaction of proprioceptive neck and vestibular motion signals thus represents an organizational feature to adequately interpret head movements and to construct respective motion percepts. If proprioceptive signals from the limbs or axial musculature are similarly integrated with vestibular signals is currently unknown.

6.7.3 MOTOR EFFERENCE COPIES

The generation of motor commands for head movements or locomotor activity also produces efference copies that represent a close spatio-temporal correlate of the motion profile (Straka et al., 2018). During orienting head movements, feedforward neck motor efference copies jointly interact with neck proprioceptive feedback signals (Cullen, 2011, 2012). In fact, in primates where most knowledge was obtained, neck proprioceptive signals and neck motor efference copies are virtually indistinguishable in their synergistic function. However, a clear role of motor efference copies in vestibular information processing has been demonstrated in amphibians during locomotor activity (Straka and Chagnaud, 2017). During rhythmic locomotion in *Xenopus* larvae and adults, ascending efference copies that derive from spinal central pattern generator activity are used to initiate gaze-stabilizing compensatory eye movements while concurrently suppressing vestibular-evoked compensatory eye movements in a plane-specific manner (Lambert et al., 2012). The suppressive effect of locomotor efference copies on vestibular signals occurs through vestibular efferent neurons at the level of the sensory periphery in the inner ear by attenuating the signal transmission at the hair cell-afferent synapse (Chagnaud et al., 2015). In addition, locomotor efference copies might influence the signal processing in central vestibular neurons through pre- and postsynaptic inhibition mostly in high-dynamic vestibulo-motor pathway elements (Straka and Chagnaud, 2017). While this predictive signaling pathway is highly effective in amphibians, studies in humans have demonstrated a comparable role of locomotor efference copies in maintaining gaze stability during walking with a concurrent attenuation of vestibular inputs (Dietrich et al., 2020; Dietrich and Wuehr, 2019). Accordingly, efference copy influences on gaze, posture, and locomotor control appears to be widespread among vertebrates, independent of locomotor style and constitutes an important component for vestibular computations with immediate consequences for the generation of motion percepts (Straka et al., 2018).

6.8 CONCLUSION

The vestibular system is renowned for its ability to initiate relatively 'simple' motor reactions that allow animals and humans to maintain stability of gaze and posture during self-induced motion or external perturbations of the head/body orientation. These reflexive motor reactions have been implemented in early vertebrate ancestors and were retained in a rather conserved organizational fashion (Straka and Gordy, 2020). The production of these reflexive motor responses is governed by a number of equally conserved functional principles. These rules guide the vectorial and dynamic decomposition of head/body motion, as well as the hierarchical transmission of corresponding signals along frequency-tuned pathways into stereotyped distributions within a topographical motor map. A more detailed knowledge about this functional organization, including multimodal convergence of motion-related signals, is more important than ever before, given the critical role of this sensory system in determining and updating the organism about the orientation and movement in space. In fact, an intact vestibular system is indispensable for orientation and navigation in any environmental niche (Cullen and Taube, 2017).

While the conceptual understanding of how the vestibular system influences navigation and orientation is still incomplete, the societal demand to better comprehend these influences becomes increasingly larger given a number of future medical and technical challenges of our society. In particular, here we highlight a few of the most striking demands: (1) Age-related deterioration of the vestibular system has a significant impact on navigation and orientation skills and thus causes complications for the cognitive capabilities of the large populations of aging humans in the modern world (Agrawal et al., 2020). Amelioration of sensory capability deficits by way of pharmacological prevention or improvement of vestibular sensitivity at different neuronal levels might prolong appropriate functionality. (2) Technological advancements in robotics and artificial intelligence/machine learning continues to provide, in part, automated vehicles and machines, which allow unsupervised motion in space using a combination of allocentric and egocentric reference frames (McBeath et al., 2018). This necessitates detailed understanding of how accelerometers and graviceptors can provide signals to stabilize gaze and posture of a running robot, including the role of locomotor efference copies, but also to update digital spatial maps following path-integration of the motion trajectory. (3) In the advent of space travel, prolonged sojourns in microgravity have a severe impact on physiological processes, often with resultant anatomical deteriorations. In the case of the vestibular system, these negative impacts heavily influence the navigational skills and orientation of astronauts under microgravitational conditions. While most fictional writings about space travel regularly neglect the negative impact of microgravity on the vestibular system and its consequent functional implications (e.g., 'The Hitchhiker's Guide to the Galaxy' by Adams, 1979), current scientific studies have begun to acknowledge the deteriorating influence of spaceflight on physiological processes including brain function (Roy-O'Reilly et al., 2021). Prolonged exposure to microgravity not only alters skeletal composition and muscle mass (Petersen et al., 2016) but also severely impairs orientation and navigational skills (Stahn et al., 2020), conditions that in fictional writing can be compensated and even improved beyond control levels by neuro-pharmacologically active spices (Herbert, 1965). While this is conceptually at present out of reach, a better understanding of how microgravity through various direct and indirect effects influences physiological parameters of astronauts during prolonged space travels is required. Based on its natural sensitivity for gravitational cues, the vestibular system plays a key role in exploring applicable conditions and technical solutions for building spaceships and orbital stations within which this sensory system can continue to operate normally and provide the positional information necessary for a life under earth-like environmental conditions.

ACKNOWLEDGEMENTS

The authors thank Clayton Gordy for critically reading the manuscript and providing valuable suggestions on key aspects of the chapter. The authors also acknowledge financial support by the NIH to KLE (R03 DC015333) and by the German Science Foundation to HS (CRC 870, B12, Ref. nr.:118803580).

REFERENCES

Adams, D. 1979. *The Hitchhiker's Guide to the Galaxy*. Pan books, London.

Agrawal, Y., Smith, P.F., Merfeld, D.M. 2020. Dizziness, imbalance and age-related vestibular loss. In: Fritzsch, B. (Editor) and Straka, H. (Volume Editor), *The Senses: A Comprehensive Reference, vol. 6*. Elsevier, Academic Press, Amsterdam, London, pp. 567–580.

Alvarez, Y., Alonso, M.T., Vendrell, V., Zelarayan, L.C., Chamero, P., Theil, T., Bösl, M.R., Kato, S., Maconochie, M., Riethmacher, D., Schimmang, T., 2003. Requirements for FGF3 and FGF10 during inner ear formation. *Development* 130:6329–6338.

Annoni, J.M., Cochran, S.L., Precht, W., 1984. Pharmacology of the vestibular hair cell-afferent fiber synapse in the frog. *J Neurosci* 4:2106–2116.

Appler, J.M., Goodrich, L.V., 2011. Connecting the ear to the brain: Molecular mechanisms of auditory circuit assembly. *Prog Neurobiol* 93:488–508.

Baird, R.A., 1992. Morphological and electrophysiological properties of hair cells in the bullfrog utriculus. *Ann NY Acad Sci* 656:12–26.

Baird, R.A., Lewis, E.R., 1986. Correspondences between afferent innervation patterns and response dynamics in the bullfrog utricle and lagena. *Brain Res* 369:48–64.

Beraneck, M., Straka, H., 2011. Vestibular signal processing by separate sets of neuronal filters. *J Vest Res* 21:5–19.

Bermingham, N.A., Hassan, B.A., Price, S.D., Vollrath, M.A., Ben Arie, N., Eatock, R.A., Bellen, H.J., Lysakowski, A., Zoghbi, H.Y., 1999. Math1: an essential gene for the generation of inner ear hair cells. *Science* 284:1837–1841.

Biesdorf, S., Malinvaud, D., Reichenberger, I., Pfanzelt, S., Straka, H., 2008. Differential inhibitory control of semicircular canal nerve afferent-evoked inputs in second-order vestibular neurons by glycinergic and GABAergic circuits. *J Neurophysiol* 99:1758–1769.

Birinyi, A., Straka, H., Matesz, C., Dieringer, N., 2001. Location of dye-coupled second order and of efferent vestibular neurons labeled from individual semicircular canal or otolith organs in the frog. *Brain Res* 921:44–59.

Birol, O., Ohyama, T., Edlund, R.K., Drakou, K., Georgiades, P., Groves, A.K., 2016. The mouse Foxi3 transcription factor is necessary for the development of posterior placodes. *Dev Biol* 409:139–151.

Blazquez, P.M., Hirata, Y. Pastor, A.M., 2020. Functional organization of cerebellar feedback loops and plasticity of influences on vestibular function. In: Fritzsch, B. (Editor) and Straka, H. (Volume Editor), *The Senses: A Comprehensive Reference, vol. 6.* Elsevier, Academic Press, Amsterdam, London, pp. 389–413.

Bok, J., Chang, W., Wu, D.K., 2007. Patterning and morphogenesis of the vertebrate inner ear. *Int J Dev Biol* 51:521–533.

Bouchard, M., de Caprona, D., Busslinger, M., Xu, P., Fritzsch, B., 2010. Pax2 and Pax8 cooperate in mouse inner ear morphogenesis and innervation. *BMC Dev Biol* 10:89.

Brandt, T., Dieterich, M., 2019. Thalamocortical network: a core structure for integrative multimodal vestibular functions. *Curr Opin Neurol* 32:154–164.

Brown, A.S., Epstein, D.J., 2011. Otic ablation of smoothened reveals direct and indirect requirements for Hedgehog signaling in inner ear development. *Development* 138:3967–3976.

Brunelli, S., Innocenzi, A., Cossu, G., 2003. Bhlhb5 is expressed in the CNS and sensory organs during mouse embryonic development. *Gene Expr Patterns* 3:755–759.

Burns, J.C., Stone, J.S., 2017. Development and regeneration of vestibular hair cells in mammals. *Semin Cell Dev Biol* 65:96–105.

Burton, Q., Cole, L.K., Mulheisen, M., Chang, W., Wu, D.K., 2004. The role of Pax2 in mouse inner ear development. *Dev Biol* 272:161–175.

Cambronero, F., Puelles, L., 2000. Rostrocaudal nuclear relationships in the avian medulla oblongata: a fate map with quail chick chimeras. *J Comp Neurol* 427:522–545.

Carney, P.R., Silver, J., 1983. Studies on cell migration and axon guidance in the developing distal auditory system of the mouse. *J Comp Neurol* 215:359–369.

Chagnaud, B.P., Banchi, R., Simmers, J., Straka, H., 2015. Spinal corollary discharge modulates motion sensing during vertebrate locomotion. *Nat Commun* 6:7982.

Chagnaud, B.P., Engelmann, J., Fritzsch, B., Glover, J.C., Straka, H., 2017. Sensing external and self-motion with hair cells, a comparison of the lateral line and vestibular systems from a developmental and evolutionary perspective. *Brain Behav Evol* 90: 98–116.

Chen, Y., Takano-Maruyama, M., Fritzsch, B., Gaufo, G.O., 2012. Hoxb1 controls anteroposterior identity of vestibular projection neurons. *PloS One* 7:e34762.

Chizhikov, V.V., Iskusnykh, I.Y., Fattakhov, N., Fritzsch, B., 2021. Lmx1a and Lmx1b are redundantly required for the development of multiple components of the mammalian auditory system. *Neuroscience* 452:247–264.

Christ, S., Biebel, U.W., Hoidis, S., Friedrichsen, S., Bauer, K., Jean, W.T.J.A., 2004. Hearing loss in athyroid pax8 knockout mice and effects of thyroxine substitution. *Audiol Neurotol* 9:88–106.

Cochran, S.L., Correia, M.J., 1995. Functional support of glutamate as a vestibular hair cell transmitter in an amniote. *Brain Res* 670:321–325.

Collewijn, H., 1969. Optokinetic eye movements in the rabbit: Input-output relations. *Vis Res* 9:117–132.

Contini, D., Holstein, G.R., Art, J.J., 2020. Synaptic transmission between hair cells and afferent fibers. In: Fritzsch, B. (Editor) and Straka, H. (Volume Editor), The Senses: A Comprehensive Reference, vol. 6. Elsevier, Academic Press, Amsterdam, London, pp. 185–210.

Corey, D.P., Hudspeth, A.J., 1983. Kinetics of the receptor current in bullfrog saccular hair cells. *J Neurosci* 3:962–976.

Cullen, K.E., 2011. The neural encoding of self-motion. *Curr Opin Neurobiol* 21:587–595.

Cullen, K.E., 2012. The vestibular system: multimodal integration and encoding of self-motion for motor control. *Trends Neurosci* 35:185–196.

Cullen, K.E., 2016. Physiology of central pathways. *Handb Clin Neurol* 137:17–40.

Cullen, K.E., 2019. Vestibular processing during natural self-motion: implications for perception and action. *Nat Rev Neurosci* 20:346–363.

Cullen, K.E., Taube, J.S., 2017. Our sense of direction: progress, controversies and challenges. *Nat Neurosci* 20:1465–1473.

Cullen, K.E., Zobeiri, O.A., 2021. Proprioception and the predictive sensing of active self-motion. *Curr Opin Physiol* 20:29–38.

Curtin, J.A., Quint, E., Tsipouri, V., Arkell, R.M., Cattanach, B., Copp, A.J., Henderson, D.J., Spurr, N., Stanier, P., Fisher, E.M., 2003. Mutation of Celsr1 disrupts planar polarity of inner ear hair cells and causes severe neural tube defects in the mouse. *Curr Biol* 13:1129–1133.

de Burlet, H.M., 1929. Zur vergleichenden Anatomie der Labyrinthinnervation. *J Comp Neurol* 47:155–169.

Deans, M.R., 2013. A balance of form and function: planar polarity and development of the vestibular maculae. *Semin Cell Dev Biol* 24:490–498.

Delacroix, L., Malgrange, B., 2015. Cochlear afferent innervation development. *Hear Res* 330:157–169.

Desai, S.S., Ali, H., Lysakowski, A., 2005b. Comparative morphology of rodent vestibular periphery. II. Cristae ampullares. *J Neurophysiol* 93:267–280.

Desai, S.S., Zeh, C., Lysakowski, A., 2005a. Comparative morphology of rodent vestibular periphery. I. Saccular and utricular maculae. *J Neurophysiol* 93:251–266.

Di Bonito, M., Boulland, J.-L., Krezel, W., Setti, E., Studer, M., Glover, J.C., 2015. Loss of projections, functional compensation, and residual deficits in the mammalian vestibulospinal system of Hoxb1-deficient mice. *eNeuro* 2:ENEURO.0096-15.2015.

Díaz, C., Puelles, L., 2019. Segmental analysis of the vestibular nerve and the efferents of the vestibular complex. *Anat Rec* 302:472–484.

Dickman, J.D., Angelaki, D.E., 2002. Vestibular convergence patterns in vestibular nuclei neurons of alert primates. *J Neurophysiol* 88:3518–3533.

Dietrich, H., Heidger, F., Schniepp, R., MacNeilage, P.R., Glasauer, S., Wuehr, M., 2020. Head motion predictability explains activity-dependent suppression of vestibular balance control. *Sci Rep* 10:668.

Dietrich, H., Wuehr, M., 2019. Selective suppression of the vestibulo-ocular reflex during human locomotion. *J Neurol* 266(Suppl 1):101–107.

Dlugaiczyk, J., Gensberger, K.D., Straka, H., 2019. Galvanic vestibular stimulation: from basic concepts to clinical applications. *J Neurophysiol* 121:2237–2255.

Duncan, J.S., Fritzsch, B., 2013. Continued expression of GATA3 is necessary for cochlear neurosensory development. *PloS One* 8:e62046.

Duncan, J.S., Fritzsch, B., Houston, D.W., Ketchum, E.M., Kersigo, J., Deans, M.R., Elliott, K.L., 2019. Topologically correct central projections of tetrapod inner ear afferents require Fzd3. *Sci Rep* 9:10298.

Duncan, J.S., Lim, K.-C., Engel, J.D., Fritzsch, B., 2011. Limited inner ear morphogenesis and neurosensory development are possible in the absence of GATA3. *Int J Dev Biol* 55:297–303.

Duncan, J.S., Stoller, M.L., Francl, A.F., Tissir, F., Devenport, D., Deans, M.R., 2017. Celsr1 coordinates the planar polarity of vestibular hair cells during inner ear development. *Dev Biol* 423:126–137.

Dvorakova, M., Macova, I., Bohuslavova, R., Anderova, M., Fritzsch, B., Pavlinkova, G., 2020. Early ear neuronal development, but not olfactory or lens development, can proceed without SOX2. *Dev Biol* 457:43–56.

Eatock, R.A., Songer, J.E., 2011. Vestibular hair cells and afferents: two channels for head motion signals. *Annu Rev Neurosci* 34:501–534.

Eatock, R.A., Xue, J., Kalluri R., 2008. Ion channels in mammalian vestibular afferents may set regularity of firing. *J Exp Biol* 211:1764–1774.

Elliott, K.L., Fritzsch, B., Duncan, J.S., 2018. Evolutionary and developmental biology provide insights into the regeneration of organ of Corti hair cells. *Front Cell Neurosci* 12:252.

Elliott, K.L., Gordy, C., 2020. Evolution and plasticity of inner ear vestibular neurosensory development. In: Fritzsch, B. (Editor) and Straka, H. (Volume Editor), The Senses: A Comprehensive Reference, vol. 6. Elsevier, Academic Press, Amsterdam, London, pp. 145–161.

Elliott, K.L., Pavlinkova, G., Chizhikov, V.V., Yamoah, E.N., Fritzsch, B., 2021. Devlopment in the mammalian auditory system depends on transcription factors. *Int J Mol Sci* 22:4189.

Fariñas, I., Jones, K.R., Tessarollo, L., Vigers, A.J., Huang, E., Kirstein, M., de Caprona, D.C., Coppola, V., Backus, C., Reichardt, L.F., Fritzsch, B., 2001. Spatial shaping of cochlear innervation by temporally regulated neurotrophin expression. *J Neurosci* 21:6170–6180.

Filova, I., Dvorakova, M., Bohuslavova, R., Pavlinek, A., Elliott, K.L., Vochyanova, S., Fritzsch, B., Pavlinkova, G., 2020. Combined Atoh1 and Neurod1 deletion reveals autonomous growth of auditory nerve fibers. *J Neurosci* 57:5307–5323.

Fritzsch, B., 2003. Development of inner ear afferent connections: forming primary neurons and connecting them to the developing sensory epithelia. *Brain Res Bull* 60:423–433.

Fritzsch, B., Beisel, K.W., Jones, K., Farinas, I., Maklad, A., Lee, J., Reichardt, L.F., 2002. Development and evolution of inner ear sensory epithelia and their innervation. *J Neurobiol* 53:143–156.

Fritzsch, B., Elliott, K.L. 2017. Gene, cell, and organ multiplication drives inner ear evolution. *Dev Biol* 431:3–15.

Fritzsch, B., Elliott, K.L., Pavlinkova, G., 2019. Primary sensory map formations reflect unique needs and molecular cues specific to each sensory system. 8:F1000 Faculty Rev-345.

Fritzsch, B., Pan, N., Jahan, I., Duncan, J.S., Kopecky, B.J., Elliott, K.L., Kersigo, J., Yang, T., 2013. Evolution and development of the tetrapod auditory system: an organ of Corti-centric perspective. *Evol Dev* 15:63–79.

Fritzsch, B., Pauley, S., Feng, F., Matei, V., Nichols, D., 2006. The molecular and developmental basis of the evolution of the vertebrate auditory system. *Int J Comp Psychol* 19:1–25.

Fritzsch, B., Signore, M., Simeone, A., 2001. Otx1 null mutant mice show partial segregation of sensory epithelia comparable to lamprey ears. *Dev Genes Evol* 211:388–396.

Fritzsch, B., Straka, H., 2014. Evolution of mechanosensory hair cells and inner ears: identifying stimuli to select altered molecular development toward new morphologies. *J Comp Physiol* 200:5–18.

Gacek, R.R., Rasmussen, G.L., 1961. Fiber analysis of the statoacoustic nerve of guinea pig, cat, and monkey. *Anat Rec* 139:455–463.

Gittis, A.H., du Lac, S., 2006. Intrinsic and synaptic plasticity in the vestibular system. *Curr Opin Neurobiol* 16:385–390.

Glasauer, S., Knorr, A.G., 2020. Physical nature of vestibular stimuli. In: Fritzsch, B. (Editor) and Straka, H. (Volume Editor), The Senses: A Comprehensive Reference, vol. 6. Elsevier, Academic Press, Amsterdam, London, pp. 6–11.

Glover, J.C., 2000. Neuroepithelial 'compartments' and the specification of vestibular projections. *Prog Brain Res* 124:3–21.

Glover, J.C., 2003. The development of vestibulo-ocular circuitry in the chicken embryo. *J Physiol (Paris)* 97:17–25.

Glover, J.C., 2020. Development and evolution of vestibulo-ocular reflex circuitry. In: Fritzsch, B. (Editor) and Straka, H. (Volume Editor), The Senses: A Comprehensive Reference, vol. 6. Elsevier, Academic Press, Amsterdam, London, pp. 309–325.

Glover, J.C., Elliott, K.L., Erives, A., Chizhikov, V.V., Fritzsch, B., 2018. Wilhelm His' lasting insights into hindbrain and cranial ganglia development and evolution. *Dev Biol* 444:S14–S24.

Goldberg, J.M., 2000. Afferent diversity and the organization of central vestibular pathways. *Exp Brain Res* 130:277–297.

Graf, W., Simpson, J.I., 1981. The relations between the semicircular canals, the optic axis, and the extraocular muscles in lateral-eyed and frontal-eyed animals. In: Fuchs, A. and Becker W (Eds.), Progress in Oculomotor Research, Developments in Neuroscience, vol. 12. Elsevier, New York, pp. 411–420.

Hans, S., Liu, D., Westerfield, M.J.D., 2004. Pax8 and Pax2a function synergistically in otic specification, downstream of the Foxi1 and Dlx3b transcription factors. *Development* 131, 5091–5102.

Herbert, F. 1965. *Dune.* Chilton Books, Boston.

Hernandez-Miranda, L.R., Müller, T., Birchmeier, C., 2017. The dorsal spinal cord and hindbrain: From developmental mechanisms to functional circuits. *Dev Biol* 432:34–42.

Hertzano,. R., Montcouquiol, M., Rashi-Elkeles, S., Elkon, R., Yucel, R., Frankel, W.N., Rechavi, G., Moroy, T., Friedman, T.B., Kelley, M.W., Avraham, K.B., 2004. Transcription profiling of inner ears from Pou4f3(ddl/ddl) identifies Gfi1 as a target of the Pou4f3 deafness gene. *Hum Mol Genet* 13:2143–2153.

Highstein, S.M., Goldberg, J.M., Moschovakis, A.K., Fernández, C., 1987. Inputs from regularly and irregularly discharging vestibular nerve afferents to secondary neurons in the vestibular nuclei of the squirrel monkey. II. Correlation with output pathways of secondary neurons. *J Neurophysiol* 58:719–738.

Higuchi, S., Sugahara, F., Pascual-Anaya, J., Takagi, W., Oisi, Y., Kuratani, S., 2019. Inner ear development in cyclostomes and evolution of the vertebrate semicircular canals. *Nature* 565:347.

Hoffman, L.R., Paulin, M.G., 2020. Peripheral innervation patterns and discharge properties of vestibular afferents in amniotes and anamniotes. In: Fritzsch, B. (Editor) and Straka, H. (Volume Editor), The Senses: A Comprehensive Reference, vol. 6. Elsevier, Academic Press, Amsterdam, London, pp. 228–255.

Holley, M., Rhodes, C., Kneebone, A., Herde, M.K., Fleming, M., Steel, K.P., 2010. Emx2 and early hair cell development in the mouse inner ear. *Dev Biol* 340:547–556.

Holstein, G.R., 2020. Morphophysiological organization of vestibulo-autonomic pathways. In: Fritzsch, B. (Editor) and Straka, H. (Volume Editor), The Senses: A Comprehensive Reference, vol. 6. Elsevier, Academic Press, Amsterdam, London, pp. 432–444.

Holstein, G.R., Martinelli, G.P., Henderson, S.C., Friedrich, V.L. Jr, Rabbitt, R.D., Highstein, S.M., 2004b. Gamma-aminobutyric acid is present in a spatially discrete subpopulation of hair cells in the crista ampullaris of the toadfish *Opsanus tau. J Comp Neurol* 471:1–10.

Holstein, G.R., Rabbitt, R.D., Martinelli, G.P., Friedrich, V.L. Jr, Boyle, R.D., Highstein, S.M., 2004. Convergence of excitatory and inhibitory hair cell transmitters shapes vestibular afferent responses. *Proc Natl Acad Sci USA* 101:15766–15771.

Holt, J.C., 2020. Synaptic and pharmacological organization of efferent influences on hair cells and vestibular afferent fibers. In: Fritzsch B. (Editor) and Straka, H. (Volume Editor), The Senses: A Comprehensive Reference, vol. 6. Elsevier, Academic Press, Amsterdam, London, pp. 526–554.

Horn, A.K.E., Straka, H., 2021. Functional organization of extraocular motoneurons and eye muscles. *Annu Rev Vis Sci* 7: 793–825. doi.org/10.1146/annurev-vision-100119-125043

Horn, A.K.E., 2020. Neuroanatomy of central vestibular connections. In: Fritzsch, B. (Editor) and Straka, H. (Volume Editor), The Senses: A Comprehensive Reference, vol. 6. Elsevier, Academic Press, Amsterdam, London, pp. 21–37.

Huang, E.J., Liu, W., Fritzsch, B., Bianchi, L.M., Reichardt, L.F., Xiang, M., 2001. Brn3a is a transcriptional regulator of soma size, target field innervation and axon pathfinding of inner ear sensory neurons. *Development* 128:2421–2432.

I Gusti Bagus, M., Gordy, C., Sanchez-Gonzalez, R., Strupp, M., Straka, H., 2019. Impact of 4-aminopyridine on vestibulo-ocular reflex performance. *J Neurol* 266 (Suppl 1):S93–S100.

Jahan, I., Kersigo, J., Pan, N., Fritzsch, B., 2010a. Neurod1 regulates survival and formation of connections in mouse ear and brain. *Cell Tiss Res* 341:95–110.

Jahan, I., Pan, N., Kersigo, J., Fritzsch, B., 2010b. Neurod1 suppresses hair cell differentiation in ear ganglia and regulates hair cell subtype development in the cochlea. *PLoS One.* 5:e11661.

Jamali, M., Carriot, J., Chacron, M.J., Cullen, K.E., 2013. Strong correlations between sensitivity and variability give rise to constant discrimination thresholds across the otolith afferent population. *J Neurosci* 33:11302–11313.

Karis, A., Pata, I., van Doorninck, J.H., Grosveld, F., de Zeeuw, C.I., de Caprona, D., Fritzsch, B., 2001. Transcription factor GATA-3 alters pathway selection of olivocochlear neurons and affects morphogenesis of the ear. *J Comp Neurol* 429:615–630.

Kashara, M., Uchino, Y., 1974. Bilateral semicircular canal inputs to neurons in cat vestibular nuclei. *Exp Brain Res* 20:285–296.

Ketchum, E.M., Sheltz-Kempf, S.N., Duncan, J.S., 2020. Molecular Basis of Vestibular Organ Formation During Ontogeny. In: Fritzsch B. (Editor) and Straka, H. (Volume Editor), The Senses: A Comprehensive Reference, vol. 6. Elsevier, Academic Press, Amsterdam, London, pp. 129–144.

Khatri, S.B., Edlund, R.K., Groves, A.K., 2014. Foxi3 is necessary for the induction of the chick otic placode in response to FGF signaling. *Dev Biol* 391:158–169.

Kiernan, A.E., Pelling, A.L., Leung, K.K., Tang, A.S., Bell, D.M., Tease, C., Lovell-Badge, R., Steel., K.P., Cheah, K.S., 2005. Sox2 is required for sensory organ development in the mammalian inner ear. *Nature* 434: 1031–1035.

Kodama, T., Gittis, A.H., Shin, M., Kelleher, K., Kolkman, K.E., McElvain, L., Lam, M., du Lac, S., 2020. Graded coexpression of ion channel, neurofilament, and synaptic genes in fast-spiking vestibular nucleus neurons. *J Neurosci* 40:496–508.

Kodama, T., Guerrero, S., Shin, M., Moghadam, S., Faulstich, M., du Lac, S., 2012. Neuronal classification and marker gene identification via single-cell expression profiling of brainstem vestibular neurons subserving cerebellar learning. *J Neurosci* 32:7819–7831.

Kopecky, B., Santi, P., Johnson, S., Schmitz, H., Fritzsch, B., 2011. Conditional deletion of N-Myc disrupts neurosensory and non-sensory development of the ear. *Dev Dyn* 240:1373–1390.

Koundakjian, E.J., Appler, J.L., Goodrich, L.V., 2007. Auditory neurons make stereotyped wiring decisions before maturation of their targets. *J Neurosci* 27:14078–14088.

Ladich, F., Schulz-Mirbach, T., 2016. Diversity in fish auditory systems: One of the riddles of sensory biology. *Front Ecol Evol* 4:28.

Lambert, F.M., Beck, J.C., Baker, R., Straka, H., 2008. Semicircular canal size determines the developmental onset of angular vestibuloocular reflexes in larval *Xenopus*. *J Neurosci* 28:8086–8096.

Lambert, F.M., Combes, D., Simmers, J., Straka, H., 2012. Gaze stabilization by efference copy signaling without sensory feedback during vertebrate locomotion. *Curr Biol* 22:1649–1658.

Lambert, F.M., Malinvaud, D., Glaunès, J., Bergot, C., Straka, H., Vidal, P.P., 2009. Vestibular asymmetry as the cause of idiopathic scoliosis: A possible answer from *Xenopus*. *J Neurosci* 29:12477–12483.

Laurens, J., Angelaki, D.E., 2018. The brain compass: A perspective on how self-motion updates the head direction cell attractor. *Neuron* 97:275–289.

Lawoko-Kerali, G., Rivolta, M.N., Lawlor, P., Cacciabue-Rivolta, D.I., Langton-Hewer, C., van Doorninck, J.H., Holley, M.C., 2004. GATA3 and NeuroD distinguish auditory and vestibular neurons during development of the mammalian inner ear. *Mech Dev* 121:287–299.

Lewis, E.R., Leverenz, E.L., Bialek, W.S., 1985. The Vertebrate Inner Ear. CRC Press, Boca Raton, FL.

Lewis, E.R., Li, C.W., 1975. Hair cell types and distributions in the otolithic and auditory organs of the bullfrog. *Brain Res* 83:35–50.

Liu, Z., Kimura, Y., Higashijima, S.I., Hildebrand, D.G.C., Morgan, J.L., Bagnall, M.W., 2020. Central vestibular tuning arises from patterned convergence of otolith afferents. *Neuron* 108:748–762.

Lunde, A., Okaty, B.W., Dymecki, S.M., Glover, J.C., 2019. Molecular profiling defines evolutionarily conserved transcription factor signatures of major vestibulospinal neuron groups. *eNeuro* 6:ENEURO.0475-18.2019.

Lysakowski, A., 2020. Anatomy and microstructural organization of vestibular hair cells. In: Fritzsch, B. (Editor) and Straka, H. (Volume Editor), The Senses: A Comprehensive Reference, vol. 6. Elsevier, Academic Press, Amsterdam, London, pp. 173–184.

Ma, Q., Anderson, D.J., Fritzsch, B., 2000. Neurogenin1 null mutant ears develop fewer, morphologically normal hair cells in smaller sensory epithelia devoid of innervation. *J Assoc Res Otolaryngol* 1:129–143.

Ma, Q.F., Chen, Z.F., Barrantes, I.D., de la Pompa, J.L., Anderson, D.J., 1998. Neurogenin1 is essential for the determination of neuronal precursors for proximal cranial sensory ganglia. *Neuron* 20:469–482.

Mackereth, M.D., Kwak, S.-J., Fritz, A., Riley, B.B., 2005. Zebrafish pax8 is required for otic placode induction and plays a redundant role with Pax2 genes in the maintenance of the otic placode. *Development* 132, 371–382.

Macova, I., Pysanenko, K., Chumak, T., Dvorakova, M., Bohuslavova, R., Syka, J., Fritzsch, B., Pavlinkova, G., 2019. Neurod1 Is Essential for the Primary Tonotopic Organization and Related Auditory Information Processing in the Midbrain. *J Neurosci* 39:984–1004.

Maklad, A., Fritzsch, B., 1999. Incomplete segregation of endorgan-specific vestibular ganglion cells in mice and rats. *J Vestib Res* 9:387–399.

Maklad, A., Kamel, S., Wong, E., Fritzsch, B., 2010. Development and organization of polarity-specific segregation of primary vestibular afferent fibers in mice. *Cell Tiss Res* 340:303–321.

Malinvaud, D., Vassias, I., Reichenberger, I., Rössert, C., Straka, H., 2010. Functional organization of vestibular commissural connections in frog. *J Neurosci* 30:3310–3325.

Marín, F., Puelles, L., 1995. Morphological fate of rhombomeres in quail/chick chimeras: a segmental analysis of hindbrain nuclei. *Eur J Neurosci* 7:1714–1738.

Martin, P., Swanson, G.J., 1993. Descriptive and experimental analysis of the epithelial remodellings that control semicircular canal formation in the developing mouse inner ear. *Dev Biol* 159, 549–558.

Matern, M.S., Milon, B., Lipford, E.L., McMurray, M., Ogawa, Y., Tkaczuk, A., Song, Y., Elkon, R., Hertzano, R., 2020. GFI1 functions to repress neuronal gene expression in the developing inner ear hair cells. *Development* 147:dev186015.

Mathews, M.A., Magnusson, A.K., Murray, A.J., Camp, A.J., 2020. The efferent vestibular and octavolateralis system: Anatomy, physiology and function. In: Fritzsch, B. (Editor) and Straka, H. (Volume Editor), The Senses: A Comprehensive Reference, vol. 6. Elsevier, Academic Press, Amsterdam, London, pp. 512–525.

May-Simera, H., Kelley, M.W., 2012. Planar cell polarity in the inner ear. *Curr Top Dev Biol* 101:111–140.

McBeath, M.K., Tang, T.Y., Shaffer, D.M., 2018. The geometry of consciousness. *Conscious Cogn* 64:207–215.

McInturff, S., Burns, J.C., Kelley, M.W., 2018. Characterization of spatial and temporal development of Type I and Type II hair cells in the mouse utricle using new cell-type-specific markers. *Biol Open* 7:bio038083.

Montcouquiol, M., Rachel, R.A., Lanford, P.J., Copeland, N.G., Jenkins, N.A., Kelley, M.W., 2003. Identification of Vangl2 and Scrb1 as planar polarity genes in mammals. *Nature* 423:173–177.

Morsli, H., Choo, D., Ryan, A., Johnson, R., Wu, D.K., 1998. Development of the mouse inner ear and origin of its sensory organs. *J Neurosci* 18:3327–3335.

Nichols, D.H., Bouma, J.E., Kopecky, B.J., Jahan, I., Beisel, K.W., He, D.Z., Liu, H., Fritzsch, B., 2020. Interaction with ectopic cochlear crista sensory epithelium disrupts basal cochlear sensory epithelium development in Lmx1a mutant mice. *Cell Tissue Res* 380:435–448.

Nichols, D.H., Pauley, S., Jahan, I., Beisel, K.W., Millen, K.J., Fritzsch, B., 2008. Lmx1a is required for segregation of sensory epithelia and normal ear histogenesis and morphogenesis. *Cell Tissue Res* 334:339–358.

Noden, D., Van De Water, T., 1986. The developing ear: tissue origins and interactions. In: Ruben, R.J., Van De Water, T.R., Rubel, E.W. (Eds.), The Biology of Change in Otolaryngology. Elsevier, Amsterdam, pp. 15–46.

Pan, N., Jahan, I., Kersigo, J., Kopecky, B., Santi, P., Johnson, S., Schmitz, H., Fritzsch, B., 2011. Conditional deletion of Atoh1 using Pax2-Cre results in viable mice without differentiated cochlear hair cells that have lost most of the organ of Corti. *Hear Res* 275:66–80.

Pasqualetti, M., Díaz, C., Renaud, J.-S., Rijli, F.M., Glover, J.C., 2007. Fate-mapping the mammalian hindbrain: segmental origins of vestibular projection neurons assessed using rhombomere-specific Hoxa2 enhancer elements in the mouse embryo. *J Neurosci* 27:9670–9681.

Paulin, M.G., Hoffman, L.F., 2019. Models of vestibular semicircular canal afferent neuron firing activity. *J Neurophysiol* 122:2548–2567.

Petersen, N., Jaekel, P., Rosenberger, A., Weber, T., Scott, J., Castrucci, F., Lambrecht, G., Ploutz-Snyder, L., Damann, V., Kozlovskaya, I., Mester, J., 2016. Exercise in space: the European Space Agency approach to in-flight exercise countermeasures for long-duration missions on ISS. *Extrem Physiol Med* 5:9.

Peusner, K.D., Giaume, C., 1997. Ontogeny of electrophysiological properties and dendritic pattern in second-order chick vestibular neurons. *J Comp Neurol* 384:621–633.

Pfanzelt, S., Rössert, C., Rohregger, M., Glasauer, S., Moore, L.E., Straka, H., 2008. Differential dynamic processing of afferent signals in frog tonic and phasic second-order vestibular neurons. *J Neurosci* 28:10349–10362.

Platt, C., Straka, H. 2020. Vestibular endorgans in vertebrates and adequate sensory stimuli. In: Fritzsch B. (Editor) and Straka H. (Volume Editor), The Senses: A Comprehensive Reference, vol. 6. Elsevier, Academic Press, Amsterdam, London, pp. 108–128.

Qian, Y., Fritzsch, B., Shirasawa, S., Chen, C.-L., Choi, Y., Ma, Q., 2001. Formation of brainstem (nor) adrenergic centers and first-order relay visceral sensory neurons is dependent on homeodomain protein Rnx/Tlx3. *Genes Dev* 15:2533–2545.

Radde-Gallwitz, K., Pan, L., Gan, L., Lin, X., Segil, N., Chen, P., 2004. Expression of Islet1 marks the sensory and neuronal lineages in the mammalian inner ear. *J Comp Neurol* 477:412–421.

Raft, S., Koundakjian, E.J., Quinones, H., Jayasena, C.S., Goodrich, L.V., Johnson, J.E., Segil, N., Groves, A.K., 2007. Cross-regulation of Ngn1 and Math1 coordinates the production of neurons and sensory hair cells during inner ear development. *Development* 134:4405–4415.

Ray, R.S., Dymecki, S.M., 2009. Rautenlippe Redux—toward a unified view of the precerebellar rhombic lip. *Curr Opin Cell Biol* 21:741–747.

Reichenberger, I., Straka, H., Ottersen, O.P., Streit, P., Gerrits, N.M., Dieringer, N., 1997. Distribution of GABA, glycine and glutamate immunoreactivities in the vestibular nuclear complex of the frog. *J Comp Neurol* 377:149–164.

Riccomagno, M.M., Martinu, L., Mulheisen, M., Wu, D.K., Epstein, D.J., 2002. Specification of the mammalian cochlea is dependent on Sonic hedgehog. *Genes Dev* 16:2365–2378.

Riccomagno, M.M., Takada, S., Epstein, D.J., 2005. Wnt-dependent regulation of inner ear morphogenesis is balanced by the opposing and supporting roles of Shh. *Genes Dev* 19:1612–1623.

Rössert, C., Straka, H., 2011. Interactions between intrinsic membrane and emerging network properties determine signal processing in central vestibular neurons. *Exp Brain Res* 210:437–449.

Roy-O'Reilly, M., Mulavara, A., Williams, T., 2021. A review of alterations to the brain during spaceflight and the potential relevance to crew in long-duration space exploration. *NPJ Microgravity* 7:5.

Ruben, R.J., 1967. Development of the inner ear of the mouse: a radioautographic study of terminal mitoses. *Acta Otolaryngol* 220:1–44.

Sadeghi, S.G., Chacron, M.J., Taylor, M.C., Cullen, K.E., 2007. Neural variability, detection thresholds, and information transmission in the vestibular system. *J Neurosci* 27:771–781.

Sato, F., Sasaki, H., 1993. Morphological correlations between spontaneously discharging primary vestibular afferents and vestibular nucleus neurons in the cat. *J Comp Neurol* 333:554–566.

Schlosser, G., 2006. Induction and specification of cranial placodes. *Dev Biol* 294: 303–351.

Schlosser, G., 2010. Making senses: development of vertebrate cranial placodes. *Int Rev Cell Mol Biol* 283:129–234.

Schlosser, G., Ahrens, K., 2004. Molecular anatomy of placode development in *Xenopus laevis*. *Dev Biol* 271:439–466.

Schmitz, J., 2020. Graviception in invertebrates. In: Fritzsch B. (Editor) and Straka H. (Volume Editor), The Senses: A Comprehensive Reference, vol. 6. Elsevier, Academic Press, Amsterdam, London, pp. 88–107.

Schneider, A.D., Jamali, M., Carriot, J., Chacron, M.J., Cullen, K.E., 2015. The increased sensitivity of irregular peripheral canal and otolith vestibular afferents optimizes their encoding of natural stimuli. *J Neurosci* 35:5522–5536.

Schultz, J.A., Zeller, U., Luo, Z.X., 2017. Inner ear labyrinth anatomy of monotremes and implications for mammalian inner ear evolution. *J Morphol* 278:236–263.

Serafin, M., de Waele, C., Khateb, A., Vidal, P.P., Mühlethaler, M., 1991. Medial vestibular nucleus in the guinea-pig. I. Intrinsic membrane properties in brainstem slices. *Exp Brain Res* 84:417–425.

Shimazu, H., Precht, W., 1965. Tonic and kinetic responses of cat's vestibular neurons to horizontal angular acceleration. *J Neurophysiol* 28:991–1013.

Shroyer, N.F., Wallis, D., Venken, K.J., Bellen, H.J., Zoghbi, H.Y., 2005. Gfi1 functions downstream of Math1 to control intestinal secretory cell subtype allocation and differentiation. *Genes Dev* 19, 2412–2417.

Simpson, J.I., Graf, W., 1985. The selection of reference frames by nature and its investigators. In: Berthoz, A., Jones, G.M. (Eds.), Adaptive Mechanisms in Gaze Control: Facts and Theories. Elsevier, Amsterdam, pp. 3–16.

Slowik, A.D., Bermingham-McDonogh, O., 2016. A central to peripheral progression of cell cycle exit and hair cell differentiation in the developing mouse cristae. *Dev Biol* 411:1–14.

Solomon, K.S., Kudoh, T., Dawid, I.B., Fritz, A., 2003. Zebrafish foxi1 mediates otic placode formation and jaw development. *Development* 130:929–940.

Solomon, K.S., Kwak, S.-J., Fritz, A., 2004. Genetic interactions underlying otic placode induction and formation. *Dev Dyn* 230:419–433.

Soupiadou, P., Branoner, F., Straka, H., 2018. Pharmacological profile of vestibular inhibitory inputs to superior oblique motoneurons. *J Neurol* 265 (Suppl 1):S18–S25.

Stahn, A.C., Riemer, M., Wolbers T., Werner, A., Brauns, K., Besnard, S., Denise, P., Kühn, S., Gunga, H.-C., 2020. Spatial updating depends on gravity. *Front Neural Circuits* 14:20.

Straka, H., 2010., Ontogenetic rules and constraints of vestibulo-ocular reflex development. *Curr Opin Neurobiol* 20:689–695.

Straka, H., 2020. Functional organization of vestibular commissural pathways. In: Fritzsch B. (Editor) and Straka H. (Volume Editor), The Senses: A Comprehensive Reference, vol. 6. Elsevier, Academic Press, Amsterdam, London, pp. 371–388.

Straka, H., Baker, R., 2013. Vestibular blueprint in early vertebrates. *Front Neural Circuits* 7:182.

Straka, H., Baker, R., Gilland, E., 2001. Rhombomeric organization of vestibular pathways in larval frogs. *J Comp Neurol* 437:42–55.

Straka, H., Beraneck, M., Rohregger, M., Moore, L.E., Vidal, P.P., Vibert, N., 2004. Second-order vestibular neurons form separate populations with different membrane and discharge properties. *J Neurophysiol* 92:845–861.

Straka, H., Biesdorf, S., Dieringer, N., 1997. Canal-specific excitation and inhibition of frog second-order vestibular neurons. *J Neurophysiol* 78:1363–1372.

Straka, H., Chagnaud, B.P., 2017. Moving or being moved: that makes a difference. *J Neurol* 264 (Suppl 1):S28–S33.

Straka, H., Dieringer, N., 2004. Basic organization principles of the VOR: lessons from frogs. *Prog Neurobiol* 73:259–309.

Straka, H., Fritzsch, B., Glover, J.C., 2014. Connecting ears to eye muscles: evolution of a 'simple' reflex arc. *Brain Behav Evol* 83:162–175.

Straka, H., Gordy, C., 2020. The vestibular system: The "Leatherman™" among sensory systems. In: Fritzsch B. (Editor) and Straka H. (Volume Editor), The Senses: A Comprehensive Reference, vol. 6. Elsevier, Academic Press, Amsterdam, London, pp. 708–720.

Straka, H., Holler, S., Goto, F., 2002. Patterns of canal and otolith afferent input convergence in frog second-order vestibular neurons. *J Neurophysiol* 88:2287–2301.

Straka, H., Lambert, F.M., Pfanzelt, S., Beraneck, M., 2009. Vestibulo-ocular signal transformation in frequency-tuned channels. *Ann NY Acad Sci* 1164:37–44.

Straka, H., Simmers, J., Chagnaud, B.P., 2018. A new perspective on predictive motor signaling. *Curr Biol* 28:R232–R243.

Straka, H., Vibert, N., Vidal, P.P., Moore, L.E., Dutia, M.B., 2005. Intrinsic properties of vertebrate vestibular neurons: function, development and plasticity. *Prog Neurobiol* 76:349–392.

Straka, H., Zwergal, A., Cullen, K.E., 2016. Vestibular animal models: contributions to understanding physiology and disease. *J Neurol* 263 (Suppl 1):S10–S23.

Sun, S.-K., Dee, C.T., Tripathi, V.B., Rengifo, A., Hirst, C.S., Scotting, P.J., 2007. Epibranchial and otic placodes are induced by a common Fgf signal, but their subsequent development is independent. *Dev Biol* 303:675–686.

Tomsa, J.M., Langeland, J.A., 1999. Otx expression during lamprey embryogenesis provides insights into the evolution of the vertebrate head and jaw. *Dev Biol* 207:26–37.

Torres, M., Giráldez, F., 1998. The development of the vertebrate inner ear. *Mech Dev* 71:5–21.

Uchino, Y., 2001. Otolith and semicircular canal inputs to single vestibular neurons in cats. *Biol Sci Space* 15:375–381.

Uchino, Y., 2004. Role of cross-striolar and commissural inhibition in the vestibulocollic reflex. *Prog Brain Res* 143:403–409.

Uchino, Y., Hirai, N., Suzuki, S., Watanabe, S., 1981. Properties of secondary vestibular neurons fired by stimulation of ampullary nerve of the vertical, anterior or posterior semicircular canals in the cat. *Brain Res* 223:273–286.

Uchino, Y., Kushiro, K., 2011. Differences between otolith- and semicircular canal-activated neural circuitry in the vestibular system. *Neurosci Res* 71:315–327.

Uchino, Y., Sato, H., Zakir, M., Kushiro, K., Imagawa, M., Ogawa, Y., Ono, S., Meng, H., Zhang, X., Katsatu, M., Isu, N., Wilson, V.J., 2001. Commissural effects in the otolith system. *Exp Brain Res* 136:421–430.

Urness, L.D., Paxton, C.N, Wang, X., Shoenwolf, C.G., Mansour, S.L., 2010. FGF signaling regulates otic placode induction and refinement by controlling both ectodermal target genes and hindbrain Wnt8a. *Dev Biol* 340:595–604.

Vidal, P.P., Cullen, K., Curthoys, I.S., du Lac, S., Holstein, G., Idoux, E., Lysakowski, A., Peusner, K., Sans, A., Smith, P., 2015. Chapter 28 - The Vestibular System, in: Paxinos, G. (Ed.), The Rat Nervous System (Fourth Edition). Academic Press, San Diego, pp. 805–864.

Walberg, F., Ottersen, O.P., Rinvik, E., 1990. GABA, glycine, aspartate, glutamate and taurine in the vestibular nuclei: an immunocytochemical investigation in the cat. *Exp Brain Res* 79:547–563.

Walls, G.L., 1962. The evolutionary history of eye movements. *Vis Res* 2:69–80.

Wang, Y., Guo, N., Nathans, J., 2006. The role of Frizzled3 and Frizzled6 in neural tube closure and in the planar polarity of inner-ear sensory hair cells. *J Neurosci* 26:2147–2156.

Wersäll, J., 1956. Studies on the structure and innervation of the sensory epithelium of the cristae ampullares in the guinea pig. *Acta Otolaryngol Suppl* 126:1–85.

Wilson, V.J., Felpel, L.P., 1972. Specificity of semicircular canal input to neurons in the pigeon vestibular nuclei. *J Neurophysiol* 35:253–254.

Wright, T.J., Mansour, S.L., 2003. Fgf3 and Fgf10 are required for mouse otic placode induction. *Development* 130:3379–3390.

Wu, D.K., Kelley, M.W., 2012. Molecular mechanisms of inner ear development. *Cold Spring Harb Perspect Biol* 4:a008409.

Xiang, M., Maklad, A., Pirvola, U., Fritzsch, B., 2003. Brn3c null mutant mice show longterm, incomplete retention of some afferent inner ear innervation. *BMC Neurosci* 4:1–16.

Xu, J., Li, J., Zhang, T., Jiang, H., Ramakrishnan, A., Fritzsch, B., Shen, L., Xu, P.X., 2021. Chromatin remodelers and lineage-specific factors interact to target enhancers to establish proneurosensory fate within otic ectoderm. *PNAS* 118(12): e2025196118.

Yang, T., Kersigo, J., Wu, S., Fritzsch, B., Bassuk, A.G., 2017. Prickle1 regulates neurite outgrowth of apical spiral ganglion neurons but not hair cell polarity in the murine cochlea. *PloS One* 12:e0183773.

Yin, H., Copley, C.O., Goodrich, L.V., Deans, M.R., 2012. Comparison of phenotypes between different vangl2 mutants demonstrates dominant effects of the Looptail mutation during hair cell development. *PloS One* 7:e31988.

Zhang, X., Zakir, M., Meng, H., Sato, H., Uchino, Y., 2001. Convergence of the horizontal semicircular canal and otolith afferents on cat single vestibular neurons. *Exp Brain Res* 140:1–11.

7 Morphological and Molecular Ontogeny of the Auditory System

Jeremy S. Duncan, Sydney N. Sheltz-Kempf, Karen L. Elliott

CONTENTS

7.1 INTRODUCTION: OVERVIEW AND PHYLOGENETIC DIFFERENCES OF THE AUDITORY SYSTEM

The vertebrate inner ear located on either side of the head is a complex labyrinth composed of caverns and tubes, and contains up to nine different sensory organs (Fritzsch and Wake, 1988; Lewis et al., 1985; van Bergeijk, 1967) for the detection of sound, self-motion, and gravity. Depending on the species, one or more of these sensory organs detects auditory information and are closely related to the vestibular sensory organs ontogenetically. Auditory sensory organs are found in every vertebrate including lampreys, elasmobranchs, bony fish, lungfish, anurans, urodeles, gymnophionans, and amniotes (Christensen et al., 2015; Fay and Popper, 2000; Ladich and Fay, 2013; Mickle et al., 2019; Mickle et al., 2020; Popper and Coombs, 1982; van Bergeijk, 1966). In teleost fish, the otolith organ of the vestibular system, the saccule, not only plays a role for balance but is also responsible for sound perception (Popper and Fay, 1999). However, in tetrapods, additional sensory organs dedicated specifically for auditory perception have evolved, such as the lagena, basilar papilla, amphibian papilla, and the organ of Corti (Figure 7.1; Fritzsch et al., 2013; Luo and Manley, 2020).

The derived auditory organization of tetrapods is shared with the coelacanth, *Latimeria*. The lagena, which responds to auditory stimulation in some species,

DOI: 10.1201/9781003092810-7

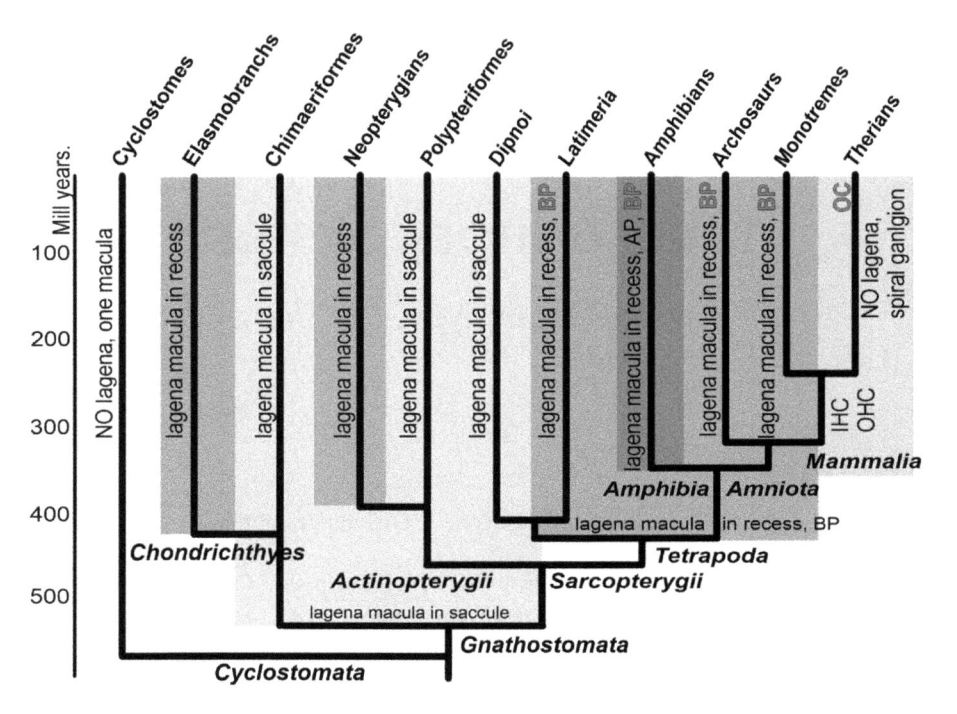

FIGURE 7.1 Cladistic analysis of the morphological changes during auditory evolution. The lagena evolved within its own recess three times: elasmobranchs, neopterygians, and tetrapods. Tetrapods and the coelacanth *Latimeria* evolved a basilar papilla (BP) for sound detection. With the exception of mammals, the basilar papilla is located within the lagena recess. Amphibians uniquely evolved an amphibian papilla (AP) in addition to the BP. The organ of Corti (OC) evolved in mammals from the BP and contains two types of hair cells: the inner and outer hair cells (IHC, OHC). The OC is located within an elongated lagenar recess referred to as the cochlea, although mammals have lost the lagena macula present in other tetrapods. (Modified from Fritzsch et al., 2013.)

first evolved in elasmobranchs, adjacent to the saccule (Khorevin, 2008). In *Latimeria* and non-mammalian tetrapods, the lagena is located in a distinct recess (Fritzsch, 1987; Fritzsch and Neary, 1998), whereas the lagena of lungfish resides with the saccule (Platt et al., 2004). The presence of the basilar papilla in *Latimeria* suggests that it may have arisen in ancestral lobe-finned fish (Basch et al., 2016; Fritzsch, 1987; Fritzsch, 1992). In *Latimeria* and many tetrapods, the basilar papilla is located adjacent to the lagena within the lagenar recess (Fritzsch et al., 2013). The amphibian papilla, a unique organ in amphibians for middle frequency hearing (Fritzsch, 1999; Fritzsch and Wake, 1988; Smotherman and Narins, 2004), likely arose from a doubling of the neglected papilla (Duncan and Fritzsch, 2012; Fritzsch, 1992; Fritzsch and Elliott, 2017; Fritzsch and Wake, 1988; Smotherman and Narins, 2004). The mammalian organ of Corti is thought to have evolved out of the basilar papilla (Fritzsch et al., 2013; Luo and Manley, 2020; Schultz et al., 2017).

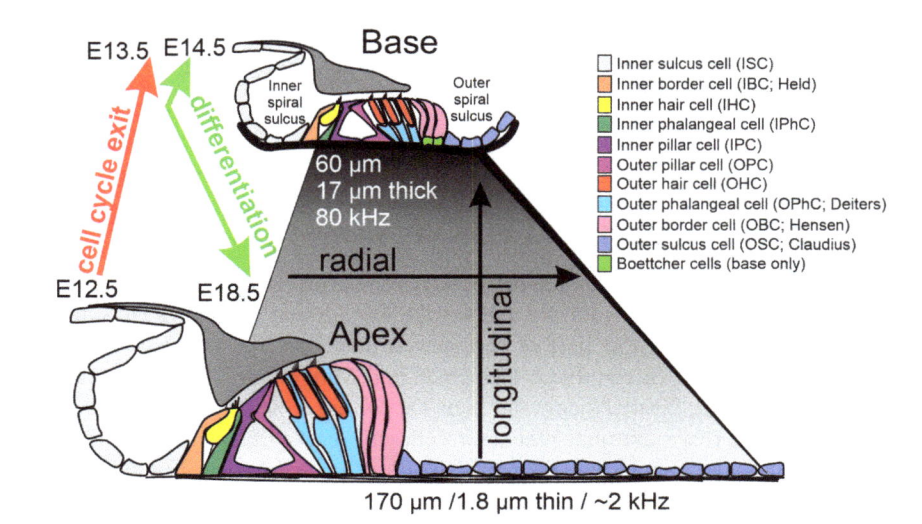

FIGURE 7.2   Hair cell development. Hair cells exit the cell cycle from apex to base between E12.5 and E14.5 (red arrow) and begin differentiation near the base and progress toward the apex between E14.5 and E18.5 (green arrow). Several morphological differences occur between the apex and base. For instance, hair cells and supporting cells are taller at the apex than they are at the base and, in addition, stereocilia are longer. An additional difference is that Boettcher cells are only found at the base. Finally, the basilar membrane is longer and thinner at the apex and shorter and thicker at the base. (Modified from Booth et al., 2018.)

In monotremes, the basilar papilla resides with the lagena, and this likely evolved into the coiled cochlea containing the organ of Corti and loss of the lagena in all therians (Fritzsch et al., 2013; Luo and Manley, 2020; Schultze et al., 2018).

Across species, sound pressure information is detected through mechanosensory hair cells (Coffin et al., 2004; Elliott et al., 2018). These cells are termed hair cells due to modified microvilli called stereocilia that extend from the apical surface. These stereocilia are polarized in a stair-like configuration (Figure 7.2; Elliott et al., 2018). The stereocilia are connected by cell surface linkages (links). One type of link are the tip links which connect to mechanosensory channels encoded by the TMC1/2 genes (Erives and Fritzsch, 2020; Kawashima et al., 2015; Marcovich and Holt, 2020; Pan et al., 2013). Hair cells synapse on auditory afferent sensory neurons/spiral ganglion neurons. Auditory afferent neurons project centrally to the cochlear nucleus within the hindbrain. Auditory nuclei neurons project bilaterally to the superior olivary complex (Kandler et al., 2020; Lohmann and Friauf, 1996). From the superior olivary complex, projections go to the inferior colliculus (Syka, 2020), the medial geniculate body (Malone et al., 2020), and finally, the auditory cortex (Rauschecker, 2020).

In this review, we provide a summary of the different components of the auditory system in the context of ontogeny, gene networks, and circuit formation. We will discuss the development of mechanosensory hair cells and the pathway of transmission through auditory neurons to reach the hindbrain. From there, we will discuss higher-order auditory processing in additional nuclei and conclude with the auditory cortex.

7.2 DEVELOPMENT OF AUDITORY NEURONS AND THEIR PERIPHERAL CONNECTIONS

The first cells of the otocyst that are restricted to a neurosensory fate are the proneuronal progenitor cells. Both auditory and vestibular neuronal progenitor cells delaminate from the anteroventral part of the otocyst and form the cochlear-vestibular ganglion (Carney and Silver, 1983; Delacroix and Malgrange, 2015; Elliott et al., 2021; Farinas et al., 2001; Noden and Van De Water, 1986). Several early genes expressed in inner ear neuronal precursor cells are responsible for their specification and development, including *Eya1/Six1, Sox2, Lfng*, and *Neurog1* (Ahmed et al., 2012b; Cole et al., 2000; Dvorakova et al., 2020; Ma et al., 2000; Ma et al., 1998). Expression of *Sox2* promotes self-renewal of the neural progenitor cells (Kiernan et al., 2005b). This period of neuronal proliferation differs depending on the species. For example, fish and amphibians have a longer period of inner ear neuronal proliferation than birds or mammals (Fritzsch et al., 1988; Katayama and Corwin, 1989; Ruben, 1967; Whitfield, 2015). Based on gene expression, some of these cells in mice may even be postmitotic while still in the otocyst (Ma et al., 1998). After various rounds of proliferation, auditory neuroblasts exit the cell cycle in a basal to apical progression in mice, and conversely in an apical to basal progression in chickens (D'amico-Martel, 1982).

Neurog1 (Ngn1, Neurod3, neurogenin1) is a basic helix-loop-helix (bHLH) transcription factor essential for neuronal differentiation and subtype specification during embryogenesis and is one of the earliest genes expressed within the developing otocyst at E8.5. It is conserved across vertebrate species and is essential for formation of all inner ear neurons, as loss of *Neurog1* results in their absence (Ma et al., 2000; Ma et al., 1998). *Neurog1* is only transiently expressed and induces the expression of additional essential genes for neuronal differentiation, such as *Neurod1* (Jahan et al., 2010a; Macova et al., 2019), *Nscl1* (Krüger et al., 2006), *Isl1* (Deng et al., 2014; Radde-Gallwitz et al., 2004), *Pou4f1* (Huang et al., 2001), and several others (Duncan and Cox, 2020; Fekete and Campero, 2007; Goodrich, 2016). As neuronal precursors delaminate from the otocyst, they downregulate expression of *Neurog1* and upregulate *Neurod1* (Kim et al., 2001; Liu et al., 2000; Ma et al., 2000). *Neurod1* is also important for the differentiation of neuroblasts into neurons. Its loss not only leads to the loss of many neurons, but also leads to the formation of 'hair-cell'-like cells within neuronal ganglia (Figure 7.3; Jahan et al., 2010b).

While auditory and vestibular neurons are collectively generated from otic neuroblasts and initially form a shared cochlear-vestibular ganglion, there is evidence that they develop from distinct neuroblast populations. For example, early deletion of *Sox2* through *Foxg1-cre* results in a complete loss of all auditory neurons, but not vestibular neurons (Dvorakova et al., 2020), suggesting there

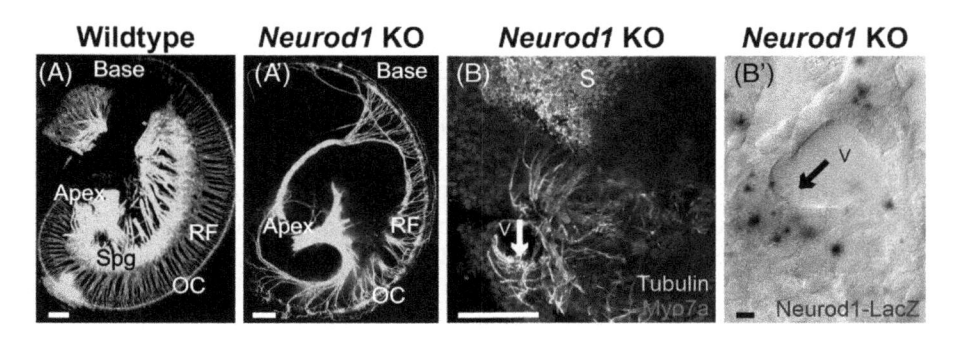

FIGURE 7.3 *Neurod1* **is important for neuronal differentiation.** Lipophilic dye reveals a normal innervation pattern to the organ of Corti (OC) by spiral ganglion neurons (A). In *Neurod1* null mice, many neurons are lost (A'). In addition, some neurons are converted into 'hair-cell'-like cells in vesicles (V) within the ganglia (B, B'; arrows). RF, radial fibers; S, saccule; Spg, spiral ganglion. Scale bar 100 μm in (A), (A'), (B); 10 μm in (B'). (Modified from Jahan et al., 2010b.)

is a clear distinction between vestibular and spiral ganglion neurons at the time of a shared cochlear-vestibular ganglion. During development, the two neuronal populations in the cochlear-vestibular ganglion subsequently segregate from each other to form separate auditory and vestibular ganglia (Delacroix and Malgrange, 2015). Beyond *Sox2*, other genes play essential roles for the development of auditory but not vestibular neurons. For instance, loss of *Lmx1a/b* results in the complete loss of the cochlea and auditory neurons, but vestibular neurons are still present (Chizhikov et al., 2021). *Gata3* is also required for formation of all auditory neurons, but only for a subset of vestibular neuron development (Duncan and Fritzsch, 2013; Duncan et al., 2011; Lawoko-Kerali et al., 2004), and downstream of *Gata3*, MAF BZIP transcription factor B (*Mafb*) further promotes auditory neuron differentiation (Yu *et al.*, 2013).

Auditory neurons/SGNs are bipolar, sending their dendrites toward the auditory epithelia at the periphery and their axons toward the brainstem (Appler and Goodrich, 2011; Rubel and Fritzsch, 2002). During development, the cochlear-vestibular ganglion forms a funnel-like shape, thought to guide auditory axons back toward the otocyst (Carney and Silver, 1983). In mammals, the earliest SGN neurites arrive at the base just prior to hair cell differentiation, and arrival of neurites to the organ of Corti progresses toward the apex (Druckenbrod and Goodrich, 2015; Rubel and Fritzsch, 2002). Proper extension of neurites from SGNs to the cochlea, and subsequently to the hair cells, depends upon the presence of Schwann cells (Mao et al., 2014; Sandell et al., 2014), as well as expression of several genes, such as *Foxg1* (He et al., 2019; Pauley et al., 2006), *Neuropilin1 and 2* (Coate et al., 2015; Zhang and Coate, 2017), *Gata3* (Appler et al., 2013), *Prox1* (Fritzsch et al., 2010), *Fgfr3* (Puligilla et al., 2007), and various Wnt/PCP genes (Ghimire and Deans, 2019; Ghimire et al., 2018; Yang et al., 2017). These genes are important both alone and in combination for proper hair cells targeting by SGNs (Coate and Kelley, 2013; Fekete and Campero, 2007; Nishimura et al., 2017).

SGNs are divided into two broad subtypes: type I and type II. These two subtypes synapse onto the two different types of hair cells, inner (IHC) and outer (OHC) hair cells, respectively (Hafidi, 1998; Maison et al., 2016; Petitpre et al., 2018; Shrestha et al., 2018; Sun et al., 2018). Each IHC is innervated by many type I fibers, whereas a single type II afferent innervates multiple OHCs (Hafidi, 1998; Rubel and Fritzsch, 2002; Simmons and Liberman, 1988). At first, the type I afferents project to both hair cell types, but later retract their processes from OHCs to selectively innervate the IHCs (Coate et al., 2015; Druckenbrod and Goodrich, 2015). Some of the molecular drivers differentiating type I from type II have been recently identified. For instance, *Epha4* and *Neuropilin2* are expressed on type I SGNs and act to prevent innervation of OHCs, which express the ligands for these two receptors (Coate et al., 2015; Defourny et al., 2013). Other genes are specifically important in type II neuronal navigation to OHCs. In mice, the Wnt/PCP genes, *Fzd3*, *Fzd6*, and *Vangl2*, cooperate together for proper guidance and turning of type II SGNs (Ghimire and Deans, 2019). Another Wnt/PCP gene, *Prickle1*, has also been shown to have an important role in type II fiber outgrowth and turning (Yang et al., 2017). In addition, the homeobox gene *Prox1*, is also essential for proper turning of type II SGNs (Fritzsch et al., 2010).

Recent work investigating SGNs has revealed that, within type I SGNs, there are differential gene expression profiles that correspond to three different type I subpopulations, each with a different spontaneous rate of firing (Petitpre et al., 2018; Shrestha et al., 2018; Sun et al., 2018). These studies showed that type Ia, which expressed higher levels of *Calb2*, corresponded to the high spontaneous rate neurons, while type Ib, which expressed higher levels of *Calb1*, corresponded to the medium spontaneous rate neurons, and type Ic, which expressed higher levels of *Pou4f1*, corresponded to low spontaneous rate neurons (Petitpre et al., 2018; Shrestha et al., 2018; Sun et al., 2018). These findings demonstrate that there is more heterogeneity among SGNs than initially understood.

Neurotrophic factors play an important role in inner ear afferent survival. SGNs express specific neurotrophin receptors (TrkB/*NtrkB* and TrkC/*Ntrk3*) for the neurotrophins present in hair cells (Bdnf/*Ntf2* and NT-3/*Ntf3*) (Green et al., 2012). Mice lacking *TrkB* and *TrkC* expression had a complete loss of SGNs (Silos-Santiago et al., 1997). Similarly, loss of both neurotrophin genes, *Bdnf* and *Ntf3*, through conditional deletion by *Pax2-cre* results in an eventual complete loss of SGNs (Kersigo and Fritzsch, 2015). Loss of *Ntf3* results in the loss of SGNs at the base and an overall reduction of neurons throughout the cochlea (Farinas et al., 2001). In contrast, loss of *Bdnf* specifically affected only type II SGNs (Ernfors et al., 1995). Interestingly, replacing one neurotrophin with the other had very little effect on SGN survival or the pattern of hair cell innervation (Agerman et al., 2003; Tessarollo et al., 2004), aside from cochlear innervation by vestibular afferents when *Ntf3* was replaced by *Bdnf* (Tessarollo et al., 2004). This suggests that these two neurotrophic factors are effectively equal in their trophic support of SGNs during development (Fritzsch et al., 2016; Green et al., 2012).

7.3 DEFINING AUDITORY HAIR CELLS AND THEIR HETEROGENEITY

Hair cells are the mechanosensory cells of the inner ear that release neurotransmitters in response to sound energy, motion, or gravity. These cells along with supporting cells constitute the sensory cells of the ear and the epithelia which contain these cell types are termed sensory epithelia. Hair cells differentiate after auditory and vestibular neurons begin sending their neurites toward the ear. *Neurog1* null mice, mentioned in above in Section 7.2, not only have neuronal defects but also have smaller overall sensory epithelia, including a shortened cochlear duct (Figure 7.4). This length reduction is attributed to an early cell cycle exit since the both IHCs and OHCs are located in the apex (Ma et al., 2000; Matei et al., 2005). Conventionally, there is one row of IHCs and three rows of OHCs; however, in the apex of these

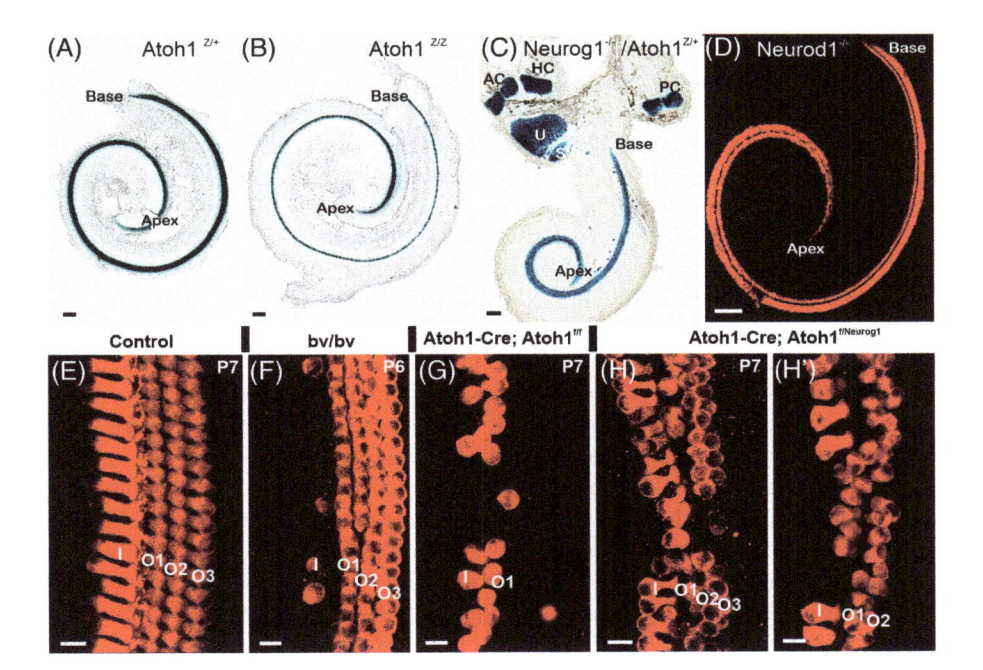

FIGURE 7.4 Cochlea and hair cell development. Compared with controls (A), loss of *Atoh1* has a limited effect on cochlea extension (B). In contrast, cochlear length is shortened in *Neurog1* (C) and *Neurod1* (D) null mice. Control organ of Corti (E) contains one row of inner hair cells (I) and three rows of outer hair cells (O1–O3). Bronx waltzer (*bv/bv*) (F) mice lack most inner hair cells. *Atoh1* 'self-terminating' (*Atoh1-Cre; Atoh1f/f*) mice have few differentiating hair cells (G); however, replacing one floxed allele of *Atoh1* with *Neurog1* results in more differentiated hair cells (H, H'). AC, anterior canal crista; HC, horizontal canal crista; PC, posterior canal crista; S, saccule; U, utricle. Scale bar is 100 µm. (Modified from Booth et al., 2018; Fritzsch et al., 2005; Jahan et al., 2010b; Matei et al., 2005.)

Neurog1 null mice, two rows of IHCs and up to five rows of OHCs were observed (Ma et al., 2000). Similar to this phenotype shown in *Neurog1* null mice, *Foxg1* CKO studies also demonstrated a short, wide cochlea with 10 or more rows of hair cells that are lost with age (Elliott et al., 2021; Ma et al., 2000; Pauley et al., 2006). The phenotypes seen in both of these CKOs are similar to normal hair cell development in other tetrapods. For example, monotremes like the platypus and echidna, develop 4–5 rows of IHCs and 6–8 rows of OHCs within a shorter cochlear duct, as compared to eutherian mammals (Ladhams and Pickles, 1996). The hair cells in frogs, turtles and lizards, and chickens do not follow the strict organization of rows seen in mammals, and in contrast, are arrayed in a single plane and surrounded by histologically similar supporting cells (Lewis et al., 1985). Chickens have 40 or more hair cells organized by height, while the hair cells in turtles and amphibians are organized based on zones and less on morphology (Christensen-Dalsgaard and Carr, 2008; Lewis et al., 1985; Tilney and Saunders, 1983; Wever, 1974). Interestingly, this lack of mammalian-type organization is comparable to the basilar papilla of *Latimeria* (Fritzsch, 1987).

Differentiation of hair cells in the developing ear across species requires the bHLH transcription factor *Atoh1*, previously known as *Math1* (Ahmed et al., 2012a; Bermingham et al., 1999). *Atoh1* null mice lack all hair cells and supporting cells (Pan et al., 2011; Woods et al., 2004) and several different *Atoh1* CKO models have demonstrated defects in other sensory cell types that are directly or indirectly affected. Furthermore, it has been suggested that the intital expression of *Atoh1* in the sensory epithelia may not be specific to hair cells since transient expression is shown in supporting cells as well (Driver et al., 2013; Matei et al., 2005). *Atoh1* is regulated by *Neurod1* expression. As mentioned above, in the absence of *Neurod1*, 'hair-cell'-like cells develop within inner ear ganglia (Jahan et al., 2010b), suggesting that *Neurod1* downregulates *Atoh1* expression to promote neuronal fates. There is no noticeable effect of nascent hair cells in the organ of Corti in the absence of *Neurod1* (Figure 7.4). *Atoh1* lends itself to manipulation because it has two different enhancers and also auto-regulates its own expression by binding to one of its own enhancers (Pan et al., 2012; Pan et al., 2011). Using *Atoh1-cre* to eliminate *Atoh1* expression in a 'self-terminating' mouse model results in an initial transient upregulation of *Atoh1* that was sufficient to initiate development of some hair cells (Pan et al., 2012). However, there was progressive hair cell loss after birth in these mice, suggesting that sustained *Atoh1* expression is necessary for maintenance of most hair cells. Replacing one allele of *Atoh1* with *Neurog1* in this 'self-terminating' mouse model partially rescued hair cells (Jahan et al., 2015; Figure 7.4), suggesting that *Neurog1* can partially substitute for a loss of *Atoh1*. In addition, unlike the shortened cochlear duct in the *Neurog1* null model, *Atoh1* deletion does not affect overall cochlear length (Matei et al., 2005; Pan et al., 2011; Figure 7.4). Taken together, these studies show the importance of *Atoh1* in hair cell formation and suggest an important cross-regulation between the bHLH genes.

Mammalian cochlear hair cells exit the cell cycle in an apical to basal progression, with an opposing gradient of differentiation that occurs from base to apex (Chen et al., 2002; Kopecky et al., 2011; Matei et al., 2005; Ruben, 1967; Figure 7.2). For cells located in the apical region of the cochlea, this results in a gap between the timing of cell cycle exit and differentiation, whereas the cells in the basal cochlea exit the cell cycle and almost immediately begin differentiating. This phenomenon may be partially responsible for the gradual shift in cytological features of hair cells from base to apex, including overall size, height and arrangement of stereocilia, as well as physiological responses. Throughout the entire length of the cochlea, there is one row of IHC on the neuronal side of the organ of Corti and three rows of OHC on the abneural side. Unlike auditory hair cells of birds, amphibians, and fish, both types of mammalian cochlear hair cells are unable to regenerate (Edge and Chen, 2008). Inactivation of the retinoblastoma (Rb) protein demonstrated that aberrant hair cell proliferation in older mice results in massive apoptotic activity, suggesting that even regulation of cell cycle exit is not sufficient for mammalian hair cell regeneration (Mantela et al., 2005). In contrast, there is a distinct, and likely continuous, proliferation of hair cells in amphibians that contributes to this ability (Corwin, 1985; Fritzsch et al., 1988). Proliferation in chick hair cells occurs in an outer to inner progression across the basilar papilla (Katayama and Corwin, 1989, 1993). Previous studies have suggested that this specific developmental pattern may also attribute to the regenerative capacities of chick hair cells (Corwin and Cotanche, 1988; Rubel et al., 2013). Therefore, if the specific patterns of hair cell development may strongly affect the ability of those hair cells to regenerate, then it stands to reason that the genes expressed early in the otocyst may directly impact the heterogeneity of the auditory hair cells specified within these tissues.

As no single gene is responsible for the differentiation of a cell, additional genes beyond *Atoh1* are necessary for hair cell differentiation. There are several genes downstream of *Atoh1* that are also required for normal hair cell development, such as *Pou4f3* (Xiang et al., 2003), *Gfi1* (Hertzano et al., 2004; Matern et al., 2020), and *Barhl1* (Chellappa et al., 2008; Li et al., 2002). Additional hair cell specification requires *Notch* signaling. *Notch* signaling plays a role in hair cell differentiation by establishing the prosensory domain and mediating lateral inhibition via ligands such as *Delta-like 1*, and *Jagged 1/2* (Doetzlhofer et al., 2009; Kiernan et al., 2001; Kiernan et al., 2005a; Kiernan et al., 2006). Expression of *Hes/Hey* transcription factors is also dependent on Notch signaling, and these transcription factors maintain supporting cell fate through *Fgf* signaling. CKO models for *Shh* also result in the absence of both types of cochlear hair cells (Bok et al., 2007; Bok et al., 2013; Hu et al., 2010). Similarly, early CKO models for *Pax2* and *Gata3* also result in the absence of all hair cells (Bouchard et al., 2010; Burton et al., 2004; Duncan and Fritzsch, 2013; Duncan et al., 2011; Hans et al., 2004; Karis et al., 2001). The *Lmx1a/b* double knockout model results in the absence of all hair cells (Chizhikov et al., 2021), while the loss of *Lmx1a* only results in the irregular segregation of sensory epithelia (Koo et al., 2009; Nichols et al., 2008).

During development, the stereocilia on the apical surface of hair cells start in the center of the cell and are polarized to one side during development. The orientation of this polarity is essential for proper function (Corey and Hudspeth, 1983a, 1983b). Genes such as *Vangl2*, *Dvl1*, *Celsr1*, and *Gal2* from the PCP pathway have been shown to be necessary for the histological development of hair cells and the orientation of their stereocilia (Duncan et al., 2017; Ghimire et al., 2018; Lu and Sipe, 2015; Montcouquiol et al., 2003; Sienknecht et al., 2014; Tarchini et al., 2013). Normal function of the hair cells results from proper stereocilia polarity forming the mechanoelectrical transduction (MET) channel near the stereocilia tips, which is dependent on *CDH23, PCDH15*, and *TMC1/2* (Elliott et al., 2018; Erives and Fritzsch, 2020; Nist-Lund et al., 2019; Pan et al., 2013; Qiu and Müller, 2018; Shibata et al., 2016).

While the specification and differentiation of all hair cells rely upon a generic hair cell genetic program (above), there are many subtypes of hair cells and some of the genetic differences driving this heterogeneity have been uncovered. For example, only IHCs depend on *Cdc42* since they are lost preferentially in the high frequency range in *Cdc42* CKO models (Ueyama et al., 2014). In addition, IHCs specifically are dependent on Fgf8 (Hayashi et al., 2008; Huh et al., 2012; Jacques et al., 2007). Conversely, loss of MANF results in the death of OHCs in the basal region soon after the onset of hearing. Additionally, *Emx2* and *Jag1* are both needed for the development of OHCs (Holley et al., 2010; Jiang et al., 2017; Kimura et al., 2005). Furthermore, both the contractile function and survival of OHC depend on motor protein Prestin.

Levels of genes may also play a role in the specification (or maintenance) of hair cell sub-types. For instance in the Bronx-Walter mice, most IHCs are lost, while numbers of OHCs remain unaffected (Figure 7.4; Nakano et al., 2012; Nakano et al., 2020). When *Atoh1* is conditionally deleted using *Atoh1-cre*, allowing for elimination of *Atoh1* shortly after differentiation begins, there is a complete loss of the second and third rows of OHCs, while some IHCs and the first row of OHCs remain (Figure 7.4).

7.4 THE AUDITORY COCHLEAR NUCLEUS

The cochlear nucleus is the first region of the brain to receive auditory input and is located on the dorsolateral surface of the brainstem where the medulla meets the pons. The cochlear nucleus develops from several rhombomeres (r2–r5) that collectively form the anteroventral cochlear nucleus (AVCN), posteroventral cochlear nucleus (PVCN), and dorsal cochlear nucleus (DCN) (Farago et al., 2006). While the AVCN is derived from r2–3 and the PVCN comes from r3–4, the DCN is composed of r3, r4, and r5 (Di Bonito and Studer, 2017). The rhombic lip is the most dorsal region of the hindbrain epithelium and is thought to generate the entire cochlear nucleus (Wang et al., 2005). *Lmx1a* and *Lmx1b* co-expression early in development defines the dorsal progenitor domain of the hindbrain, which will give rise to the roof plate of the brainstem, and eventually, the choroid plexus (Glover et al., 2018). Without the roof plate of the fourth ventricle and choroid plexus, no cochlear nuclei develop, and there is also a reduced formation of the vestibular nuclei (Chizhikov et al., 2021). This is consistent with the gene expression network in this region, since previous studies have shown that *Lmx1a* and *Lmx1b* specifically regulate *Gdf7, Atoh1, Wnt1*, and *Wnt3a*, and also play a role in the reduction of *BMP4* and *BMP7* (Chizhikov et al., 2021; Chung et al., 2009; Huang et al., 2018; Mishima et al., 2009; Yan et al., 2011). An interaction between *Shh* (ventral) and *BMPs* and *Wnts* (dorsal) define a longitudinal expression of *Atoh1* that extends from the spinal cord to the cerebellum (Bermingham et al., 2001; Bok et al., 2005; Farago et al., 2006). Both *Atoh1* and *Olig3* are needed for the derivative cells within the cochlear nuclei (Wang et al., 2005). *Olig3* is also expressed in early vestibular nuclei neurons at various rostro-caudal expression, alongside *Ascl1, Ptf1a*,

Neurog2, and *Neurog1* (Bermingham et al., 2001). *Ptf1a* is expressed ventral to *Atoh1* in the hindbrain (Fujiyama et al., 2009). Inhibitory cochlear nucleus neurons are derived from the *Ptf1a*-expressing regions of the hindbrain, whereas excitatory cochlear nucleus neurons are derived from the *Atoh1*-expressing regions (Fujiyama et al., 2009).

There are five major cell types in the cochlear nucleus based on Nissl preparations. Fusiform, or pyramidal, cells are unique to the DCN and it has been suggested that the patches of *Atoh1* expression in the DCN form a specific population of these cells that are calbindin positive (Nothwang, 2016). Spherical cells are limited to the AVCN, and octopus cells are exclusive to the PVCN. Globular and multipolar cells are unique to the AVCN and PVCN, and are distributed around the location where SGNs enter the cochlear nucleus. The cochlear nucleus is entirely encapsulated by WNT-dependent granule cells (Hutson and Morest, 1996; Nichols and Bruce, 2006).

Both type I and type II SGN central projections can be traced to the cochlear nucleus from the apical and basal peripheral projections in the organ of Corti while maintaining their tonotopic map (Figure 7.5) as early as E12.5 (Fritzsch et al., 2015;

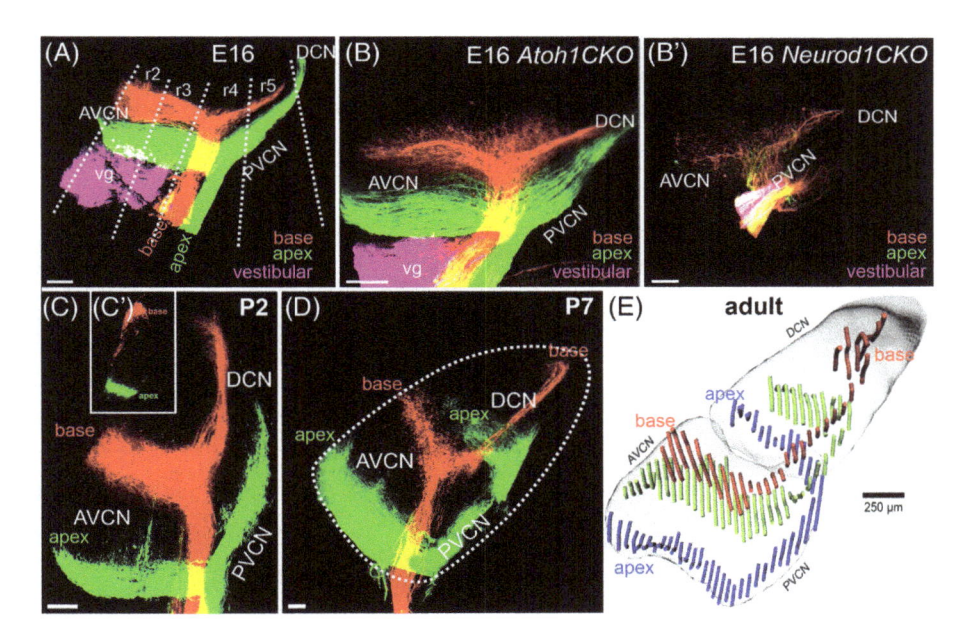

FIGURE 7.5 Spiral Ganglion Central Projections. Spiral ganglion neurons enter the hindbrain at rhombomere (r)4 and extend within the cochlear nuclei from r2–r5 (A). Fibers from the base are labeled with lipophilic dye in red and from the apex in green. Nearby vestibular central projections can be seen in purple. A similar central projection is present in the absence of all hair cells following conditional deletion of *Atoh1* in neurons with *Islet1-cre* (B). However, conditional deletion of *Neurod1* with *Iselt1-cre* resulted in aberrant central projections (B'). The angle\shape of the DCN changes as the pontine flexure develops (C, C'). The basal fibers end up projecting more dorsally in the DCN than the apical fibers (D, E). (Modified from Filova et al., 2020; Muniak, 2016; Schmidt and Fritzsch, 2019.)

Nayagam et al., 2011; Schmidt and Fritzsch, 2019). The segregation of these projections develops before the peripheral and central targets differentiate, suggesting that the dorso-ventral gradients defined by *Shh*, *BMP4*, and *Wnts* may control the temporal progression of development (Bermingham et al., 2001; Bok et al., 2005; Farago et al., 2006). Unlike the AVCN and PVCN, the three-dimensional shape of the DCN changes from a simple longitudinal extension to an obtuse angle arranged perpendicular to the AVCN. This change in shape correlates with the morphological development of the pontine angle. While projections from the apical organ of Corti can reach the ventral region of the DCN, the basal region cannot extend to the tip of the DCN. *Atoh1* expression is high in the AVCN compared to the PCVN, and *Atoh1* expression is only in certain regions of the DCN (Fritzsch et al,, 2006; Figure 7.6). Interestingly, *Atoh1* CKOs and *Atoh1* null mice models result in near normal central projection (Figure 7.5) despite the loss of hair cells and/or most cochlear nuclei (Elliott et al., 2017; Filova et al., 2020; Fritzsch et al., 2005). However, mutations in *Neurod1*, *Gata3*, *Prickle1*, *Fzd3*, and *Npr2* can dramatically change the organization of these central projections (Duncan and Fritzsch, 2013; Duncan et al., 2019; Jahan et al., 2010a; Schmidt and Fritzsch, 2019; Yang et al., 2017). For example, in *Neurod1* null mice, the central projections are severely reduced, and the projections that remain are strongly disorganized (Figure 7.5; Filova et al., 2020; Jahan et al., 2010a; Macova et al., 2019). Mutations in NT-3 and Eph complex also affect central projections (Fritzsch et al,, 1997; Cramer and Gabriele, 2014; Milinkeviciute and Cramer, 2020).

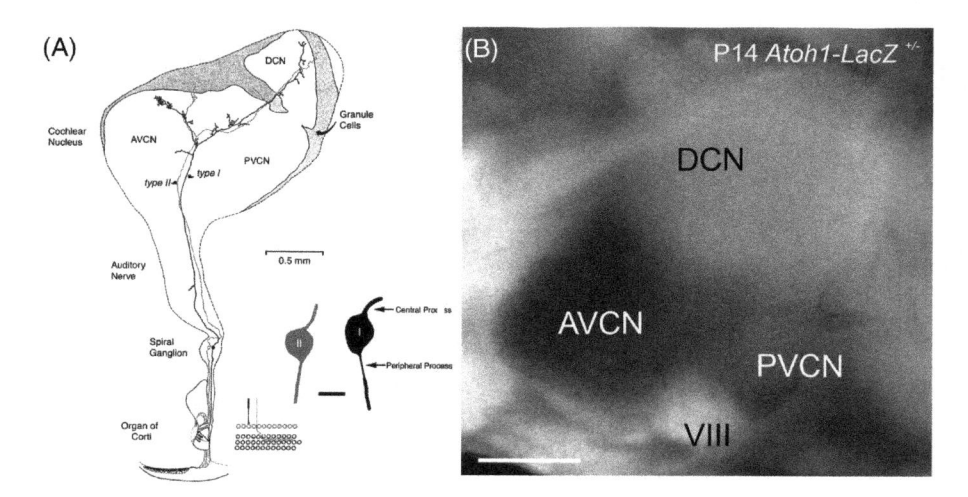

FIGURE 7.6   Cochlear nuclei. Cochlear fibers extend from two different types (type 1, type 2) to project to all three cochlear nuclei, the AVCN, PVCN, and DCN (A). *Atoh1* is expressed throughout the cochlear nucleus; however, the expression varies from high in the AVCN, intermediate in the PVCN, and low in the DCN (B). (Modified from Fritzsch et al., 2006; Muniak, 2016.)

7.5 HIGHER-ORDER PROJECTIONS TO SOC, IC, AND MGB

Axons leaving the AVCN and PVCN will converge to form the ventral acoustic stria and intermediate acoustic stria as they approach the superior olivary complex (SOC) (Figure 7.7). Axons projecting from the DCN will form the dorsal acoustic stria to project to both the lateral lemniscus and inferior colliculus (Grothe et al., 2004; Marrs and Spirou, 2012). The level of the SOC is the first time that auditory information from both ears is integrated for sound localization and coincidence detection. The SOC is composed of four sub-nuclei embedded within the trapezoid body: lateral superior olive (LSO), the medial superior olive (MSO), the medial nucleus of the trapezoid body (MNTB), and a group of cells collectively termed the peri-olivary nuclei that surround the other three sub-nuclei (Di Bonito et al., 2013; Lipovsek and Wingate, 2018; Maricich et al., 2009; Nothwang, 2016). The SOC sub-nuclei have

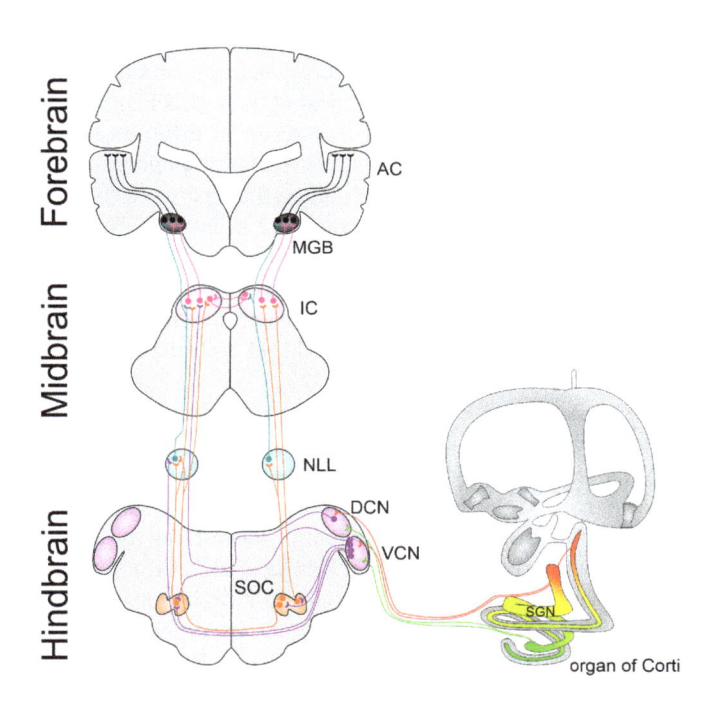

FIGURE 7.7   Auditory pathways from the inner ear to the auditory cortex. Spiral ganglion neurons (SGNs, red-base and green-apex) relay sound information from auditory hair cells in the organ of Corti to the ventral cochlear nuclei (VCN), the anteroventral cochlear nucleus, and the posteroventral cochlear nucleus, as well as the dorsal cochlear nucleus (DCN), maintaining tonotopic organization. DCN neurons (purple) project contralaterally to the inferior colliculus (IC). VCN neurons (purple) project to ipsilateral and contralateral superior olivary complex (SOC) neurons. From the SOC, projections reach the nucleus of the lateral lemniscus (NLL) and IC. Neurons from the NLL (cyan) and IC (pink) project to the medial geniculate body (MGB) in the thalamus and from there, thalamic relays (black) to the primary auditory cortex (AC).

distinct origins, the LSO and MSO being derived from the rhombic lip, while the trapezoid body is derived from the basal plate (Maricich et al., 2009; Rose et al., 2009).

The cells of the MSO are arranged in an easily recognizable dorsal-ventral column, with two dendrites oriented in opposite directions following the medio-lateral axis. These cells receive direct input from the AVCN and PVCN of both ears in order to preserve precise time information about a sound stimulus in space (Fischl et al., 2016). In contrast, the LSO is distinguished by its overall sigmoid shape, and fusiform cells of the LSO have large, polar dendrites that extend to the margin of the nucleus. These cells receive both ipsilateral direct excitatory input and contralateral indirect inhibitory input from an interneuron circuit through the MNTB (Fischl et al., 2016). Previous studies have shown that reduced expression of *Egfr* in r3 and r5 results in a smaller LSO (Maricich et al., 2009; Nothwang, 2016). *Atoh1* deletion in r3 and r5 also causes glutamatergic neuronal death in the LSO and MSO, and there is an additional reduction of neurons in the entire trapezoid body (Maricich et al., 2009). *En1* CKO models in r3 and r5 also result in the absence of the MNTB in addition to the loss of glycinergic/cholinergic neurons and the GABAergic neurons located in the lateral nucleus of the trapezoid body (LNTB) (Altieri et al., 2015; Altieri et al., 2014). Because *Atoh1* and *En1* result in different types of neuronal cells, it has been suggested that r3–r5 have two different populations that are determined by the expression of these genes compared to the r4-derived tissues. There is also weak expression of *Calretinin* around P12 in both SGNs and the cochlear nuclei, with strong expression within the PVCN that is able to be detected by antibodies in recent studies. There is limited expression of *Calretinin* in the SOC complex early on, but after P8, expression can be visualized in the octopus cells of the AVCN, fusiform cells in the DCN, and in the MNTB (Friauf, 1994; Kandler et al., 2020; Rubio and Juiz, 2004).

The lateral lemniscus is located where the midbrain meets the pons, and the nuclei are divided into dorsal (DLL) and ventral (VLL) regions, whose functions are the determination of sound localization and coincidence detection, respectively. The VLL is the larger of the two nuclei and is further divided into three zones: dorsal, middle, and ventral. The VLL contains several cell types organized in a dorso-ventral gradient derived from r4 (Di Bonito et al., 2013). Large oval cells are located ventrally, while the middle zone contains a high density of elongate cells, which are less densely organized in the dorsal zone. In comparison, most of the cells in the DLL are elongate cells with horizontal dendrites. While the DLL is a binaural system because it receives input from both ears and projects to both inferior colliculi (IC), the VLL is considered a monoaural system because it only receives input from one ear and connects to the ipsilateral IC (Brunso-Bechtold et al., 1981; Kudo, 1981; Strutz, 1987). Both of these projections will form the lateral lemniscus tract that will approach the IC, branch to the superior colliculus and ultimately project into the medial geniculate body in the midbrain (Di Bonito et al., 2013). The IC itself is organized into four sub-nuclei: the central nucleus, the dorsomedial nucleus, the pericentral nucleus, and the external nucleus. The cells in the central nucleus are primarily responsible for the projection to the medial geniculate body (MGB) and can be characterized into two types: primary cells with oval dendritic fields and stellate cells with sphere-shaped dendritic fields. The MGB is divided into ventral (vMGB), medial (mMGB), and dorsal (dMGB) regions based anatomically and functionally. The vMGB only receives input from the IC in order to maintain the tonotopic map, while the dMGB does not respond strongly to auditory signals. The vMGB is characterized by small, densely organized neurons, whereas the dMGB is formed by parallel neurons forming oriented across the dorsolateral to ventromedial axis, and the neurons in the mMGB are organized in concentric circles (Davis, 2005; Kudo and Nakamura, 1988).

Studies have shown that the earliest central projections from the CN to the SOC and the IC, and the fibers that project from the IC to the MGB both develop prior to the onset of auditory function (Gurung and Fritzsch, 2004; Kandler and Friauf, 1993).

7.6 AUDITORY CORTEX PROJECTIONS

Information sent from the cochlea to the primary auditory cortex (A1) has equal and reciprocal information sent from A1 back to the MGB in the thalamus, the midbrain, and the other CN. The A1 is distinguished from other auditory cortical fields by receiving its main ascending input from the ventral nucleus of the MGB. A1 significantly expands the magnitude and variety of its functional processing repertoire over subcortical stations due to its wide convergence of inputs, sophistication of local circuitry, and high degree of functional plasticity (Malone et al., 2020; Rauschecker, 2020). A1 is thus a hub for the convergence of subcortical and corticocortical information streams, located at a pivotal position within the auditory system between sensing, perception, and interpretation of the acoustic environment. Progress in understanding the functional and perceptual contributions of primary auditory cortex depends on an expanded range of approaches that identify and characterize the strong modulatory influences inherent in the different tasks that are contributed by cortical processing (Malone et al., 2020).

A1 neurons and adjacent core areas respond preferentially to tones of a single frequency, whereas distinct pathways participate in the decoding of complex sounds. While the ventral stream supports auditory patterning or object recognition, the dorsal stream supports sensorimotor integration and control in the production of sounds, as well as processing of auditory space and motion (Rauschecker, 2020). The process is particularly complex in humans.

7.7 SUMMARY AND CONCLUSION

Directional hearing evolved in the common tetrapod ancestor. Although the auditory system has been studied in many tetrapods, the mammalian auditory system is the most highly studied and the most derived of all the tetrapods. The mammalian auditory system is composed of a peripheral sensory organ, the organ of Corti that is evolved from the basilar papilla. The organ of Corti is connected to a chain of highly interconnected nuclei within the central nervous system by a heterogeneous set of spiral ganglion neurons. We are starting to understand the how these spiral ganglion neurons develop and diversify into two broad but distinct subtypes to reach the two types of hair cells, and more recently, the heterogeneity of these subtypes. This mammalian diversity within auditory neurons is unique compared to non-mammalian tetrapods, and the evolutionary novelties that lead to this diversity remain to be answered. One unifying theme of the mammalian auditory system is that all cochlear nuclei, spiral ganglion neurons, and cochlear hair cells depend on the Lmx1a/b genes. Higher cortical inputs influence the momentary state of cortical processing including fast and slow adaptive processes, plasticity, stimulus context, attentional states, and behavioral contingencies on decision-making processes. Nonhuman primates could serve as excellent models for studies not only of higher auditory perception and cognition, but also in examining the evolution of these abilities.

REFERENCES

Agerman, K., Hjerling-Leffler, J., Blanchard, M.P., Scarfone, E., Canlon, B., Nosrat, C., Ernfors, P.J.D., 2003. BDNF gene replacement reveals multiple mechanisms for establishing neurotrophin specificity during sensory nervous system. Development 130, 1479–1491.

Ahmed, M., Wong, E.Y., Sun, J., Xu, J., Wang, F., Xu, P.X., 2012a. Eya1-Six1 interaction is sufficient to induce hair cell fate in the cochlea by activating Atoh1 expression in cooperation with Sox2. Developmental Cell 22, 377–390.

Ahmed, M., Xu, J., Xu, P.X., 2012b. EYA1 and SIX1 drive the neuronal developmental program in cooperation with the SWI/SNF chromatin-remodeling complex and SOX2 in the mammalian inner ear. Development 139, 1965–1977.

Altieri, S.C., Jalabi, W., Zhao, T., Romito-DiGiacomo, R.R., Maricich, S.M.M., 2015. En1 directs superior olivary complex neuron positioning, survival, and expression of FoxP1. Developmental Biology 408, 99–108.

Altieri, S.C., Zhao, T., Jalabi, W., Maricich, S.M.M., 2014. Development of glycinergic innervation to the murine LSO and SPN in the presence and absence of the MNTB. Frontiers in Neural Circuits 8, 109.

Appler, J.M., Goodrich, L.V., 2011. Connecting the ear to the brain: Molecular mechanisms of auditory circuit assembly. Progress in Neurobiology 93, 488–508.

Appler, J.M., Lu, C.C., Druckenbrod, N.R., Yu, W.M., Koundakjian, E.J., Goodrich, L.V., 2013. Gata3 is a critical regulator of cochlear wiring. The Journal of Neuroscience: The Official Journal of the Society for Neuroscience 33, 3679–3691.

Basch, M.L., Brown, R.M., 2nd, Jen, H.I., Groves, A.K., 2016. Where hearing starts: the development of the mammalian cochlea. Journal of Anatomy 228, 233–254.

Bermingham, N.A., Hassan, B.A., Price, S.D., Vollrath, M.A., Ben-Arie, N., Eatock, R.A., Bellen, H.J., Lysakowski, A., Zoghbi, H.Y., 1999. Math1: An essential gene for the generation of inner ear hair cells. Science 284, 1837–1841.

Bermingham, N.A., Hassan, B.A., Wang, V.Y., Fernandez, M., Banfi, S., Bellen, H.J., Fritzsch, B., Zoghbi, H.Y., 2001. Proprioceptor pathway development is dependent on Math1. Neuron 30, 411–422.

Bok, J., Bronner-Fraser, M., Wu, D.K., 2005. Role of the hindbrain in dorsoventral but not anteroposterior axial specification of the inner ear. Development 132, 2115–2124.

Bok, J., Dolson, D.K., Hill, P., Ruther, U., Epstein, D.J., Wu, D.K., 2007. Opposing gradients of Gli repressor and activators mediate Shh signaling along the dorsoventral axis of the inner ear. Development 134, 1713–1722.

Bok, J., Zenczak, C., Hwang, C.H., Wu, D.K., 2013. Auditory ganglion source of Sonic hedgehog regulates timing of cell cycle exit and differentiation of mammalian cochlear hair cells. Proceedings of the National Academy of Sciences of the United States of America 110, 13869–13874.

Booth, K.T., Azaiez, H., Jahan, I., Smith, R.J., Fritzsch, B., 2018. Intracellular regulome variability along the organ of corti: evidence, approaches, challenges, and perspective. Frontiers in Genetics 9, 156.

Bouchard, M., Busslinger, M., Xu, P., De Caprona, D., Fritzsch, B., 2010. PAX2 and PAX8 cooperate in mouse inner ear morphogenesis and innervation. BMC Developmental Biology 10, 89.

Brunso-Bechtold, J., Thompson, G., Masterton, R., 1981. HRP study of the organization of auditory afferents ascending to central nucleus of inferior colliculus in cat. Journal of Comparative Neurology 197, 705–722.

Burton, Q., Cole, L.K., Mulheisen, M., Chang, W., Wu, D.K., 2004. The role of Pax2 in mouse inner ear development. Dev Biol 272, 161–175.

Carney, P.R., Silver, J., 1983. Studies on cell migration and axon guidance in the developing distal auditory system of the mouse. Journal of Comparative Neurology 215, 359–369.

Chellappa, R., Li, S., Pauley, S., Jahan, I., Jin, K., Xiang, M., 2008. Barhl1 regulatory sequences required for cell-specific gene expression and autoregulation in the inner ear and central nervous system. Molecular and Cellular Biology 28, 1905–1914.

Chen, P., Johnson, J.E., Zoghbi, H.Y., Segil, N., 2002. The role of Math1 in inner ear development: Uncoupling the establishment of the sensory primordium from hair cell fate determination. Development 129, 2495–2505.

Chizhikov, V.V., Iskusnykh, I.Y., Fattakhov, N., Fritzsch, B., 2021. Lmx1a and Lmx1b are Redundantly Required for the Development of Multiple Components of the Mammalian Auditory System. Neuroscience 452, 247–264.

Christensen-Dalsgaard, J., Carr, C.E., 2008. Evolution of a sensory novelty: tympanic ears and the associated neural processing. Brain Research Bulletin 75, 365–370.

Christensen, C.B., Christensen-Dalsgaard, J., Madsen, P.T., 2015. Hearing of the African lungfish (Protopterus annectens) suggests underwater pressure detection and rudimentary aerial hearing in early tetrapods. Journal of Experimental Biology 218, 381–387.

Chung, S., Leung, A., Han, B.-S., Chang, M.-Y., Moon, J.-I., Kim, C.-H., Hong, S., Pruszak, J., Isacson, O., Kim, K.-S.J.C.s.c., 2009. Wnt1-lmx1a forms a novel autoregulatory loop and controls midbrain dopaminergic differentiation synergistically with the SHH-FoxA2 pathway. Cell Stem Cell 5, 646–658.

Coate, T.M., Kelley, M.W., 2013. Making connections in the inner ear: recent insights into the development of spiral ganglion neurons and their connectivity with sensory hair cells. Seminars in Cell & Developmental Biology 24, 460–469.

Coate, T.M., Spita, N.A., Zhang, K.D., Isgrig, K.T., Kelley, M.W., 2015. Neuropilin-2/Semaphorin-3F-mediated repulsion promotes inner hair cell innervation by spiral ganglion neurons. Elife 4, e07830.

Coffin, A., Kelley, M., Manley, G.A., Popper, A.N., 2004. Evolution of sensory hair cells, Evolution of the Vertebrate Auditory System. Springer, New York, pp. 55–94.

Cole, L.K., Le Roux, I., Nunes, F., Laufer, E., Lewis, J., Wu, D.K., 2000. Sensory organ generation in the chicken inner ear: contributions of bone morphogenetic protein 4, serrate1, and lunatic fringe. Journal of Comparative Neurology 424, 509–520.

Corey, D.P., Hudspeth, A.J., 1983a. Analysis of the microphonic potential of the bullfrog's sacculus. The Journal of Neuroscience 3, 942–961.

Corey, D.P., Hudspeth, A.J., 1983b. Kinetics of the receptor current in bullfrog saccular hair cells. The Journal of Neuroscience 3, 962–976.

Corwin, J.T., 1985. Perpetual production of hair cells and maturational changes in hair cell ultrastructure accompany postembryonic growth in an amphibian ear. Proceedings of the National Academy of Sciences 82, 3911–3915.

Corwin, J.T., Cotanche, D.A.J.S., 1988. Regeneration of sensory hair cells after acoustic trauma. Science 240, 1772–1774.

Cramer, K.S., Gabriele, M.L., 2014. Axon guidance in the auditory system: multiple functions of Eph receptors. Neuroscience 277, 152–162.

D'amico-Martel, A., 1982. Temporal patterns of neurogenesis in avian cranial sensory and autonomic ganglia. American Journal of Anatomy 163, 351–372.

Davis, K.A., 2005. Spectral processing in the inferior colliculus. International Review of Neurobiology 70, 169–205.

Defourny, J., Poirrier, A.L., Lallemend, F., Mateo Sanchez, S., Neef, J., Vanderhaeghen, P., Soriano, E., Peuckert, C., Kullander, K., Fritzsch, B., Nguyen, L., Moonen, G., Moser, T., Malgrange, B., 2013. Ephrin-A5/EphA4 signalling controls specific afferent targeting to cochlear hair cells. Nature Communications 4, 1438.

Delacroix, L., Malgrange, B., 2015. Cochlear afferent innervation development. Hearing Research 330, 157–169.

Deng, M., Yang, H., Xie, X., Liang, G., Gan, L., 2014. Comparative expression analysis of POU4F1, POU4F2 and ISL1 in developing mouse cochleovestibular ganglion neurons. Gene Expression Patterns: GEP 15, 31–37.

Di Bonito, M., Narita, Y., Avallone, B., Sequino, L., Mancuso, M., Andolfi, G., Franze, A.M., Puelles, L., Rijli, F.M., Studer, M., 2013. Assembly of the auditory circuitry by a Hox genetic network in the mouse brainstem. PLOS Genetics 9, e1003249.

Di Bonito, M., Studer, M., 2017. Cellular and Molecular Underpinnings of Neuronal Assembly in the Central Auditory System during Mouse Development. Frontiers in Neural Circuits 11, 18.

Doetzlhofer, A., Basch, M.L., Ohyama, T., Gessler, M., Groves, A.K., Segil, N., 2009. Hey2 regulation by FGF provides a Notch-independent mechanism for maintaining pillar cell fate in the organ of Corti. Developmental Cell 16, 58–69.

Driver, E.C., Sillers, L., Coate, T.M., Rose, M.F., Kelley, M.W., 2013. The Atoh1-lineage gives rise to hair cells and supporting cells within the mammalian cochlea. Developmental Biology 376, 86–98.

Druckenbrod, N.R., Goodrich, L.V., 2015. Sequential retraction segregates SGN processes during target selection in the cochlea. Journal of Neuroscience 35, 16221–16235.

Duncan, J.S., Cox, B.C., 2020. Anatomy and development of the inner ear, in: Fritzsch, B. (Ed.), The Senses. Elsevier, Cambridge, pp. 253–275.

Duncan, J.S., Fritzsch, B., 2012. Transforming the vestibular system one molecule at a time: the molecular and developmental basis of vertebrate auditory evolution, Sensing in Nature. Springer, New York, pp. 173–186.

Duncan, J.S., Fritzsch, B., 2013. Continued expression of GATA3 is necessary for cochlear neurosensory development. PloS One 8, e62046.

Duncan, J.S., Fritzsch, B., Houston, D.W., Ketchum, E.M., Kersigo, J., Deans, M.R., Elliott, K.L., 2019. Topologically correct central projections of tetrapod inner ear afferents require Fzd3. Scientific Reports 9, 10298.

Duncan, J.S., Lim, K.C., Engel, J.D., Fritzsch, B., 2011. Limited inner ear morphogenesis and neurosensory development are possible in the absence of GATA3. The International Journal of Developmental Biology 55, 297–303.

Duncan, J.S., Stoller, M.L., Francl, A.F., Tissir, F., Devenport, D., Deans, M.R., 2017. Celsr1 coordinates the planar polarity of vestibular hair cells during inner ear development. Developmental Biology 423, 126–137.

Dvorakova, M., Macova, I., Bohuslavova, R., Anderova, M., Fritzsch, B., Pavlinkova, G., 2020. Early ear neuronal development, but not olfactory or lens development, can proceed without SOX2. Developmental Biology 457, 43–56.

Edge, A.S., Chen, Z.-Y.C., 2008. Hair cell regeneration. Elsevier 18, 377–382.

Elliott, K.L., Fritzsch, B., Duncan, J.S., 2018. Evolutionary and developmental biology provide insights into the regeneration of organ of Corti hair cells. Frontiers in Cellular Neuroscience 12, 252.

Elliott, K.L., Kersigo, J., Pan, N., Jahan, I., Fritzsch, B., 2017. Spiral ganglion neuron projection development to the hindbrain in mice lacking peripheral and/or central target differentiation. Front Neural Circuits 11, 25.

Elliott, K.L., Pavlínková, G., Chizhikov, V.V., Yamoah, E.N., Fritzsch, B., 2021. Development in the Mammalian Auditory System Depends on Transcription Factors. International Journal of Molecular Sciences 22, 4189.

Erives, A., Fritzsch, B., 2020. A screen for gene paralogies delineating evolutionary branching order of early Metazoa. G3: Genes, Genomes, Genetics 10, 811–826.

Ernfors, P., Van De Water, T., Loring, J., Jaenisch, R.J.N., 1995. Complementary roles of BDNF and NT-3 in vestibular and auditory development. Neuron 14, 1153–1164.

Farago, A.F., Awatramani, R.B., Dymecki, S.M., 2006. Assembly of the brainstem cochlear nuclear complex is revealed by intersectional and subtractive genetic fate maps. Neuron 50, 205–218.

Farinas, I., Jones, K.R., Tessarollo, L., Vigers, A.J., Huang, E., Kirstein, M., de Caprona, D.C., Coppola, V., Backus, C., Reichardt, L.F., Fritzsch, B., 2001. Spatial shaping of cochlear innervation by temporally regulated neurotrophin expression. The Journal of Neuroscience: The Official Journal of the Society for Neuroscience 21, 6170–6180.

Fay, R., Popper, A., 2000. Evolution of hearing in vertebrates: the inner ears and processing. Hear Research 149, 1–10.

Fekete, D.M., Campero, A.M., 2007. Axon guidance in the inner ear. International Journal of Developmental Biology 51, 549–556.

Filova, I., Dvorakova, M., Bohuslavova, R., Pavlinek, A., Elliott, K.L., Vochyanova, S., Fritzsch, B., Pavlinkova, G., 2020. Combined Atoh1 and Neurod1 Deletion Reveals Autonomous Growth of Auditory Nerve Fibers. Molecular Neurobiology 57, 5307–5323.

Fischl, M.J., Burger, R.M., Schmidt-Pauly, M., Alexandrova, O., Sinclair, J.L., Grothe, B., Forsythe, I.D., Kopp-Scheinpflug, C.J.J.o.n., 2016. Physiology and anatomy of neurons in the medial superior olive of the mouse. Journal of Neurophysiology 116, 2676–2688.

Friauf, E., 1994. Distribution of calcium-binding protein calbindin-D28k in the auditory system of adult and developing rats. Journal of Comparative Neurology 349, 193–211.

Fritzsch, B., 1987. The inner ear of the coelacanth fish *Latimeria* has tetrapod affinities. Nature 327, 153–154.

Fritzsch, B., 1992. The water-to-land transition: evolution of the tetrapod basilar papilla, middle ear, and auditory nuclei. The Evolutionary Biology of Hearing. Springer, New York, pp. 351–375.

Fritzsch, B., 1999. Hearing in two worlds: theoretical and actual adaptive changes of the aquatic and terrestrial ear for sound reception, Comparative Hearing: Fish and Amphibians. Springer, New York, pp. 15–42.

Fritzsch, B., Dillard, M., Lavado, A., Harvey, N.L., Jahan, I., 2010. Canal cristae growth and fiber extension to the outer hair cells of the mouse ear require Prox1 activity. PloS One 5, e9377.

Fritzsch, B., Elliott, K.L., 2017. Gene, cell, and organ multiplication drives inner ear evolution. Developmental Biology 431, 3–15.

Fritzsch, B., Kersigo, J., Yang, T., Jahan, I., Pan, N., 2016. Neurotrophic factor function during ear development: expression changes define critical phases for neuronal viability. The Primary Auditory Neurons of the Mammalian Cochlea. Springer, New York, pp. 49–84.

Fritzsch, B., Matei, V., Nichols, D., Bermingham, N., Jones, K., Beisel, K., Wang, V., 2005. Atoh1 null mice show directed afferent fiber growth to undifferentiated ear sensory epithelia followed by incomplete fiber retention. Developmental Dynamics: An Official Publication of the American Association of Anatomists 233, 570–583.

Fritzsch, B., Neary, T., 1998. The octavolateralis system of mechanosensory and electrosensory organs. Amphibian Biology 3, 878–922.

Fritzsch, B., Pan, N., Jahan, I., Duncan, J.S., Kopecky, B.J., Elliott, K.L., Kersigo, J., Yang, T., 2013. Evolution and development of the tetrapod auditory system: an organ of Corti-centric perspective. Evolution & Development 15, 63–79.

Fritzsch, B., Pan, N., Jahan, I., Elliott, K.L., 2015. Inner ear development: building a spiral ganglion and an organ of Corti out of unspecified ectoderm. Cell Tissue Research 361, 7–24.

Fritzsch, B., Pauley, S., Feng, F., Matei, V., Nichols, D.H., 2006. The molecular and developmental basis of the evolution of the vertebrate auditory system. International Journal of Comparative Psychology 19(1), 1–25.

Fritzsch, B., Wahnschaffe, U., Bartsch, U., 1988. Metamorphic changes in the octavolateralis system of amphibians. The Evolution of the Amphibian Auditory System. Wiley, New York, 359–376.

Fritzsch, B., Wake, M., 1988. The inner ear of gymnophione amphibians and its nerve supply: a comparative study of regressive events in a complex sensory system (Amphibia, Gymnophiona). Zoomorphology 108, 201–217.

Fritzsch, B., Fariñas, I., Reichardt, L.F., 1997. Lack of neurotrophin 3 causes losses of both classes of spiral ganglion neurons in the cochlea in a region-specific fashion. Journal of Neuroscience 17(16), 6213–6225.

Fujiyama, T., Yamada, M., Terao, M., Terashima, T., Hioki, H., Inoue, Y.U., Inoue, T., Masuyama, N., Obata, K., Yanagawa, Y., 2009. Inhibitory and excitatory subtypes of cochlear nucleus neurons are defined by distinct bHLH transcription factors, Ptf1a and Atoh1. Development 136, 2049–2058.

Ghimire, S.R., Deans, M.R., 2019. Frizzled3 and Frizzled6 Cooperate with Vangl2 to Direct Cochlear Innervation by Type II Spiral Ganglion Neurons. Journal of Neuroscience 39, 8013–8023.

Ghimire, S.R., Ratzan, E.M., Deans, M.R., 2018. A non-autonomous function of the core PCP protein VANGL2 directs peripheral axon turning in the developing cochlea. Development. 145 (12)

Glover, J.C., Elliott, K.L., Erives, A., Chizhikov, V.V., Fritzsch, B., 2018. Wilhelm His' lasting insights into hindbrain and cranial ganglia development and evolution. Developmental biology.

Goodrich, L.V., 2016. Early development of the spiral ganglion, in: Dabdoub, A., Fritzsch, B., Popper, A.N., Fay, R.R. (Eds.), The Primary Auditory Neurons of the Mammalian Cochlea. Springer, New York, pp. 11–48.

Green, S.H., Bailey, E., Wang, Q., Davis, R.L., 2012. The Trk A, B, C's of neurotrophins in the cochlea. The Anatomical Record 295, 1877–1895.

Grothe, B., Carr, C.E., Casseday, J.H., Fritzsch, B., Köppl, C., 2004. The evolution of central pathways and their neural processing patterns. Evolution of the Vertebrate Auditory System. Springer, pp. 289–359.

Gurung, B., Fritzsch, B., 2004. Time course of embryonic midbrain and thalamic auditory connection development in mice as revealed by carbocyanine dye tracing. Journal of Comparative Neurology 479, 309–327.

Hafidi, A., 1998. Peripherin-like immunoreactivity in type II spiral ganglion cell body and projections. Brain Research 805, 181–190.

Hans, S., Liu, D., Westerfield, M., 2004. Pax8 and Pax2a function synergistically in otic specification, downstream of the Foxi1 and Dlx3b transcription factors. Development 131, 5091–5102.

Hayashi, T., Ray, C.A., Bermingham-McDonogh, O., 2008. Fgf20 is required for sensory epithelial specification in the developing cochlea. The Journal of Neuroscience 28, 5991–5999.

He, Z., Fang, Q., Li, H., Shao, B., Zhang, Y., Zhang, Y., Han, X., Guo, R., Cheng, C., Guo, L., 2019. The role of FOXG1 in the postnatal development and survival of mouse cochlear hair cells. Neuropharmacology 144, 43–57.

Hertzano, R., Montcouquiol, M., Rashi-Elkeles, S., Elkon, R., Yucel, R., Frankel, W.N., Rechavi, G., Moroy, T., Friedman, T.B., Kelley, M.W., Avraham, K.B., 2004. Transcription profiling of inner ears from Pou4f3(ddl/ddl) identifies Gfi1 as a target of the Pou4f3 deafness gene. Human Molecular Genetics 13, 2143–2153.

Holley, M., Rhodes, C., Kneebone, A., Herde, M.K., Fleming, M., Steel, K.P., 2010. Emx2 and early hair cell development in the mouse inner ear. Developmental Biology 340, 547–556.

Hu, X., Huang, J., Feng, L., Fukudome, S., Hamajima, Y., Lin, J., 2010. Sonic hedgehog (SHH) promotes the differentiation of mouse cochlear neural progenitors via the Math1-Brn3.1 signaling pathway in vitro. Journal of Neuroscience Research 88, 927–935.

Huang, E.J., Liu, W., Fritzsch, B., Bianchi, L.M., Reichardt, L.F., Xiang, M., 2001. Brn3a is a transcriptional regulator of soma size, target field innervation and axon pathfinding of inner ear sensory neurons. Development 128, 2421–2432.

Huang, Y., Hill, J., Yatteau, A., Wong, L., Jiang, T., Petrovic, J., Gan, L., Dong, L., Wu, D.K., 2018. Reciprocal negative regulation between Lmx1a and Lmo4 is required for inner ear formation. Journal of Neuroscience 38, 5429–5440.

Huh, S.H., Jones, J., Warchol, M.E., Ornitz, D.M., 2012. Differentiation of the lateral compartment of the cochlea requires a temporally restricted FGF20 signal. PLoS Biology 10, e1001231.

Hutson, K.A., Morest, D.K., 1996. Fine structure of the cell clusters in the cochlear nerve root: stellate, granule, and mitt cells offer insights into the synaptic organization of local circuit neurons. Journal of Comparative Neurology 371, 397–414.

Jacques, B.E., Montcouquiol, M.E., Layman, E.M., Lewandoski, M., Kelley, M.W., 2007. Fgf8 induces pillar cell fate and regulates cellular patterning in the mammalian cochlea. Development 134, 3021–3029.

Jahan, I., Kersigo, J., Pan, N., Fritzsch, B., 2010a. Neurod1 regulates survival and formation of connections in mouse ear and brain. Cell and Tissue Research 341, 95–110.

Jahan, I., Pan, N., Kersigo, J., Fritzsch, B., 2010b. Neurod1 suppresses hair cell differentiation in ear ganglia and regulates hair cell subtype development in the cochlea. PLoS One. 5, e11661.

Jahan, I., Pan, N., Kersigo, J., Fritzsch, B., 2015. Neurog1 can partially substitute for Atoh1 function in hair cell differentiation and maintenance during organ of Corti development. Development 142, 2810–2821.

Jiang, T., Kindt, K., Wu, D.K., 2017. Transcription factor Emx2 controls stereociliary bundle orientation of sensory hair cells. Elife 6, e23661.

Kandler, K., Friauf, E., 1993. Pre-and postnatal development of efferent connections of the cochlear nucleus in the rat. Journal of Comparative Neurology 328, 161–184.

Kandler, K., Lee, J., Pecka, M., 2020. The superior olivary complex, in: Fritzsch, B. (Ed.), The Senses. Elsevier, pp. 533–555.

Karis, A., Pata, I., van Doorninck, J.H., Grosveld, F., de Zeeuw, C.I., de Caprona, D., Fritzsch, B., 2001. Transcription factor GATA-3 alters pathway selection of olivocochlear neurons and affects morphogenesis of the ear. The Journal of Comparative Neurology 429, 615–630.

Katayama, A., Corwin, J.T., 1989. Cell production in the chicken cochlea. Journal of Comparative Neurology 281, 129–135.

Katayama, A., Corwin, J.T., 1993. Cochlear cytogenesis visualized through pulse labeling of chick embryos in culture. Journal of Comparative Neurology 333, 28–40.

Kawashima, Y., Kurima, K., Pan, B., Griffith, A.J., Holt., 2015. Transmembrane channel-like (TMC) genes are required for auditory and vestibular mechanosensation. Pflügers Archiv - European Journal of Physiology 467, 85–94.

Kersigo, J., Fritzsch, B., 2015. Inner ear hair cells deteriorate in mice engineered to have no or diminished innervation. Frontiers in Aging Neuroscience 7, 33.

Khorevin, V. I. 2008. The lagena (the third otolith endorgan in vertebrates). Neurophysiology, 40(2), 142–159.

Kiernan, A.E., Ahituv, N., Fuchs, H., Balling, R., Avraham, K.B., Steel, K.P., Hrabe de Angelis, M., 2001. The Notch ligand Jagged1 is required for inner ear sensory development. Proceedings of the National Academy of Sciences of the United States of America 98, 3873–3878.

Kiernan, A.E., Cordes, R., Kopan, R., Gossler, A., Gridley, T., 2005a. The Notch ligands DLL1 and JAG2 act synergistically to regulate hair cell development in the mammalian inner ear. Development 132, 4353–4362.

Kiernan, A.E., Pelling, A.L., Leung, K.K., Tang, A.S., Bell, D.M., Tease, C., Lovell-Badge, R., Steel, K.P., Cheah, K.S., 2005b. Sox2 is required for sensory organ development in the mammalian inner ear. Nature 434, 1031–1035.

Kiernan, A.E., Xu, J., Gridley, T., 2006. The Notch ligand JAG1 is required for sensory progenitor development in the mammalian inner ear. PLOS Genetics 2, e4.

Kim, W.Y., Fritzsch, B., Serls, A., Bakel, L.A., Huang, E.J., Reichardt, L.F., Barth, D.S., Lee, J.E., 2001. NeuroD-null mice are deaf due to a severe loss of the inner ear sensory neurons during development. Development 128, 417–426.

Kimura, J., Suda, Y., Kurokawa, D., Hossain, Z.M., Nakamura, M., Takahashi, M., Hara, A., Aizawa, S., 2005. Emx2 and Pax6 function in cooperation with Otx2 and Otx1 to develop caudal forebrain primordium that includes future archipallium. The Journal of Neuroscience 25, 5097–5108.

Koo, S.K., Hill, J.K., Hwang, C.H., Lin, Z.S., Millen, K.J., Wu, D.K., 2009. Lmx1a maintains proper neurogenic, sensory, and non-sensory domains in the mammalian inner ear. Developmental Biology 333, 14–25.

Kopecky, B., Santi, P., Johnson, S., Schmitz, H., Fritzsch, B., 2011. Conditional deletion of N-Myc disrupts neurosensory and non-sensory development of the ear. Developmental Dynamics 240, 1373–1390.

Krüger, M., Schmid, T., Krüger, S., Bober, E., Braun, T., 2006. Functional redundancy of NSCL-1 and NeuroD during development of the petrosal and vestibulocochlear ganglia. European Journal of Neuroscience 24, 1581–1590.

Kudo, M., Nakamura, Y., 1988. Organization of the lateral lemniscal fibers converging onto the inferior colliculus in the cat: an anatomical review. Auditory Pathway. Springer, New York, pp. 171–183.

Kudo, M.J.B.r., 1981. Projections of the nuclei of the lateral lemniscus in the cat: An autoradiographic study. Brain Research 221, 57–69.

Ladhams, A., Pickles, J.O., 1996. Morphology of the monotreme organ of Corti and macula lagena. The Journal of Comparative Neurology 366, 335–347.

Ladich, F., Fay, R.R., 2013. Auditory evoked potential audiometry in fish. Reviews in Fish Biology and Fisheries 23, 317–364.

Lawoko-Kerali, G., Rivolta, M.N., Lawlor, P., Cacciabue-Rivolta, D.I., Langton-Hewer, C., van Doorninck, J.H., Holley, M.C., 2004. GATA3 and NeuroD distinguish auditory and vestibular neurons during development of the mammalian inner ear. Mechanisms of Development 121, 287–299.

Lewis, E.R., Leverenz, E.L., Bialek, W.S., 1985. The vertebrate inner ear. CRC Press, Boca Raton.

Li, S., Price, S.M., Cahill, H., Ryugo, D.K., Shen, M.M., Xiang, M., 2002. Hearing loss caused by progressive degeneration of cochlear hair cells in mice deficient for the Barhl1 homeobox gene. Development 129, 3523–3532.

Lipovsek, M., Wingate, R.J., 2018. Conserved and divergent development of brainstem vestibular and auditory nuclei. Elife 7, e40232.

Liu, M., Pereira, F.A., Price, S.D., Chu, M.-j., Shope, C., Himes, D., Eatock, R.A., Brownell, W.E., Lysakowski, A., Tsai, M.-J., 2000. Essential role of BETA2/NeuroD1 in development of the vestibular and auditory systems. Genes & Development 14, 2839–2854.

Lohmann, C., Friauf, E., 1996. Distribution of the calcium-binding proteins parvalbumin and calretinin in the auditory brainstem of adult and developing rats. Journal of Comparative Neurology 367, 90–109.

Lu, X., Sipe, C.W., 2015. Developmental regulation of planar cell polarity and hair-bundle morphogenesis in auditory hair cells: lessons from human and mouse genetics. Wiley interdisciplinary reviews. Developmental Biology 5, 85–101.

Luo, Z.-X., Manley, G.A., 2021. Origins and early evolution of mammalian ears and hearing function, in: Fritzsch, B. (Ed.), The Senses. Elsevier, pp. 207–252.

Ma, Q., Anderson, D.J., Fritzsch, B., 2000. Neurogenin 1 null mutant ears develop fewer, morphologically normal hair cells in smaller sensory epithelia devoid of innervation. Journal of the Association for Research in Otolaryngology: JARO 1, 129–143.

Ma, Q.F., Chen, Z.F., Barrantes, I.D., de la Pompa, J.L., Anderson, D.J., 1998. Neurogenin1 is essential for the determination of neuronal precursors for proximal cranial sensory ganglia. Neuron 20, 469–482.

Macova, I., Pysanenko, K., Chumak, T., Dvorakova, M., Bohuslavova, R., Syka, J., Fritzsch, B., Pavlinkova, G., 2019. Neurod1 is essential for the primary tonotopic organization and related auditory information processing in the midbrain. The Journal of Neuroscience: The Official Journal of the Society for Neuroscience 39, 984–1004.

Maison, S., Liberman, L.D., Liberman, M.C., 2016. Type II cochlear ganglion neurons do not drive the olivocochlear reflex: re-examination of the cochlear phenotype in peripherin knock-out mice. Eneuro 3, 207–216.

Malone, B.J., Hasenstaub, A.R., Schreiner, C.E., 2020. Primary auditory cortex II. Some functional considerations, in: Fritzsch, B. (Ed.), The Senses. Elsevier, pp. 657–680.

Mantela, J., Jiang, Z., Ylikoski, J., Fritzsch, B., Zacksenhaus, E., Pirvola, U., 2005. The retinoblastoma gene pathway regulates the postmitotic state of hair cells of the mouse inner ear. Development 132, 2377–2388.

Mao, Y., Reiprich, S., Wegner, M., Fritzsch, B., 2014. Targeted deletion of Sox10 by Wnt1-cre defects neuronal migration and projection in the mouse inner ear. PloS One 9, e94580.

Marcovich, I., Holt, 2020. Evolution and function of Tmc genes in mammalian hearing. Current Opinion in Physiology 18, 11–19.

Maricich, S.M., Xia, A., Mathes, E.L., Wang, V.Y., Oghalai, J.S., Fritzsch, B., Zoghbi, H.Y., 2009. Atoh1-lineal neurons are required for hearing and for the survival of neurons in the spiral ganglion and brainstem accessory auditory nuclei. The Journal of Neuroscience: The Official Journal of the Society for Neuroscience 29, 11123–11133.

Marrs, G.S., Spirou, G.A., 2012. Embryonic assembly of auditory circuits: spiral ganglion and brainstem. The Journal of Physiology 590, 2391–2408.

Matei, V., Pauley, S., Kaing, S., Rowitch, D., Beisel, K.W., Morris, K., Feng, F., Jones, K., Lee, J., Fritzsch, B., 2005. Smaller inner ear sensory epithelia in Neurog 1 null mice are related to earlier hair cell cycle exit. Developmental Dynamics 234, 633–650.

Matern, M.S., Milon, B., Lipford, E.L., McMurray, M., Ogawa, Y., Tkaczuk, A., Song, Y., Elkon, R., Hertzano, R., 2020. GFI1 functions to repress neuronal gene expression in the developing inner ear hair cells. Development 147, dev186015.

Mickle, M.F., Miehls, S.M., Johnson, N.S., Higgs, D.M., 2019. Hearing capabilities and behavioural response of sea lamprey (Petromyzon marinus) to low-frequency sounds. Canadian Journal of Fisheries and Aquatic Sciences 76, 1541–1548.

Mickle, M.F., Pieniazek, R.H., Higgs, D.M., 2020. Field assessment of behavioural responses of southern stingrays (Hypanus americanus) to acoustic stimuli. Royal Society Open Science 7, 191544.

Milinkeviciute, G., Cramer, K.S., 2020. Development of the ascending auditory pathway, in: Fritzsch, B. (Ed.), The Senses. Elsevier, Cambridge, pp. 337–353.

Mishima, Y., Lindgren, A.G., Chizhikov, V.V., Johnson, R.L., Millen, K.J., 2009. Overlapping function of Lmx1a and Lmx1b in anterior hindbrain roof plate formation and cerebellar growth. Journal of Neuroscience 29, 11377–11384.

Montcouquiol, M., Rachel, R.A., Lanford, P.J., Copeland, N.G., Jenkins, N.A., Kelley, M.W., 2003. Identification of Vangl2 and Scrb1 as planar polarity genes in mammals. Nature 423, 173–177.

Muniak, M.A., Connelly, C.J., Suthakar, K., Milinkeviciute, G., Ayeni, F.E., Ryugo, D.K, 2016. Central Projections of Spiral Ganglion Neurons. Springer, New York.

Nakano, Y., Jahan, I., Bonde, G., Sun, X., Hildebrand, M.S., Engelhardt, J.F., Smith, R.J., Cornell, R.A., Fritzsch, B., Bánfi, B.J.P.G., 2012. A mutation in the Srrm4 gene causes alternative splicing defects and deafness in the Bronx Waltzer mouse. Plos Genetics 8, e1002966.

Nakano, Y., Wiechert, S., Fritzsch, B., Bánfi, B., 2020. Inhibition of a transcriptional repressor rescues hearing in a splicing factor–deficient mouse. Life Science Alliance 3(12): e202000841.

Nayagam, B.A., Muniak, M.A., Ryugo, D.K., 2011. The spiral ganglion: connecting the peripheral and central auditory systems. Hearing Research 278, 2–20.

Nichols, D.H., Bruce., 2006. Migratory routes and fates of cells transcribing the Wnt-1 gene in the murine hindbrain. Developmental Dynamics 235, 285–300.

Nichols, D.H., Pauley, S., Jahan, I., Beisel, K.W., Millen, K.J., Fritzsch, B., 2008. Lmx1a is required for segregation of sensory epithelia and normal ear histogenesis and morphogenesis. Cell Tissue Research 334, 339–358.

Nishimura, K., Noda, T., Dabdoub, A., 2017. Dynamic expression of Sox2, Gata3, and Prox1 during primary auditory neuron development in the mammalian cochlea. PloS One 12, e0170568.

Nist-Lund, C.A., Pan, B., Patterson, A., Asai, Y., Chen, T., Zhou, W., Zhu, H., Romero, S., Resnik, J., Polley, D.B., 2019. Improved TMC1 gene therapy restores hearing and balance in mice with genetic inner ear disorders. Nature Communications 10, 1–14.

Noden, D., Van De Water, T., 1986. The developing ear: tissue origins and interactions. The Biology of Change in Otolaryngology, 15–46.

Nothwang, H.G., 2016. Evolution of mammalian sound localization circuits: A developmental perspective. Progress in Neurobiology 141, 1–24.

Pan, B., Geleoc, G.S., Asai, Y., Horwitz, G.C., Kurima, K., Ishikawa, K., Kawashima, Y., Griffith, A.J., Holt, J.R., 2013. TMC1 and TMC2 are components of the mechanotransduction channel in hair cells of the mammalian inner ear. Neuron 79, 504–515.

Pan, N., Jahan, I., Kersigo, J., Duncan, J., Kopecky, B., Fritzsch, B., 2012. A novel Atoh1 'self-terminating' mouse model reveals the necessity of proper Atoh1 expression level and duration for inner ear hair cell differentiation and viability. PLoS One 7, e30358.

Pan, N., Jahan, I., Kersigo, J., Kopecky, B., Santi, P., Johnson, S., Schmitz, H., Fritzsch, B., 2011. Conditional deletion of Atoh1 using Pax2-Cre results in viable mice without differentiated cochlear hair cells that have lost most of the organ of Corti. Hearing Research 275, 66–80.

Pauley, S., Lai, E., Fritzsch, B., 2006. Foxg1 is required for morphogenesis and histogenesis of the mammalian inner ear. Developmental Dynamics. 235, 2470–2482.

Petitpre, C., Wu, H., Sharma, A., Tokarska, A., Fontanet, P., Wang, Y., Helmbacher, F., Yackle, K., Silberberg, G., Hadjab, S., Lallemend, F., 2018. Neuronal heterogeneity and stereotyped connectivity in the auditory afferent system. Nature Communications 9, 3691.

Platt, C., Jørgensen, J.M., Popper, A.N., 2004. The inner ear of the lungfish Protopterus. Journal of Comparative Neurology 471, 277–288.

Popper, A., Fay, R., 1999. Comparative hearing: fish and amphibians. Springer Handbook of Auditory Research. Springer, New York.

Popper, A.N., Coombs, S., 1982. The morphology and evolution of the ear in actinopterygian fishes. American Zoologist 22, 311–328.

Puligilla, C., Feng, F., Ishikawa, K., Bertuzzi, S., Dabdoub, A., Griffith, A.J., Fritzsch, B., Kelley, M.W., 2007. Disruption of fibroblast growth factor receptor 3 signaling results in defects in cellular differentiation, neuronal patterning, and hearing impairment. Developmental Dynamics: An Official Publication of the American Association of Anatomists 236, 1905–1917.

Qiu, X., Müller, U., 2018. Mechanically gated ion channels in mammalian hair cells. Frontiers in Cellular Neuroscience 12, 100.

Radde-Gallwitz, K., Pan, L., Gan, L., Lin, X., Segil, N., Chen, P., 2004. Expression of Islet1 marks the sensory and neuronal lineages in the mammalian inner ear. The Journal of Comparative Neurology. 477, 412–421.

Rauschecker, J.P., 2020. The auditory cortex of primates including man with reference to speech, in: Fritzsch, B. (Ed.), The Senses. Elsevier, Cambridge, pp. 791–811.

Rose, M.F., Ahmad, K.A., Thaller, C., Zoghbi, H.Y., 2009. Excitatory neurons of the proprioceptive, interoceptive, and arousal hindbrain networks share a developmental requirement for Math1. Proceedings of the National Academy of Sciences of the United States of America 106, 22462–22467.

Rubel, E.W., Fritzsch, B., 2002. Auditory system development: primary auditory neurons and their targets. Annual Review of Neuroscience 25, 51–101.

Rubel, E.W., Furrer, S.A., Stone, J.S., 2013. A brief history of hair cell regeneration research and speculations on the future. Hearing Research 297, 42–51.

Ruben, R.J., 1967. Development of the inner ear of the mouse: a radioautographic study of terminal mitoses. Acta Oto-laryngologica Supplementum 220:221–244.

Rubio, M.E., Juiz, J.M., 2004. Differential distribution of synaptic endings containing glutamate, glycine, and GABA in the rat dorsal cochlear nucleus. The Journal of Comparative Neurology 477, 253–272.

Sandell, L.L., Tjaden, N.E.B., Barlow, A.J., Trainor, P.A., 2014. Cochleovestibular nerve development is integrated with migratory neural crest cells. Developmental Biology 385, 200–210.

Schmidt, H., Fritzsch, B., 2019. Npr2 null mutants show initial overshooting followed by reduction of spiral ganglion axon projections combined with near-normal cochleotopic projection. Cell and Tissue Research 378, 15–32.

Schultz, J.A., Zeller, U., Luo, Z.X., 2017. Inner ear labyrinth anatomy of monotremes and implications for mammalian inner ear evolution. Journal of Morphology 278, 236–263.

Schultze, H.-P., 2018. Hard tissues in fish evolution: history and current issues. Cybium 42, 29–39.

Shibata, S.B., Ranum, P.T., Moteki, H., Pan, B., Goodwin, A.T., Goodman, S.S., Abbas, P.J., Holt, J.R., Smith, R.J., 2016. RNA interference prevents autosomal-dominant hearing loss. The American Journal of Human Genetics 98, 1101–1113.

Shrestha, B.R., Chia, C., Wu, L., Kujawa, S.G., Liberman, M.C., Goodrich, L.V., 2018. Sensory Neuron Diversity in the Inner Ear Is Shaped by Activity. Cell 174, 1229–1246 e1217.

Sienknecht, U.J., Koppl, C., Fritzsch, B., 2014. Evolution and development of hair cell polarity and efferent function in the inner ear. Brain, Behavior and Evolution 83, 150–161.

Silos-Santiago, I., Fagan, A.M., Garber, M., Fritzsch, B., Barbacid, M., 1997. Severe sensory deficits but normal CNS development in newborn mice lacking TrkB and TrkC tyrosine protein kinase receptors. European Journal of Neuroscience 9, 2045–2056.

Simmons, D.D., Liberman, M.C., 1988. Afferent innervation of outer hair cells in adult cats: I. Light microscopic analysis of fibers labeled with horseradish peroxidase. Journal of Comparative Neurology 270, 132–144.

Smotherman, M., Narins, P., 2004. Evolution of the amphibian ear, Evolution of the Vertebrate Auditory System. Springer, New York, pp. 164–199.

Strutz, J.J.H., 1987. Anatomy of the central auditory pathway. Demonstration with Horseradish Peroxidase in the Guinea Pig. 35, 407–415.

Sun, S., Babola, T., Pregernig, G., So, K.S., Nguyen, M., Su, S.M., Palermo, A.T., Bergles, D.E., Burns, J.C., Muller, U., 2018. Hair Cell mechanotransduction regulates spontaneous activity and spiral ganglion subtype specification in the auditory system. Cell 174, 1247–1263 e1215.

Syka, J., 2020. Age-related changes in the auditory brainstem and inferior colliculus, Aging and Hearing. Springer, pp. 67–96.

Tarchini, B., Jolicoeur, C., Cayouette, M., 2013. A molecular blueprint at the apical surface establishes planar asymmetry in cochlear hair cells. Developmental Cell 27, 88–102.

Tessarollo, L., Coppola, V., Fritzsch, B., 2004. NT-3 replacement with brain-derived neurotrophic factor redirects vestibular nerve fibers to the cochlea. Journal of Neuroscience 24, 2575–2584.

Tilney, L.G., Saunders, J.C., 1983. Actin filaments, stereocilia, and hair cells of the bird cochlea. I. Length, number, width, and distribution of stereocilia of each hair cell are related to the position of the hair cell on the cochlea. Journal of Cell Biology 96(3): 807–821.

Ueyama, T., Sakaguchi, H., Nakamura, T., Goto, A., Morioka, S., Shimizu, A., Nakao, K., Hishikawa, Y., Ninoyu, Y., Kassai, H., 2014. Maintenance of stereocilia and apical junctional complexes by Cdc42 in cochlear hair cells. Journal of Cell Science 127, 2040–2052.

van Bergeijk, W.A., 1966. Evolution of the sense of hearing in vertebrates. American Zoologist 6, 371–377.

van Bergeijk, W.A., 1967. The evolution of vertebrate hearing, in: Neff, W.D. (Ed.), Contributions to Sensory Physiology. Springer-Verlag, Berlin and New York, pp. 1–49.

Wang, V.Y., Rose, M.F., Zoghbi, H.Y., 2005. Math1 expression redefines the rhombic lip derivatives and reveals novel lineages within the brainstem and cerebellum. Neuron 48, 31–43.

Wever, E.G., 1974. The evolution of vertebrate hearing, in: Keidel, W.D., Neff, W.D. (Eds.), Auditory System, Springer Verlag, Berlin, Germany, pp. 423–454.

White, P.M., Doetzlhofer, A., Lee, Y.S., Groves, A.K., Segil, N., 2006. Mammalian cochlear supporting cells can divide and trans-differentiate into hair cells. Nature 441, 984–987.

Whitfield, T.T., 2015. Development of the inner ear. Current Opinion in Genetics & Development 32, 112–118.

Woods, C., Montcouquiol, M., Kelley, M.W., 2004. Math1 regulates development of the sensory epithelium in the mammalian cochlea. Nature Neuroscience 7, 1310–1318.

Xiang, M., Maklad, A., Pirvola, U., Fritzsch, B., 2003. Brn3c null mutant mice show long-term, incomplete retention of some afferent inner ear innervation. BMC Neuroscience 4, 2.

Yan, C.H., Levesque, M., Claxton, S., Johnson, R.L., Ang, S.-L., 2011. Lmx1a and lmx1b function cooperatively to regulate proliferation, specification, and differentiation of midbrain dopaminergic progenitors. Journal of Neuroscience 31, 12413–12425.

Yang, T., Kersigo, J., Wu, S., Fritzsch, B., Bassuk, A.G., 2017. Prickle1 regulates neurite outgrowth of apical spiral ganglion neurons but not hair cell polarity in the murine cochlea. PLoS One 12, e0183773.

Yu, W.M., Appler, J.M., Kim, Y.H., Ishitani, A.M., Holt, J.R., Goodrich, L.V., 2013. A Gata3-Mafb transcriptional network directs post-synaptic differentiation in synapses specialized for hearing. Elife, PMC3851837.

Zhang, K.D., Coate, T.M., 2017. Recent advances in the development and function of type II spiral ganglion neurons in the mammalian inner ear. Seminars in Cell & Developmental Biology 65, 80–87.

8 Lateral Line Input to 'Almost' All Vertebrates Shares a Common Organization with Different Distinct Connections

Jacob Engelmann, Bernd Fritzsch

CONTENTS

8.1 THREE LEVELS OF LATERAL LINE PERCEPTION: SENSORY NEURONS, HAIR CELLS, AND THE INTERMEDIATE NUCLEI OF THE BRAINSTEM

The lateral line is a sensory system that allows vertebrates to sense the pattern of water flow over their body surface via mechanosensory organs, called neuromasts, which are distributed in a characteristic pattern over the body surface (Webb, 2020, Bleckmann, 2020). While the lateral line system is shared by all craniates (Nieuwenhuys et al., 1998, Striedter and Northcutt, 2019), amniotes only retain some rudimental organs during development (Fritzsch, 1989, Schlosser, 2002b, Washausen and Knabe, 2018). A unique hair-cell population is found in several gnathostomes,

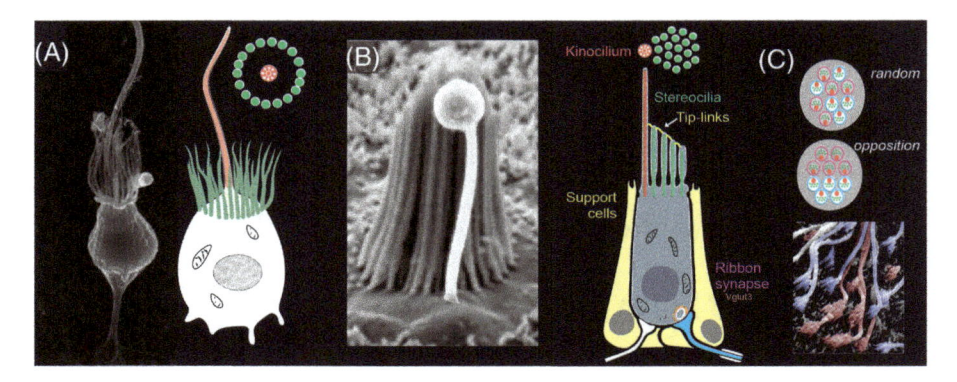

FIGURE 8.1   Hair cells evolve from choanoflagellates. (A) The single cells of choanoflagellates have a single central kinocilium (red) surrounded by a collar of microvilli (green) to filter-feed. (B). In contrast, all mechanosensory hair cells of the lateral line and the inner ear are characterized by an asymmetric position of the kinocilium and the stereocilia that are connected through tip-link proteins. The hair cells receive both afferent (blue) and efferent (white) innervation, except in cyclostomes, which lack the efferent innervation. (C). Hair cells are grouped in functional units, the so-called neuromasts. Within these, the two opposing hair-cell polarities can occur either in a random distribution (top), or ordered (bottom, schematic, and colored REM image). (Modified from (Fritzsch and Straka, 2014, Nicolson, 2017, Chagnaud et al., 2017.)

but both function and origin of these spiracular organs are unresolved (O'Neill et al., 2012; Figure 8.1).

Within vertebrates, the CNS connects to the hair cells through glutamatergic afferent neurons that connect the hair cells with their specific central lateral line nuclei within the alar plate (Elliott and Fritzsch, 2020), while cholinergic fibers (Carpaneto Freixas et al., 2021) provide a separate efferent innervation (Fritzsch and Elliott, 2017a; see Figure 8.1A,B, schematic). These sensory lateral line neurons develop from anterior and posterior neurons comparable to the development of the facialis (N VII) and glossopharyngeal (N IX) sensory neurons. Sensory neurons connect peripheral hair cells with the central nuclei of the alar plate. The precise organization of the innervation of the hair cells in the anterior and posterior parts of the body depends on *Atoh1* and *Neurog1*, respectively. The nuclei in the alar plate of the hindbrain require *Atoh1* genes and their development and maturation involve neuronal inputs (Matsuda and Chitnis, 2010).

The lateral line nuclei in the alar plate of the hindbrain, the mechanosensory hair cells at the periphery, and the afferent and efferent neurons connecting them to the CNS are referred to as the octavolateralis system (Baker, 2019, O'Neill et al., 2012, McCormick, 1999). This comprises the anterior and posterior lateral line nerves (aLL, pLL, see Figure 8.3) that innervate groups of hair cells within the functional units (neuromasts) on the head and trunk, respectively (López-Schier et al., 2004, Fritzsch and López-Schier, 2014). Each neuromast contains hair cells of opposing polarity. These are either randomly distributed within a neuromast, or they occur in a highly ordered fashion (Figures 8.1C and 8.5). Each polarity is innervated separately (Figure 8.4B) (Chagnaud et al., 2017, Lu and Sipe, 2016, Bleckmann et al., 2014).

Both lateral line nerves project to lateral line nuclei in the intermediate nucleus (IN, rhombomere r1–r7/8).

We here focus on understanding the evolutionary addition of the lateral line sensory systems and their unique projections to distinct rhombomere targets of vertebrates (Fritzsch and Elliott, 2017b, Glover et al., 2018, McCormick, 1999, Wullimann and Grothe, 2013). We will emphasize the comparative aspect of adult stages of the interlinked parts of the latera line system (sensory neurons, hair cells, intermediate nuclei) and briefly explain the higher-order connections (Dean and Claas, 2020, Chagnaud and Coombs, 2014).

8.2 A BRIEF SUMMARY OF THE EVOLUTION OF LATERAL LINE SYSTEMS

The inner ear and the mechano- and electrosensory lateral line systems arose at or before the origin of vertebrates (Figure 8.2A, see Chapter 6) and the ubiquitous ciliated mechanosensory cells probably are a relic of the first animals (Manley and

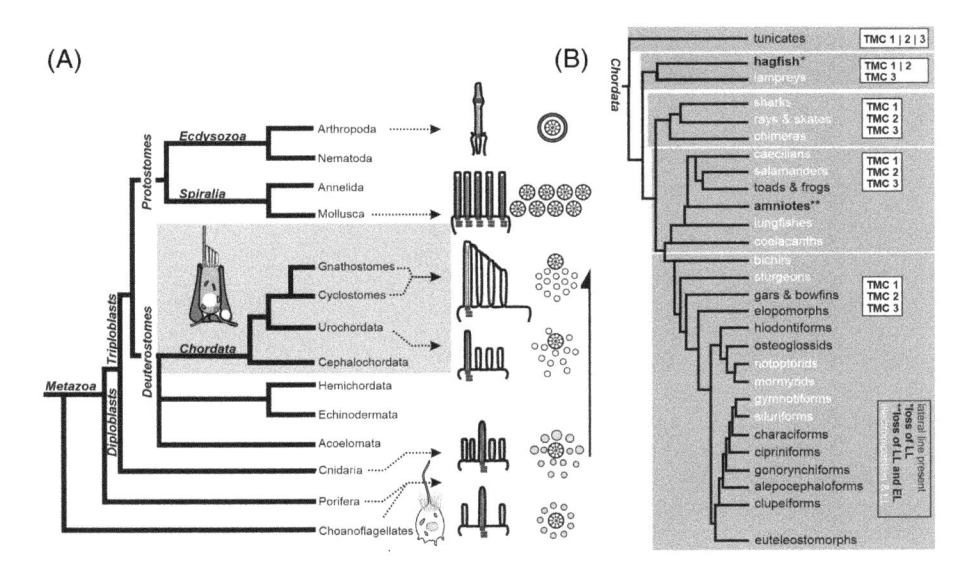

FIGURE 8.2 The evolution of superficial sensory organs. (A) Starting with choanoflagellates hair cells have a central kinocilium surrounded by stereocilia. From cnidarians to ascidians and vertebrates the hair cells progressively become more asymmetric. Protostome mechanoreceptors consist of kinocilia only. (B) Cladogram with details on presence and absence of elements: black for species with lateral line organs, white for species with lateral line and electroreceptors, bold black one asterisk for a secondary loss of the lateral line and bold black with two asterisks for a loss of both the lateral line and electroreceptors. Lateral line hair cells require TMC for their function and show a unique distribution ranging from a single TMC of tunicates, two TMCs (TMC1/2 and TMC3) among agnathans, and three distinct TMCs in all gnathostomes. In parallel, lateral line organizations started with an asymmetric stereocilia in tunicates and shows increase in size in lampreys, sharks, and rays. Length of kinocilia can be short or can be extremely long, as in zebrafish. ([A]: Modified from Fritzsch et al., 2007; [B]: Modified from Baker, 2019, Elliott and Fritzsch, 2020, Erives and Fritzsch, 2020.)

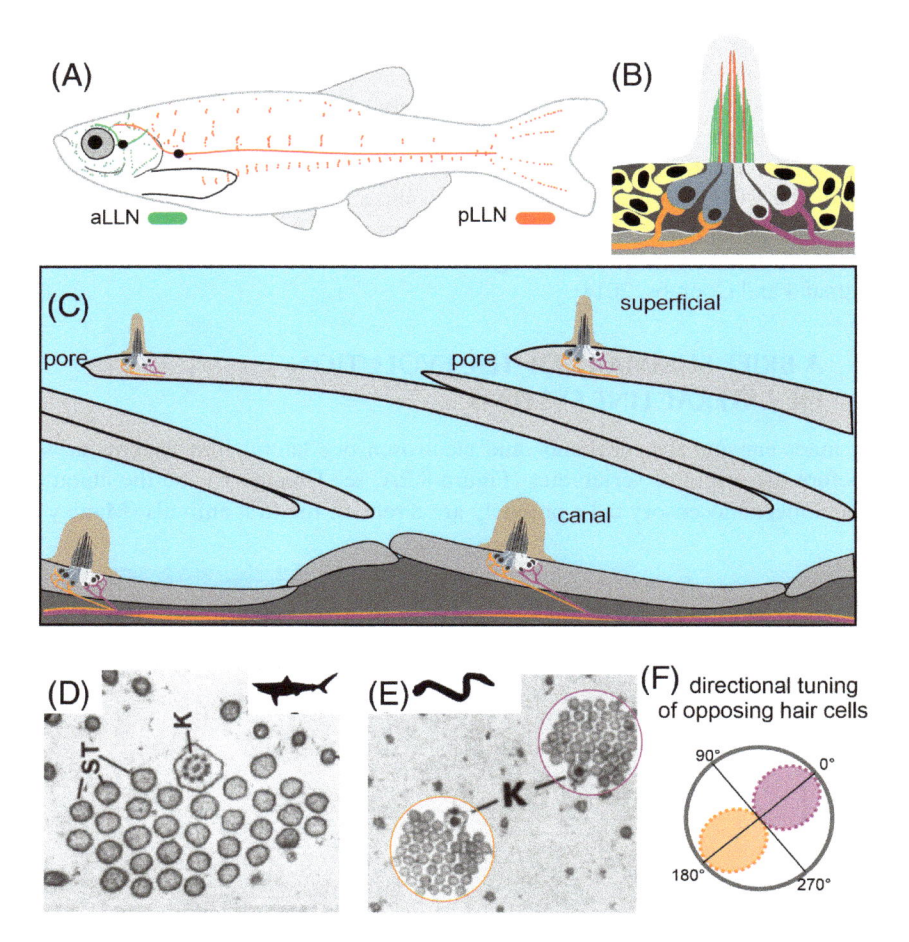

FIGURE 8.3   Hair cells are polarized. (A) In fish, the aLLN and pLLN provide the inner-vation of neuromasts on the head and trunk, respectively. (B) Within these neuromasts, both populations of opposingly oriented hair cells receive separate innervations. (C). Neuromasts occur either recessed in canals or freestanding as superficial neuromasts. Where investi-gated, these groups are independently innervated. (D, E) Thirty to forty stereocilia are typically found in sharks (D), but as many as 60 or more stereocilia are found in eels (E). The opposing polarity is shown in color in (F). (Modified from Yamada and Hama, 1972, Hama and Yamada, 1977, Weber and Schiewe, 1976.)

Ladher, 2008, Fritzsch et al., 2020). These mechanosensory cells likely evolved through a stepwise transformation, starting with choanoflagellates, the sister group of the metazoa (Figures 8.1A and 8.2A; Fritzsch and Straka, 2014, Brunet and King, 2017). The asymmetric position of the true cilia so characteristic for vertebrate hair cells first occurs in tunicates (Gasparini et al., 2013, Manni et al., 2018). Vertebrate hair cells are innervated by *Atoh* expressing sensory neurons (Tang et al., 2013) that form transcription factor *Neurogenin*-dependent 'ganglia' adjacent to the spinal cord

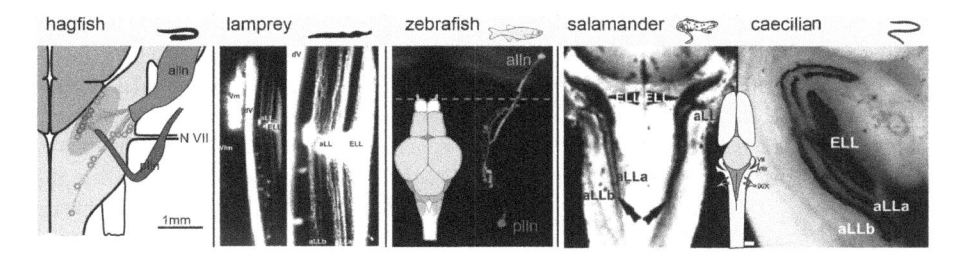

FIGURE 8.4 Lateral line projections for hagfish, lampreys, zebrafish, salamander, and caecilians. A discrete central projection forms two branches, presenting the two hair cell polarities that overlap in hagfish. Note that in some cases (lamprey, salamander, caecilian) we document independent bundles, likely associated with the hair cell polarity. (Modified from Elliott and Fritzsch, 2020, Valera et al., 2021.)

(Stolfi et al., 2015, Kim et al., 2020). In contrast, innervating neurons in hemichordates (lancelet) run inside the spinal cords and thus do not form ganglia (Fritzsch and Northcutt, 1993, Fritzsch, 1996, Ferran and Puelles, 2019, Baatrup, 1983) and don't express *Atoh* and *Neurogenin* is only found in the caudal part of the spinal cord (Holland, 2020). Both factors thus are essential in the development of the vertebrate lateral line system.

Within the cyclostomes a lateral line is present only in one of the two families of hagfish, the Eptatretida, as well as in lampreys. Hair cells of hagfish neuromasts lack the opposing polarity organization while it is present in lampreys and all other gnathostomes (Figures 8.1C, 8.3A, and 8.4E; Yamada, 1973, Elliott and Fritzsch, 2020). Neuromasts of cyclostomes lack a cupula and have no efferent innervation, both traits characteristic of all other vertebrates with neuromasts (Kozak et al., 2020, Fritzsch and López-Schier, 2014).

The lateral line system has been lost several times during vertebrate evolution, including all amniotes and derived amphibians, of which only adult urodeles may have a lateral line system (Schlosser, 2002b, Fritzsch, 1989) (Figure 8.2B). Interestingly, the hypothesis that the lateral line is completely lost in amniotes no longer stands, as it was demonstrated that mice can generate vestigial lateral lines that are otherwise suppressed by apoptosis during regular development in vertebrates (Washausen and Knabe, 2018, Washausen and Knabe, 2019).

8.3 LATERAL LINE SENSORY NEURONS DEPEND ON MOLECULAR EXPRESSION OF THE LATERAL LINE PLACODE

Lateral line nerve fibers are unmyelinated in cyclostomes and myelinated in gnathostomes (Gelman et al., 2007, Daghfous et al., 2020, Kishida et al., 1987, Braun and Northcutt, 1997). Within the neuromasts the hair cells are innervated by at least two afferents, one per hair cell orientation (Dean and Claas, 2020). These form 'bar' or 'ribbon' synapses with the hair cells (Figures 8.1B, 8.4D, and 8.4E)

(Wahnschaffe et al., 1985, Schwander et al., 2010, Jørgensen, 2005, Maxeiner et al., 2016, Chou et al., 2017).

In general, anterior neurons project in the anterior lateral line nerve (aLLN) that runs along the three major trigeminal branches (ophthalmic, maxillary, and mandibular branches) and enters the CNS at the level of the facial nerve. The posterior lateral line nerve (pLLN) starts from the glossopharyngeal nerve (Wullimann and Grothe, 2013, Northcutt, 1989, Webb, 2021) from where it diverts into several branches. The central projections of aLLN and pLLN are discrete from facial and glossopharyngeal nerve fibers and enter the hindbrain near r4 and r6, respectively. Here they form distinct fascicles that provide a very broad termination field in the central nuclei (Figures 8.3 and 8.5).

An efferent innervation is unique to gnathostomes (Hellmann and Fritzsch, 1996, Sienknecht et al., 2014, Fritzsch and Elliott, 2017a, Roberts and Meredith, 1992), and is absent in cyclostomes (Yamada, 1973, Whitear and Lane, 1983, Braun and Northcutt, 1997). The efferent innervation relies on acetylcholine and innervated hair cells that express $\alpha9$ and $\alpha10$ nicotinic acetylcholine in receptors to innervation for hair cells (Elgoyhen et al., 2001, Pisciottano et al., 2019, Zuo et al., 1999, Carpaneto Freixas et al., 2021).

The development of sensory neurons, nuclei, and the continuous adding of hair cells show a completely different developmental progression compared to most vertebrates (Vetschera et al., 2019, Wahnschaffe et al., 1987, Schlosser, 2002b). This critically depends on basic helix-loop-helix (bHLH) genes (Fritzsch et al., 2010). Historically, the first insights into early bHLH genes were from the mammalian ear development. We here assume the following bHLH genes to all other neurosensory and rhombomere nuclei in vertebrates (Fritzsch and Elliott, 2017b, Glover et al., 2018, Riddiford and Schlosser, 2016, Baker and Modrell, 2018, Baker, 2019, Sugahara et al., 2016, Wullimann and Grothe, 2013):

1. *Sox2* expression provides the precursors to initiate differentiation (Kageyama et al., 2018, Dvorakova et al., 2020).
2. *Neurog1* has a role in sensory neuron differentiation, likely including all lateral line afferents (Ma et al., 1998, Millimaki et al., 2010).
3. *Neurod1* acts downstream of *Neurog1* and is needed by neurons and hair cells (Riddiford and Schlosser, 2016, Filova et al., 2020).
4. *Atoh1* is essential for lateral line hair cells and lateral line nuclei (Bermingham et al., 1999, Wang et al., 2005, Mishima et al., 2009, Chizhikov et al., 2021, Wullimann and Grothe, 2013, Millimaki et al., 2007).

We distinguish three interrelated issues of sensory formation, 'hair cell' development, and hindbrain alar plates, consistent with the molecular biology of different vertebrates (Vetschera et al., 2019, Undurraga et al., 2019, Riddiford and Schlosser, 2016). We will describe lateral line neurons followed by 'hair cells' (lateral line 'neuromasts' of mechanosensory hair cells), and by the loss and gain of lateral line nuclei (intermediate nuclei).

The lateral-line placodal development of larval lamprey has been detailed elsewhere (Modrell et al., 2014). Evidence suggests an early development of lateral line innervation (Whitear and Lane, 1981, Yamada, 1973). Information for chondrichthyes and teleosts is limited to sturgeons, knife fish, and zebrafish. Interesting data have been obtained in the developing knife fish that suggests sensory neurons may induce the formation of the mechanosensory cells at the periphery (Vischer et al., 1989, Vischer, 1989). Ghysen and Alexandre (Alexandre and Ghysen, 1999) provided the first detailed description of posterior lateral line projections for zebrafish, a work that was followed by others (Lozano-Ortega et al., 2018, Vetschera et al., 2019, Undurraga et al., 2019, López-Schier and Pujol-Martí, 2013, Zecca et al., 2015). The organization of the central projection of lateral line sensory afferents recapitulates the temporal order of the differentiation of sensory ganglia, starting with the trigeminal ganglion, followed by the inner ear and lateral line ganglia (Zecca et al., 2015). As a consequence of this temporal succession, central projections of the trigeminal system terminate more medially than inner ear or latera line projections (Zecca et al., 2015) in parallel to the sequence of lateral line development (Vischer, 1989, Fritzsch et al., 2005).

During development the lateral line afferents of amphibians follow the path of hair cell precursors to both aLL and pLL (Northcutt, 1992, Harrison, 1903, Harrison, 1910, Baker, 2019, Winklbauer and Hausen, 1983). Likewise, the pLLN afferents of teleost fish elongate as the neuromasts but as well understood for the aLL (Ghysen and Dambly-Chaudière, 2007, Schuster and Ghysen, 2013, Vetschera et al., 2019, López-Schier et al., 2004).

8.4 MOLECULAR BASE OF HAIR CELLS AND THEIR POLARITIES

As introduced earlier, lateral line hair cells are ubiquitous in craniates. Here we review hair cell morphology, their innervation and the factors controlling their development in vertebrates. For cyclostomes as the least derived vertebrates data on the organization of hair cells and supporting cells is only available for lampreys, which share the general vertebrate *Bauplan* (Yamada, 1973). However, further work is needed to describe hair cells of the evolutionary earlier branch of hagfish (Fernholm et al., 2013).

The conserved organization of the lateral line hair cells is as follows: The functional unit of the lateral line system is the neuromasts, which physically couples hair cells to the surrounding medium (Bleckmann, 2020). Within a neuromast the hair cells are organized in two opposing polarities that are either randomly distributed within a neuromast (Xenopus), or occur in a regularized counter-organization (Figure 8.1C) (Fritzsch and López-Schier, 2014, Wahnschaffe et al., 1985, Flock and Jørgensen, 1974, Yamada and Hama, 1972, Yamada, 1973, Hama and Yamada, 1977, Ghysen and Dambly-Chaudière, 2007). The transduction from the mechanical stimulus into an electrical signal (mechano-electrical transduction, MET) requires an eccentric kinocilium and shorter stereocilia (Figures 8.1 and 8.2) (Nicolson, 2017, Schwander et al., 2010, Qiu and Muller, 2018, Lu and Sipe, 2016, Maeda et al., 2014, Krey et al., 2020, Chou et al., 2017). Similar to the situation in mammals (Qiu and Muller, 2018), these stereocilia are connected through threads of *Cdh23* and *Pcdh15* (Nicolson, 2017). Additional proteins are required for normal function, including *TMC1/2*, *TMIE*, *Clarin*, *Myo7* and *Myo6*. TMC1 and TMC2 are the likely candidates for the mechanotransduction channel (Pacentine and Nicolson, 2019, Pan et al., 2018, Shibata et al., 2016, Krey et al., 2020, Chou et al., 2017, Fritzsch et al., 2020). The absence of TMC1, TMC2, or TMIE disrupts stereocilia development (Krey et al., 2020).

Deviating from the above, only one TMC1/2 gene is known from cyclostomes, potentially explaining the distinct differences of lamprey mechanosensors already introduced above, which also include fewer and very short stereocilia (Yamada, 1973, Chagnaud et al., 2017). Combined, these features set aside lamprey and hagfish from 'typical' mechanosensory lateral line hair cells (Nicolson, 2017, Maeda et al., 2014, Krey et al., 2020, Kozak et al., 2020, Ji et al., 2018).

While the kinocilium and stereocilia are the common elements of craniate hair cells, they differ across species. Gnathostomes typically have 50–60 stereocilia (Wahnschaffe et al., 1985, Yamada and Hama, 1972), thus more than lampreys (~18–20 very short stereocilia (Yamada, 1973) (Figure. 8.4D and E). Interestingly enough, chondrichthyes have about 30–40 short stereocilia (Hama and Yamada, 1977, Roberts and Ryan, 1971) with shorter kinocilia (Figure 8.2) than those reported in teleost fish (Ghysen and Dambly-Chaudière, 2007, Webb, 2013, Puzdrowski, 1989).

Most gnathostomes share a distinct division of neuromasts recessed in groves or canals, and those found superficially (Figure 8.4A and C) (Webb, 2013, Northcutt, 1989). This results in different sensitivities of the sensory organs, with superficial neuromasts responding in proportion to the water velocity, while the neuromasts in canals respond to the acceleration (Denton and Gray, 1988). Recessed systems are unknown in lampreys and amphibians (Yamada, 1973, Braun and Northcutt, 1997, Wahnschaffe et al., 1985).

As mentioned above, hair cells occur in opposing orientations within neuromasts. Electrophysiological work in some species have each hair cells of similar orientation that are innervated by a single afferent (Figure 8.4B) (Münz et al., 1984, Vetschera et al., 2019) and it has been hypothesized that each hair cell of a given polarity is presented in a single fascicle (Fritzsch, 1989, Lozano-Ortega et al., 2018). This has been confirmed in salamanders where afferents contributes to only one of the two central fascicles (Figure 8.3) such that hair cells of opposing polarity are kept separate in the afferent stream of the peripheral nerves (Fritzsch, 1989).

It seems possible that the neurons giving rise to the two afferents, and possibly also the two opposing hair cell populations, are separated by different birthdates in teleosts (Zecca et al., 2015, Vischer, 1989). In zebrafish it was further shown that while early born afferent neurons connect hair cells to the Mauthner cell, those occurring later only project to the central nucleus (Pujol-Martí et al., 2012).

The opposing polarity of the hair cells and their selective innervation by afferent nerves is determined through the combined action of a transcription factor *Emx2* and the Notch signaling pathway (Jacobo et al., 2019, Ji et al., 2018, Jiang et al., 2017, Kozak et al., 2020, Ohta et al., 2020). Ectopic expression of *Emx2* drives all hair cells to have their kinocilia in a caudal position, while broadly activating the *Notch* pathway results in the inhibition of *Emx2* expression and thus all kinocilia are positioned rostrally (Kozak et al., 2020, Ohta et al., 2020, Chitnis, 2020). In normal development, the division of a non-sensory progenitor cell results in initially unpolarized pairs of hair cells. While details of the regulatory network remain to be revealed it was found that the *Notch* ligands expressed in the rostral cell activate *Notch* in its caudal sibling, where the expression of *Emx2* is inhibited (Jacobo et al., 2019). It appears that a bi-stable situation then determines an upregulation of *Emx2* in the rostral sibling through *Notch*-mediated lateral inhibition, that then determines the caudal position for the kinocilium of the rostral sibling and the emergence of the opposing polarity (Elliott and Fritzsch, 2020).

In summary, craniates have distinct mechanosensory hair cells that depend on *Atoh1* and other factors (Chitnis, 1995, Chitnis et al., 2012, Matsuda and Chitnis, 2010) that receive lateral line innervation by at least two afferent fibers per neuromast, with opposing hair cell orientations that is developmentally determined by *Notch1* and *Emx2*. With some variation the placement of the stereocilia and the asymmetry of the kinocilium are conserved. However, the organization of the mechano-electrical transduction apparatus by a single *TMC* in hagfish and lampreys clearly sets cyclostomes apart from the three *TMCs* in gnathostomes.

8.5 CENTRAL LATERAL LINE PROJECTIONS

The central projection of lateral line afferents is identical in hagfish (Kishida et al., 1987), lampreys (Koyama et al., 1990, Ronan and Northcutt, 1987, Fritzsch et al., 1984) and most aquatic gnathostomes (McCormick, 1999, McCormick, 1989), except some amphibians (Fritzsch, 1989, Fritzsch and Neary, 1998, Schlosser, 2002b) and all amniotes (Baker, 2019, Wullimann and Grothe, 2013). The nuclei in the alar plate receiving lateral line fibers are known as the intermediate nuclei (IN; [Nieuwenhuys et al., 1998]) or the medial octavolateralis nuclei (MON; [McCormick, 1989]). A separate projection to the dorsal descending auditory nucleus has recently been described (McCormick et al., 2016). This provides a late justification for the term octavolateral system as it shows an interaction between the acoustic and lateralis system.

Where investigated in fish and frogs, lateral line projections show a unique feature of interdigitating aLLN and pLLN fibers in the IN (e.g. zebrafish in Figure 8.3), with specific cases where these fibers were found to remain segregated (Fritzsch et al., 1984, Fritzsch and Elliott, 2017b). A distinct projection of aLLN and pLLN fibers is found in salamanders and caecilians (Fritzsch, 1989), whereas aLLN an pLLN fibers mix in Xenopus (Will, 1989, Dean and Claas, 2020).

Difference in the range of projections has been reported, but no distinct pattern has emerged thus far. For example the pLLN fibers of lampreys cross into r1 (Ronan and Northcutt, 1987), whereas projections beyond r2 are only reported for zebrafish and goldfish in gnathostomes (Puzdrowski, 1989). This distinction of more rostral projection of the pLLN is reminiscent of the pattern in the vestibular system where posterior fibers projecting somewhat more rostral and anterior vestibular fibers predominantly project more caudally (Fritzsch, 1998).

In summary, the IN are the common first order central nuclei of lateral line input and the segregation of afferents within this nucleus varies across species.

8.6 INTERMEDIATE LATERAL LINE NUCLEI DEVELOP FROM THE RHOMBENCEPHALON

Since His (His, 1890) described the development of the rhombencephalon of mammals, novel ideas concerning the roles of central nuclei have been posited (Herrick, 1948). How are these mechanosensory systems organized to connect hair cells to central nuclei via sensory afferent neurons (Glover et al., 2018)? We need molecular information to generate discrete and overlapping projections to provide a common theme of distinct losses and gains of genes across vertebrates (Fritzsch and Elliott, 2017b). We will start with the molecular background of defining the various genes that could play a role for alar plate nuclei formation, including dorsal relevant genes such as *Lmx1a* and *Lmx1b*, *Wnts*, *Sox2*, *Atoh1*, *Ptf1a*, and micro RNAs, among others.

Specifically, early expression of genes, such as *Atoh1*, develop from dorsal alar plate nuclei, such as in lampreys, bony fish, and mammals (Sugahara et al., 2016, Fritzsch et al., 2006, Millimaki et al., 2007). *Atoh1* is expressed in all intermediate nuclei, playing a similarly important role as in the cochlear nuclei of mice (Bermingham et al., 2001, Wang et al., 2005, Chizhikov et al., 2021). Other genes are dependent on *Sox2* and *Sox3* (Undurraga et al., 2019) and interact with *Lmx1a/b*, which is needed for *Atoh1* expression (Mishima et al., 2009), and all may be involved with the earliest lamprey projections (Sugahara et al., 2016).

Recent work in mice showed that deletion of both *Lmx1a* and *Lmx1b* causes complete loss of *Atoh1* and certain *Wnts* including *Wnt1* and *Wnt3a* (Mishima et al., 2009, Iskusnykh et al., 2016). In *Lmx1a/b* null mice, there is a single dorsal expression of *Wnt3a* instead of two parallel Wnt expressions between the choroid plexus (Chizhikov et al., 2021). The choroid plexus is highly positive for *Lmx1a* and *Lmx1b* (Glover et al., 2018) and loss of *Lmx1a/b* abolishes *Atoh1* in the dorsal hindbrain of mice (Mishima et al., 2009). The loss of auditory nuclei in *Lmx1a/b* null mice is likely due to the loss of the downstream gene, *Atoh1* (Mishima et al., 2009), which is comparable to the loss of nuclei described in *Atoh1* mutants (Wang et al., 2005, Fritzsch et al., 2006). In the absence of *Atoh1*, auditory afferents project nearly identical to that of wild type mice (Elliott et al., 2017, Filova et al., 2020). However, in *Lmx1a/b* null mice, central projections are aberrant. Due to the fusion of the dorsal roof plates in *Lmx1a/b* null mice, fibers of the inner ear project bilaterally at the roof plate, showing incomplete segregation of inner ear fibers as they intertwine (Chizhikov et al., 2020, Elliott et al., 2021). With this clear outline we can compare the early deletions across craniates, including other genes such as *Ptf1a* (Iskusnykh et al., 2016).

In lamprey and hagfish, some relevant genes are expressed, including *Atoh1* and *Ptf1a* (Sugahara et al., 2016, Higuchi et al., 2019). As described by others in the development of mice (Iskusnykh et al., 2016), *Atoh1* is expressed dorsally along the spinal cord and hindbrain to reach the cerebellum in mice (Bermingham et al., 2001). Given the similar gene expression in auditory nuclei in mammals that replaces the equivalent nuclei of lampreys, we can assume that loss of *Atoh1* expression will eliminate lateral line nuclei in lamprey. However, lampreys and hagfish have a single *Lmx1a/b* gene, as compared to separate *Lmx1a* and *Lmx1b* in all gnathostomes (Glover et al., 2018, Mishima et al., 2009, Chizhikov et al., 2021). Following the data generated by *Lmx1a/b* double null mice suggests that neither lamprey nor hagfish would develop lateral line nuclei following loss of the *Lmx1a/b* gene. Whether *Lmx1a* and *Lmx1b* play a role in bony fish has not yet been analyzed. There is incomplete data for any additional *Wnts* that may be needed in hagfish and lampreys (Erives and Fritzsch, 2020).

Among bony fish, only zebrafish data have provided some clues of slightly different genes in development. First, early gene duplication evolved into an *Atoh1a* and an *Atoh1b* that switches from early and late expression in the ear and brain (Millimaki et al., 2007, Millimaki et al., 2010). In addition, the ubiquitous early gene expression of *Eya1* and *Sox2*, necessary for early proneural bHLH gene expression (Kageyama et al., 2018, Xu et al., 2021), show interesting interactions of *Sox2* and *Sox3* and with downstream genes of *Atoh1*, *Neurod1*, and *Neurog1* (Gou et al., 2018, Undurraga et al., 2019). Expression of *Lmx1a/b* can relate to the loss or gain of mechanoreception of different kinds of bony fish, hagfish, lampreys, and amphibians. *Lmx1a/b* and downstream *Atoh1* and *Wnt3a* could explain the alar plate formation.

Preliminary data for salamanders and frogs show little expression of *Atoh1* and *Sox2* (Riddiford and Schlosser, 2016). The central projections were not analyzed in later developmental stages beyond earliest tracing in salamanders (Fritzsch et al., 2005) that has not fully described in the lateral line of frogs (Gordy et al., 2018). Different genes that could be eliminated and/or miss-expressed are needed to fully understand the molecular basis for brainstem development of amphibians, including salamanders (Schlosser, 2002b, Schlosser, 2002a, Fritzsch and Neary, 1998).

Dye tracings of various central projections of Axolotl showed a temporal progression starting with trigeminal projections that are followed by the inner ear, lateral line and electroreception projections (Fritzsch et al., 2005). Polarity among mammals in the vestibular system (Maklad et al., 2010, Chagnaud et al., 2017) is known for polarity projection in lateral line projections (Fritzsch and López-Schier, 2014). Albeit limited, the central development of lateral line projections of anurans suggests a similar temporal sequence or delay takes place in anurans and caecilians (Fritzsch et al., 1985).

Salamanders, caecilians and anurans have various degeneration of different lateral lines (Wahnschaffe et al., 1987, Fritzsch, 1989, Fritzsch and Neary, 1998, Schlosser, 2002b, Schlosser, 2002a). Lateral line remains during aestivation, only to emerge again to swim in water and engage in reproduction (Schlosser, 2002b, Fritzsch, 1990). Transplantation of ears to the trunk of *Xenopus laevis* embryos demonstrated unique interactions between the ear and the lateral line afferents (Gordy et al., 2018). Lateral line afferents were observed navigating along inner ear afferents to project into the inner ear and randomly project to the anterior or posterior expansion along lateral line fibers. These data suggest that the segregated projections of afferents develop with a delay to avoid cross-innervation of different sensory systems.

In summary, we are beginning a novel insight driven by newly partially characterized genes that are essential for the development of the dorsal part of the hindbrain, including the alar plate and the choroid plexus formation. The common theme might be that these genes are driving development of the dorsal parts of the alar plate, including intermediate nucleus and auditory nuclei.

8.7 LATERAL LINE REACHES THE TORUS

Second-order projections have not yet been studied in lampreys and hagfish (Ronan and Northcutt, 1987, Braun and Northcutt, 1997). In gnathostomes the projections from the IN predominantly reach the contralateral torus semicircularis and the preeminential nucleus (Figure 8.5A) and readers are referred to recent reviews for more details (Wullimann and Grothe, 2013, Will, 1989, Chagnaud et al., 2017). Parallel fiber feedback through the eminentia granularis (EGp) of the cerebellum acts as an adaptive filter at the IN that assists in the cancellation of reafferent hydrodynamic noise (Montgomery et al., 2012, Perks et al., 2020). The torus semicircularis provides input to the tectum opticum and the preglomerular nuclei, from where it reaches the telencephalon. In some gnathostomes, second-order projections also reach caudal areas to the spinal cord in *Xenopus* (Will, 1989).

With respect to encoding the hydrodynamic environment hair cells and their primary afferents can, due to the hair cell's directional sensitivity, represent the local flow amplitude and direction. Furthermore, the separation in a superficial and a canal system results in two complementary channels of information with different filter characteristics. However, to which degree flow directionality, topography and velocity or acceleration channels are kept separate within the IN is unclear. As presented earlier, the projection of the aLL and pLL within the IN is well studied mainly in zebrafish (Chagnaud et al., 2017, Ghysen and Dambly-Chaudière, 2007, Alexandre and Ghysen, 1999, Metcalfe et al., 1985). Here, afferents bifurcate

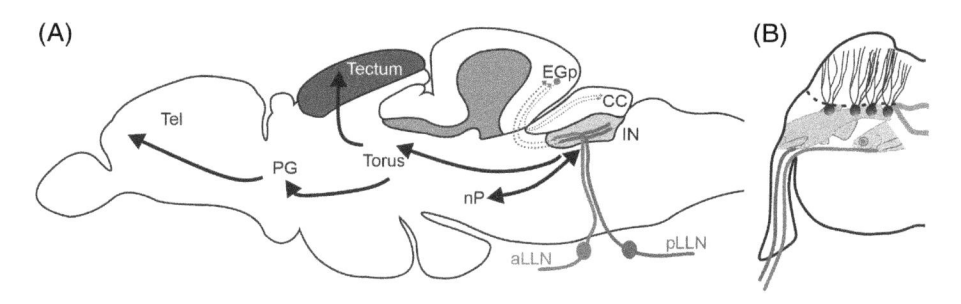

FIGURE 8.5 Schematic of the ascending central lateral line pathway of a fish. (A) Sagittal section. The anterior (light grey) and posterior (black) afferents form discrete central projections along the IN from where ascending information reaches various brain areas, including a cerebellar feedback loop. (B) Cross section through the IN, showing that fibers from the anterior lateral line nerve (aLLN) run medio ventral to the fibers of the pLLN. Presumably, this results in a separation of terminals from the trunk and the head as schematically indicated by the fish silhouettes. CC, cerebellar crest; EG, eminentia granularis; PG, preglomerula nucleus; nP, nucleus preeminetials.

along the anterior-posterior length of the IN where aLL afferents course more medially and more ventrally than those of the pLLN (Figure 8.5B). However, synaptic terminals of aLL and pLL fibers appear to largely overlap showing that the projections of the afferents maintain topography, this appears to be lost in the actual synaptic terminations in the IN (Valera et al., 2021). Contrary to this anatomical finding, physiological data suggest that a topographic representation of lateral line input exists at the level of the torus semicircularis, suggesting the presence of a weak, albeit yet anatomically unresolved topography of afferent termination within the IN.

For anurans a mapping between lateral line and visual input has been suggested for the tectum opticum (Dean and Claas, 2020) (Figure 8.6): Tectal lateral line units form a topographic map of surface wave direction in register with an overlying visual map. Deficits following tectal lesions are consistent with this map, but not all investigators agree on its organization and location. At present comparable data is lacking in teleosts.

The mechanosensory lateral line of gnathostomes also maintains a distinct lemniscal pathway. Target areas are at present not well studied, despite a series of physiological studies in different species (Bleckmann, 2020). To which extent lateral line input contributes to spatial learning, for which the lateral pallium of fish plays a crucial role (Fuss et al., 2014) remains to be investigated.

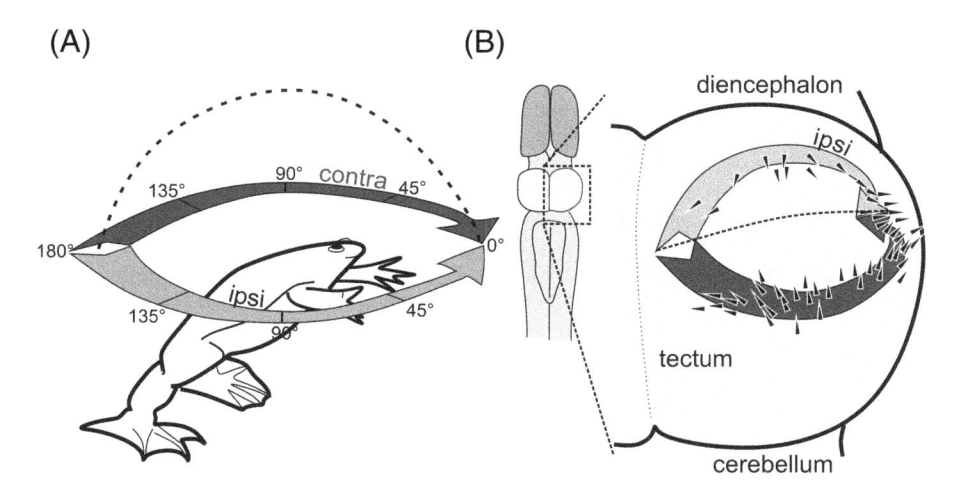

FIGURE 8.6 Tectal map of best directions for responses to surface waves. (A) Definition of azimuthal directions on the water surface surrounding a frog in a natural resting posture. The stippled line indicates the meridian that separates the ipsi- and contralateral sides. (B) Best directions and locations of tectal units (arrowheads). Thick colored arrows indicate azimuthal directions on the water surface and in the tectum, where 0° is toward the rostrolateral margin. Best directions are in register with the visual map of the hemispheric, largely dorsal visual field of the contralateral eye. No lateral line responses were recorded in areas, where visual receptive fields presumably corresponded to elevations above and below the water surface. (Modified from Dean and Claas, 2020.)

8.8 SUMMARY AND CONCLUSION

The lateral line system is common in vertebrates but is absent in certain amphibians and all amniotes. The fiber innervation of the hair cell system is provided by two or more fibers that can be subdivided into an anterior and a posterior lateral line nerve. Separate afferents innervate hair cells of opposing polarity and polarities are segregated with respect to their central projections in some vertebrates. The detailed organization of neuromasts and their innervation is driven by genetic interaction that allows the topographical polarity to be studied. Mechanotransduction requires TMC to provide connections between stereocilia that open ion channels upon proper mechanical stimulation. Central organization follows the lateral line afferents and provides the information processing for direction of sensory information. Secondary and tertiary projections reach the torus semicircularis and the tectum opticum as well as the preglomerular nuclei which in turn provide input to the pallium. These projections are currently unclear in cyclostomes.

REFERENCES

Alexandre, D. & Ghysen, A. 1999. Somatotopy of the lateral line projection in larval zebrafish. Proceedings of the National Academy of Sciences, 96, 7558–7562.

Baatrup, E. 1983. Ciliated receptors in the pharyngeal terminal buds of larval *Lampetra planeri* (Bloch)(Cyclostomata). *Acta Zoologica*, 64, 67–75.

Baker, C. V. & Modrell, M. S. 2018. Insights into electroreceptor development and evolution from molecular comparisons with hair cells. *Integrative and Comparative Biology*, 58, 329–340.

Baker, C. V. 2019. The Development and Evolution of Lateral Line Electroreceptors: Insights from Comparative Molecular Approaches. *Electroreception: Fundamental Insights from Comparative Approaches* (pp. 25–62). Cham: Springer.

Bermingham, N. A., Hassan, B. A., Price, S. D., Vollrath, M. A., Ben-Arie, N., Eatock, R. A., Bellen, H. J., Lysakowski, A. & Zoghbi, H. Y. 1999. Math1: An essential gene for the generation of inner ear hair cells. *Science*, 284, 1837–1841.

Bermingham, N. A., Hassan, B. A., Wang, V. Y., Fernandez, M., Banfi, S., Bellen, H. J., Fritzsch, B. & Zoghbi, H. Y. 2001. Proprioceptor pathway development is dependent on Math1. *Neuron*, 30, 411–422.

Bleckmann, H. 2020. Central lateral line pathways and central integration of lateral line information. *In:* Fritzsch, B. (ed.) *The Senses* (pp. 163–184). Elsevier, Academic Press.

Bleckmann, H., Mogdans, J. & Coombs, S. L. 2014. *Flow Sensing in Air and Water*. Springer, Berlin, Germany, 976.

Braun, C. B. & Northcutt, R. G. 1997. The lateral line system of hagfishes (Craniata: Myxinoidea). *Acta Zoologica*, 78, 247–268.

Brunet, T. & King, N. 2017. The origin of animal multicellularity and cell differentiation. *Developmental Cell* 43, 124–140.

Carpaneto Freixas, A. E., Moglie, M. J., Castagnola, T., Salatino, L., Domene, S., Marcovich, I., Gallino, S., Wedemeyer, C., Goutman, J. D., Plazas, P. V. & Elgoyhen, A. B. 2021. Unraveling the Molecular Players at the Cholinergic Efferent Synapse of the Zebrafish Lateral Line. *The Journal of Neuroscience*, 41, 47–60.

Chagnaud, B. P. & Coombs, S. 2014. Information Encoding and Processing by the Peripheral Lateral Line System. *In:* Coombs, S., Bleckmann, H., Fay, R. R. & Popper, A. N. (eds.) *The Lateral Line System*. New York, NY: Springer New York.

Chagnaud, B. P., Engelmann, J., Fritzsch, B., Glover, J. C. & Straka, H. 2017. Sensing external and self-motion with hair cells: a comparison of the lateral line and vestibular systems from a developmental and evolutionary perspective. *Brain, Behavior and Evolution*, 90, 98–116.

Chitnis, A. B. 1995. The role of Notch in lateral inhibition and cell fate specification. *Molecular and Cellular Neuroscience* 6, 311–321.

Chitnis, A. B. 2020. Development of the zebrafish posterior lateral line system. *In:* Fritzsch, B. (ed.) *The Senses* (pp. 66–84). Elsevier, Academic Press.

Chitnis, A. B., Nogare, D. D. & Matsuda, M. 2012. Building the posterior lateral line system in zebrafish. *Developmental Neurobiology* 72, 234–55.

Chizhikov, V. V., Iskusnykh, I. Y., Fattakhov, N. & Fritzsch, B. 2021. Lmx1a and Lmx1b are Redundantly Required for the Development of Multiple Components of the Mammalian Auditory System. *Neuroscience*, 452, 247–264.

Chou, S.-W., Chen, Z., Zhu, S., Davis, R. W., Hu, J., Liu, L., Fernando, C. A., Kindig, K., Brown, W. C. & Stepanyan, R. 2017. A molecular basis for water motion detection by the mechanosensory lateral line of zebrafish. *Nature Communications*, 8, 2234.

Daghfous, G., Auclair, F., Blumenthal, F., Suntres, T., Lamarre-Bourret, J., Mansouri, M., Zielinski, B. & Dubuc, R. 2020. Sensory cutaneous papillae in the sea lamprey (Petromyzon marinus L.): I. Neuroanatomy and physiology. *Journal of Comparative Neurology*, 528, 664–686.

Dean, J. & Claas, B. 2020. Hydrodynamic sensing by the African clawed frog, *Xenopus laevis*. *In:* Fritzsch, B. (ed.) *The Senses* (pp. 185–214). Elsevier, Academic Press.

Denton, E. J. & Gray, J. A. 1988. Mechanical factors in the excitation of the lateral lines of fishes. *Sensory Biology of Aquatic Animals*. Springer.

Dvorakova, M., Macova, I., Bohuslavova, R., Anderova, M., Fritzsch, B. & Pavlinkova, G. 2020. Early ear neuronal development, but not olfactory or lens development, can proceed without SOX2. *Developmental Biology*, 457, 43–56.

Elgoyhen, A. B., Vetter, D. E., Katz, E., Rothlin, C. V., Heinemann, S. F. & Boulter, J. 2001. α10: A determinant of nicotinic cholinergic receptor function in mammalian vestibular and cochlear mechanosensory hair cells. Proceedings of the National Academy of Sciences, 98, 3501–3506.

Elliott, K. L. & Fritzsch, B. 2020. Evolution and development of lateral line and electroreception: an integrated perception of neurons, hair cells and brainstem nuclei. *In:* Fritzsch, B. (ed.) *The Senses* (pp. 95–115). Elsevier, Academic Press.

Elliott, K. L., Kersigo, J., Pan, N., Jahan, I. & Fritzsch, B. 2017. Spiral ganglion neuron projection development to the hindbrain in mice lacking peripheral and/or central target differentiation. *Frontiers in Neural Circuits, p. 25*, 11.

Elliott, K. L., Pavlinkova, G., Chzhikov, V. V., Yamoah, E. N. & Fritzsch, B. 2021. Development in the mammalian auditory system dependent on transcription factors. *International Journal of Developmental Neuroscience*, 22, 3189.

Erives, A. & Fritzsch, B. 2020. A screen for gene paralogies delineating evolutionary branching order of early Metazoa. *G3: Genes, Genomes, Genetics*, 10, 811–826.

Fernholm, B., Norén, M., Kullander, S. O., Quattrini, A. M., Zintzen, V., Roberts, C. D., Mok, H. K. & Kuo, C. H. 2013. Hagfish phylogeny and taxonomy, with description of the new genus Rubicundus (Craniata, Myxinidae). *Journal of Zoological Systematics and Evolutionary Research*, 51, 296–307.

Ferran, J. L. & Puelles, L. 2019. Lessons from amphioxus bauplan about origin of cranial nerves of vertebrates that innervates extrinsic eye muscles. *The Anatomical Record*, 302, 452–462.

Filova, I., Dvorakova, M., Bohuslavova, R., Pavlinek, A., Vochyanova, S., Fritzsch, B. & Pavlinkova, G. 2020. Combined Atoh1 and Neurod1 deletion reveals autonomous growth of auditory nerve fibers. Molecular *Neurobiology*, 57(12), 5307–5323.

Flock, Å. & Jørgensen, J. M. 1974. The ultrastructure of lateral line sense organs in the juvenile salamander Ambystoma mexicanum. *Cell and Tissue Research*, 152, 283–292.

Fritzsch, B. & Elliott, K. L. 2017a. Evolution and development of the inner ear efferent system: transforming a motor neuron population to connect to the most unusual motor protein via ancient nicotinic receptors. *Frontiers in Cellular Neuroscience*, 11, 114.

Fritzsch, B. & Elliott, K. L. 2017b. Gene, cell, and organ multiplication drives inner ear evolution. *Developmental Biology*, 431, 3–15.

Fritzsch, B. & López-Schier, H. 2014. Evolution of polarized hair cells in aquatic vertebrates and their connection to directionally sensitive neurons. *Flow Sensing in Air and Water* (pp. 271–294). Springer, Berlin, Heidelberg.

Fritzsch, B. & Neary, T. 1998. The octavolateralis system of mechanosensory and electrosensory organs. *Amphibian Biology*, 3, 878–922.

Fritzsch, B. & Northcutt, R. G. 1993. Cranial and spinal nerve organization in amphioxus and lampreys: evidence for an ancestral craniate pattern. *Cells Tissues Organs*, 148, 96–109.

Fritzsch, B. & Straka, H. 2014. Evolution of vertebrate mechanosensory hair cells and inner ears: toward identifying stimuli that select mutation driven altered morphologies. *Journal of Comparative Physiology A*, 200, 5–18.

Fritzsch, B. 1989. Diversity and regression in the amphibian lateral line and electrosensory system. *The Mechanosensory Lateral Line*. Springer.

Fritzsch, B. 1990. The evolution of metamorphosis in amphibians. *Journal of Neurobiology*, 21, 1011–1021.

Fritzsch, B. 1996. Similarities and differences in lancelet and craniate nervous systems. *Israel Journal of Zoology*, 42, S147–S160.

Fritzsch, B. 1998. Evolution of the vestibulo-ocular system. *Otolaryngology—Head and Neck Surgery*, 119, 182–192.

Fritzsch, B., Beisel, K. W., Pauley, S. & Soukup, G. 2007. Molecular evolution of the vertebrate mechanosensory cell and ear. *The International Journal of Developmental Biology*, 51, 663.

Fritzsch, B., De Caprona, M.-D. C., Wächtler, K. & Körtje, K.-H. 1984. Neuroanatomical evidence for electroreception in lampreys. *Zeitschrift für Naturforschung C*, 39, 856–858.

Fritzsch, B., Eberl, D. F. & Beisel, K. W. 2010. The role of bHLH genes in ear development and evolution: revisiting a 10-year-old hypothesis. *Cellular and Molecular Life Sciences*, 67, 3089–99.

Fritzsch, B., Erives, A., Eberl, D. F. & Yamoah, E. N. 2021. Genetics of mechanoreceptor evolution and development. *In:* Fritzsch, B. (ed.) *The Senses*. New York: Elsevier.

Fritzsch, B., Gregory, D. & Rosa-Molinar, E. 2005. The development of the hindbrain afferent projections in the axolotl: evidence for timing as a specific mechanism of afferent fiber sorting. *Zoology*, 108, 297–306.

Fritzsch, B., Pauley, S., Feng, F., Matei, V. & Nichols, D. 2006. The molecular and developmental basis of the evolution of the vertebrate auditory system. *International Journal of Comparative Psychology*, 19, 1–25.

Fritzsch, B., Wahnschaffe, U., Caprona, M.-D. C. & Himstedt, W. 1985. Anatomical evidence for electroreception in larval *Ichthyophis kohtaoensis*. *Naturwissenschaften*, 72, 102–104.

Fuss, T., Bleckmann, H. & Schluessel, V. 2014. Place learning prior to and after telencephalon ablation in bamboo and coral cat sharks (*Chiloscyllium griseum* and *Atelomycterus marmoratus*). *Journal of Comparative Physiology A*, 200, 37–52.

Gasparini, F., Degasperi, V., Shimeld, S. M., Burighel, P. & Manni, L. 2013. Evolutionary conservation of the placodal transcriptional network during sexual and asexual development in chordates. *Developmental Dynamics*, 242, 752–66.

Gelman, S., Ayali, A., Tytell, E. & Cohen, A. 2007. Larval lampreys possess a functional lateral line system. *Journal of Comparative Physiology A*, 193, 271–277.

Ghysen, A. & Dambly-Chaudière, C. 2007. The lateral line microcosmos. *Genes & Development*, 21, 2118–2130.

Glover, J. C., Elliott, K. L., Erives, A., Chizhikov, V. V. & Fritzsch, B. 2018. Wilhelm His' lasting insights into hindbrain and cranial ganglia development and evolution. *Developmental Biology*.

Gordy, C., Straka, H., Houston, D. W., Fritzsch, B. & Elliott, K. L. 2018. Transplantation of ears provides insights into inner ear afferent pathfinding properties. *Developmental Neurobiology*, 78, 1064–1080.

Gou, Y., Vemaraju, S., Sweet, E. M., Kwon, H.-J. & Riley, B. B. 2018. sox2 and sox3 play unique roles in development of hair cells and neurons in the zebrafish inner ear. *Developmental Biology*, 435, 73–83.

Hama, K. & Yamada, Y. 1977. Fine structure of the ordinary lateral line organ. *Cell and Tissue Research*, 176, 23–36.

Harrison, R. G. 1903. Experimentelle Untersuchungen Über die Entwicklung der Sinnesorgane der Seitenlinie bei den Ampkibien. *Archiv für mikroskopische Anatomie*, 63, 35–149.

Harrison, R. G. 1910. The outgrowth of the nerve fiber as a mode of protoplasmic movement. *Journal of Experimental Zoology*, 9, 787–846.

Hellmann, B. & Fritzsch, B. 1996. Neuroanatomical and histochemical evidence for the presence of common lateral line and inner ear efferents and of efferents to the basilar papilla in a frog, Xenopus laevis. *Brain, Behavior and Evolution*, 47, 185–194.

Herrick, C. J. 1948. *The Brain of the Tiger Salamander*. Chicago: University of Chicago Press.

Higuchi, S., Sugahara, F., Pascual-Anaya, J., Takagi, W., Oisi, Y. & Kuratani, S. 2019. Inner ear development in cyclostomes and evolution of the vertebrate semicircular canals. *Nature*, 565, 347–350.

His, W. 1890. Die Entwickelung des menschlichen Rautenhirns vom Ende des ersten bis zum Beginn des dritten Monats. *Abhandlungen der mathematisch-physischen Classe der koeniglich saechsischen Gesellschaft der Wissenschaftern*, 17, 4–74.

Holland, L. 2020. Invertebrate origins of vertebrate nervous systems. *Evolutionary Neuroscience* (pp. 51–73). Elsevier, Academic Press.

Iskusnykh, I. Y., Steshina, E. Y. & Chizhikov, V. V. 2016. Loss of Ptf1a leads to a widespread cell-fate misspecification in the brainstem, affecting the development of somatosensory and viscerosensory nuclei. *The Journal of Neuroscience*, 36, 2691–2710.

Jacobo, A., Dasgupta, A., Erzberger, A., Siletti, K. & Hudspeth, A. J. 2019. Notch-mediated determination of hair-bundle polarity in mechanosensory hair cells of the zebrafish lateral line. *Current Biology*, 29, 3579–3587 e7.

Ji, Y. R., Warrier, S., Jiang, T., Wu, D. K. & Kindt, K. S. 2018. Directional selectivity of afferent neurons in zebrafish neuromasts is regulated by Emx2 in presynaptic hair cells. *Elife*, 7, e35796.

Jiang, T., Kindt, K. & Wu, D. K. 2017. Transcription factor Emx2 controls stereociliary bundle orientation of sensory hair cells. *Elife*, 6, e23661.

Jørgensen, J. M. 2005. Morphology of electroreceptive sensory organs. *Electroreception*. New York, NY: Springer.

Kageyama, R., Shimojo, H. & Ohtsuka, T. 2018. Dynamic control of neural stem cells by bHLH factors. *Neuroscience Research*, 138, 12–18.

Kim, K., Gibboney, S., Razy-Krajka, F., Lowe, E. K., Wang, W. & Stolfi, A. 2020. Regulation of neurogenesis by FGF signaling and neurogenin in the invertebrate chordate ciona. *Frontiers in Cell and Developmental Biology*, 8, 477.

Kishida, R., Goris, R., Nishizawa, H., Koyama, H., Kadota, T. & Amemiya, F. 1987. Primary neurons of the lateral line nerves and their central projections in hagfishes. *Journal of Comparative Neurology*, 264, 303–310.

Koyama, H., Kishida, R., Goris, R. C. & Kusunoki, T. 1990. Organization of the primary projections of the lateral line nerves in the lamprey *Lampetra japonica*. *Journal of Comparative Neurology*, 295, 277–289.

Kozak, E. L., Palit, S., Miranda-Rodriguez, J. R., Janjic, A., Bottcher, A., Lickert, H., Enard, W., Theis, F. J. & López-Schier, H. 2020. Epithelial planar bipolarity emerges from notch-mediated asymmetric inhibition of Emx2. *Current Biology*, 30, 1142–1151 e6.

Krey, J. F., Chatterjee, P., Dumont, R. A., O'sullivan, M., Choi, D., Bird, J. E. & Barr-Gillespie, P. G. 2020. Mechanotransduction-dependent control of stereocilia dimensions and row identity in inner hair cells. *Current Biology*, 30(3), 442–454.

López-Schier, H. & Pujol-Martí, J. 2013. Developmental and architectural principles of the lateral-line neural map. *Frontiers in Neural Circuits*, 7, 47.

López-Schier, H., Starr, C. J., Kappler, J. A., Kollmar, R. & Hudspeth, A. J. 2004. Directional cell migration establishes the axes of planar polarity in the posterior lateral-line organ of the zebrafish. *Dev Cell*, 7, 401–12.

Lozano-Ortega, M., Valera, G., Xiao, Y., Faucherre, A. & López-Schier, H. 2018. Hair cell identity establishes labeled lines of directional mechanosensation. *PLoS Biology*, 16, e2004404.

Lu, X. & Sipe, C. W. 2016. Developmental regulation of planar cell polarity and hair-bundle morphogenesis in auditory hair cells: lessons from human and mouse genetics. *Wiley Interdisciplinary Reviews: Developmental Biology*, 5, 85–101.

Ma, Q., Chen, Z., del Barco Barrantes, I., de la Pompa, J. L. & Anderson, D. J. 1998. neurogenin1 is essential for the determination of neuronal precursors for proximal cranial sensory ganglia. *Neuron*, 20, 469–482.

Maeda, R., Kindt, K. S., Mo, W., Morgan, C. P., Erickson, T., Zhao, H., Clemens-Grisham, R., Barr-Gillespie, P. G. & Nicolson, T. 2014. Tip-link protein protocadherin 15 interacts with transmembrane channel-like proteins TMC1 and TMC2. Proceedings of the National Academy of Sciences, 111, 12907–12912.

Maklad, A., Kamel, S., Wong, E. & Fritzsch, B. 2010. Development and organization of polarity-specific segregation of primary vestibular afferent fibers in mice. *Cell and Tissue Research*, 340, 303–321.

Manley, G. & Ladher, R. 2008. Phylogeny and evolution of ciliated mechanoreceptor cells. *In:* Dallos, P. & Oertel, D. (eds.) *The Senses*. New York: Elsevier.

Manni, L., Anselmi, C., Burighel, P., Martini, M. & Gasparini, F. 2018. Differentiation and induced sensorial alteration of the coronal organ in the asexual life of a tunicate. *Integrative and Comparative Biology*, 58, 317–328.

Matsuda, M. & Chitnis, A. B. 2010. Atoh1a expression must be restricted by Notch signaling for effective morphogenesis of the posterior lateral line primordium in zebrafish. *Development*, 137, 3477–3487.

Maxeiner, S., Luo, F., Tan, A., Schmitz, F. & Südhof, T. C. 2016. How to make a synaptic ribbon: RIBEYE deletion abolishes ribbons in retinal synapses and disrupts neurotransmitter release. *The EMBO Journal*, 35, 1098–1114.

Mccormick, C. A. 1989. Central lateral line mechanosensory pathways in bony fish. *The Mechanosensory Lateral Line* (pp. 341–364). New York, NY: Springer.

Mccormick, C. A. 1999. Anatomy of the central auditory pathways of fish and amphibians. *In:* Fay, R. R. & Popper, A. N. (eds.) *Comparative Hearing: Fish and Amphibians*. New York: Springer-Verlag.

Mccormick, C. A., Gallagher, S., Cantu-Hertzler, E. & Woodrick, S. 2016. Mechanosensory lateral line nerve projections to auditory neurons in the dorsal descending octaval nucleus in the goldfish, *Carassius auratus*. *Brain Behavior Evolution*, 88, 68–80.

Metcalfe, W. K., Kimmel, C. B. & Schabtach, E. 1985. Anatomy of the posterior lateral line system in young larvae of the zebrafish. *Journal of Comparative Neurology*, 233, 377–389.

Millimaki, B. B., Sweet, E. M. & Riley, B. B. 2010. Sox2 is required for maintenance and regeneration, but not initial development, of hair cells in the zebrafish inner ear. *Developmental Biology*, 338, 262–269.

Millimaki, B. B., Sweet, E. M., Dhason, M. S. & Riley, B. B. 2007. Zebrafish atoh1 genes: classic proneural activity in the inner ear and regulation by Fgf and Notch. *Development*, 134, 295–305.

Mishima, Y., Lindgren, A. G., Chizhikov, V. V., Johnson, R. L. & Millen, K. J. 2009. Overlapping function of Lmx1a and Lmx1b in anterior hindbrain roof plate formation and cerebellar growth. *Journal of Neuroscience*, 29, 11377–11384.

Modrell, M. S., Hockman, D., Uy, B., Buckley, D., Sauka-Spengler, T., Bronner, M. E. & Baker, C. V. 2014. A fate-map for cranial sensory ganglia in the sea lamprey. *Developmental Biology*, 385, 405–416.

Montgomery, J. C., Bodznick, D. & Yopak, K. E. 2012. The Cerebellum and Cerebellum-Like Structures of Cartilaginous Fishes. *Brain, Behavior and Evolution*, 80, 152–165.

Münz, H., Claas, B. & Fritzsch, B. 1984. Electroreceptive and mechanoreceptive units in the lateral line of the axolotl Ambystoma mexicanum. *Journal of Comparative Physiology A*, 154, 33–44.

Nicolson, T. 2017. The genetics of hair-cell function in zebrafish. *Journal of Neurogenetics*, 31, 102–112.

Nieuwenhuys, R., Donkelaar, H.J. & Nicholson, C. 1998. *The Central Nervous System of Vertebrates*. Berlin: Springer.

Northcutt, R. G. 1989. The phylogenetic distribution and innervation of craniate mechanoreceptive lateral lines. *The Mechanosensory Lateral Line* (pp. 17–78). New York, NY: Springer.

Northcutt, R. G. 1992. Distribution and innervation of lateral line organs in the axolotl. *Journal of Comparative Neurology*, 325, 95–123.

O'Neill, P., Mak, S.-S., Fritzsch, B., Ladher, R. K. & Baker, C. V. 2012. The amniote paratympanic organ develops from a previously undiscovered sensory placode. *Nature Communications*, 3, 1041.

Ohta, S., Ji, Y. R., Martin, D. & Wu, D. K. 2020. Emx2 regulates hair cell rearrangement but not positional identity within neuromasts. *Elife*, 9, e60432.

Pacentine, I. V. & Nicolson, T. 2019. Subunits of the mechano-electrical transduction channel, Tmc1/2b, require Tmie to localize in zebrafish sensory hair cells. *PLoS Genetics*, 15, e1007635.

Pan, B., Akyuz, N., Liu, X. P., Asai, Y., Nist-Lund, C., Kurima, K., Derfler, B. H., Gyorgy, B., Limapichat, W., Walujkar, S., Wimalasena, L. N., Sotomayor, M., Corey, D. P. & Holt, J. R. 2018. TMC1 forms the pore of mechanosensory transduction channels in vertebrate inner ear hair cells. *Neuron*, 99, 736–753 e6.

Perks, K. E., Krotinger, A. & Bodznick, D. 2020. A cerebellum-like circuit in the lateral line system of fish cancels mechanosensory input associated with its own movements. *The Journal of Experimental Biology*, 223(4), jeb204438.

Pisciottano, F., Cinalli, A. R., Stopiello, M., Castagna, V. C., Elgoyhen, A. B., Rubinstein, M., Gómez-Casati, M. E. & Franchini, L. F. 2019. Inner ear genes underwent positive selection and adaptation in the mammalian lineage. *Molecular Biology and Evolution*, 36(8), 1653–1670.

Pujol-Martí, J., Zecca, A., Baudoin, J. P., Faucherre, A., Asakawa, K., Kawakami, K. & López-Schier, H. 2012. Neuronal birth order identifies a dimorphic sensorineural map. *Journal of Neuroscience*, 32, 2976–87.

Puzdrowski, R. L. 1989. Peripheral distribution and central projections of the lateral-line nerves in goldfish, *Carasius auratus* (Part 2 of 2). *Brain, Behavior and Evolution*, 34, 121–131.

Qiu, X. & Muller, U. 2018. Mechanically gated ion channels in mammalian hair cells. *Front Cell Neuroscience*, 12, 100.

Riddiford, N. & Schlosser, G. 2016. Dissecting the pre-placodal transcriptome to reveal presumptive direct targets of Six1 and Eya1 in cranial placodes. *Elife*, 5, e17666.

Roberts, B. & Ryan, K. 1971. The fine structure of the lateral-line sense organs of dogfish. Proceedings of the Royal Society of London. Series B. Biological Sciences, 179, 157–169.

Roberts, B. L. & Meredith, G. E. 1992. The efferent innervation of the ear: variations on an enigma. *The Evolutionary Biology of Hearing*. Springer.

Ronan, M. & Northcutt, R. G. 1987. Primary projections of the lateral line nerves in adult lampreys. *Brain, Behavior and Evolution*, 30, 62–81.

Schlosser, G. 2002a. Development and evolution of lateral line placodes in amphibians I. Development. *Zoology (Jena)*, 105, 119–46.

Schlosser, G. 2002b. Development and evolution of lateral line placodes in amphibians.–II. Evolutionary diversification. *Zoology*, 105, 177–193.

Schuster, K. & Ghysen, A. 2013. Time-lapse analysis of primordium migration during the development of the fish lateral line. *Cold Spring Harb Protoc*.

Schwander, M., Kachar, B. & Müller, U. 2010. The cell biology of hearing. *The Journal of Cell Biology*, 190, 9–20.

Shibata, S. B., Ranum, P. T., Moteki, H., Pan, B., Goodwin, A. T., Goodman, S. S., Abbas, P. J., Holt, J. R. & Smith, R. J. 2016. RNA interference prevents autosomal-dominant hearing loss. *American Journal of Human Genetics*, 98, 1101–13.

Sienknecht, U. J., Köppl, C. & Fritzsch, B. 2014. Evolution and development of hair cell polarity and efferent function in the inner ear. *Brain, Behavior and Evolution*, 83, 150–161.

Stolfi, A., Ryan, K., Meinertzhagen, I. A. & Christiaen, L. 2015. Migratory neuronal progenitors arise from the neural plate borders in tunicates. *Nature*, 527, 371–374.

Striedter, G. F. & Northcutt, R. G. 2019. *Brains through Time: A Natural History of Vertebrates*, New York, NY: Oxford University Press.

Sugahara, F., Pascual-Anaya, J., Oisi, Y., Kuraku, S., Aota, S., Adachi, N., Takagi, W., Hirai, T., Sato, N., Murakami, Y. & Kuratani, S. 2016. Evidence from cyclostomes for complex regionalization of the ancestral vertebrate brain. *Nature*, 531, 97–100.

Tang, W. J., Chen, J. S. & Zeller, R. W. 2013. Transcriptional regulation of the peripheral nervous system in Ciona intestinalis. *Developmental Biology*, 378, 183–193.

Undurraga, C. A., Gou, Y., Sandoval, P. C., Nuñez, V. A., Allende, M. L., Riley, B. B., Hernández, P. P. & Sarrazin, A. F. 2019. Sox2 and Sox3 are essential for development and regeneration of the zebrafish lateral line. *bioRxiv*, 856088.

Valera, G., Markov, D. A., Bijari, K., Randlett, O., Asgharsharghi, A., Baudoin, J.-P., Ascoli, G. A., Portugues, R. & López-Schier, H. 2021. A neuronal blueprint for directional mechanosensation in larval zebrafish. *Current Biology*, 31(7), 1463–1475.

Vetschera, P., Koberstein-Schwarz, B., Schmitt-Manderbach, T., Dietrich, C., Hellmich, W., Chekkoury, A., Symvoulidis, P., Reber, J., Westmeyer, G. & López-Schier, H. 2019. Beyond early development: observing zebrafish over 6 weeks with hybrid optical and optoacoustic imaging. *bioRxiv*, 586933.

Vischer, H. A. 1989. The development of lateral-line receptors in *Eigenmannia* (Teleostei, Gymnotiformes). I. The mechanoreceptive lateral-line system (Part 1 of 2). *Brain, Behavior and Evolution*, 33, 205–213.

Vischer, H., Lannoo, M. & Heiligenberg, W. 1989. Development of the electrosensory nervous system in Eigenmannia (Gymnotiformes): I. The peripheral nervous system. *Journal of Comparative Neurology*, 290, 16–40.

Wahnschaffe, U., Bartsch, U. & Fritzsch, B. 1987. Metamorphic changes within the lateral-line system of Anura. *Anatomy and Embryology*, 175, 431–442.

Wahnschaffe, U., Fritzsch, B. & Himstedt, W. 1985. The fine structure of the lateral-line organs of larval Ichthyophis (Amphibia: Gymnophiona). *Journal of Morphology*, 186, 369–377.

Wang, V. Y., Rose, M. F. & Zoghbi, H. Y. 2005. Math1 expression redefines the rhombic lip derivatives and reveals novel lineages within the brainstem and cerebellum. *Neuron*, 48, 31–43.

Washausen, S. & Knabe, W. 2018. Lateral line placodes of aquatic vertebrates are evolutionarily conserved in mammals. *Biology Open*, 7(6), bio031815.

Washausen, S. & Knabe, W. 2019. Chicken embryos share mammalian patterns of apoptosis in the posterior placodal area. *Journal of Anatomy*, 234, 551–563.

Webb, J. F. 2013. Morphological diversity, development, and evolution of the mechanosensory lateral line system. *The Lateral Line System*. Springer.

Webb, J. F. 2020. Morphology of the mechanosensory lateral lien system of fishes. *In*: Fritzsch, B. (ed.) *The Senses* (pp. 29–46). Elsevier, Academic Press.

Weber, D. D. & Schiewe, M. H. 1976. Morphology and function of the lateral line of juvenile steelhead trout in relation to gas-bubble disease. *Journal of Fish Biology*, 9, 217–233.

Whitear, M. & Lane, E. 1983. Multivillous cells: epidermal sensory cells of unknown function in lamprey skin. *Journal of Zoology*, 201, 259–272.

Whitear, M. & Lane, E. B. 1981. Bar synapses in the end buds of lamprey skin. *Cell and Tissue Research*, 216, 445–448.

Will, U. 1989. Central mechanosensory lateral line system in amphibians. *The Mechanosensory Lateral Line System* (pp. 365–386). New York, NY: Springer.

Winklbauer, R. & Hausen, P. 1983. Development of the lateral line system in Xenopus laevis: II. Cell multiplication and organ formation in the supraorbital system. *Development*, 76, 283–296.

Wullimann, M. F. & Grothe, B. 2013. The central nervous organization of the lateral line system. *The Lateral Line System* (pp. 195–251). New York, NY: Springer.

Xu, J., Li, J., Zhang, T., Jiang, H., Ramakrishnan, A., Fritzsch, B., Shen, L. & Xu, P. X. 2021. Chromatin remodelers and lineage-specific factors interact to target enhancers to establish proneurosensory fate within otic ectoderm. *Proceedings of the National Academy of Sciences of the United States of America*, 118(12).

Yamada, Y. & Hama, K. 1972. Fine structure of the lateral-line organ of the common eel, Anguilla Japonica. *Zeitschrift für Zellforschung und mikroskopische Anatomie*, 124, 454–464.

Yamada, Y. 1973. Fine structure of the ordinary lateral line organ: I. The neuromast of lamprey, Entosphenus japonicus. *Journal of Ultrastructure Research*, 43, 1–17.

Zecca, A., Dyballa, S., Voltes, A., Bradley, R. & Pujades, C. 2015. The order and place of neuronal differentiation establish the topography of sensory projections and the entry points within the hindbrain. *Journal of Neuroscience*, 35, 7475–7486.

Zuo, J., Treadaway, J., Buckner, T. W. & Fritzsch, B. 1999. Visualization of α9 acetylcholine receptor expression in hair cells of transgenic mice containing a modified bacterial artificial chromosome. Proceedings of the National Academy of Sciences, 96, 14100–14105.

9 Electroreception Depends on Hair Cell-Derived Senses in Some Vertebrates

Sarah Nicola Jung, Valerie Lucks,
Karen L. Elliott, Bernd Fritzsch

CONTENTS

9.1 INTRODUCTION OF ELECTRORECEPTION

It took a long time to understand the mechanosensory lateral line system that has so many similarities with the inner ear and was typically referred to as the 'sixth sense'. Electric fish developed a unique 'seventh' sense that fascinated but confounded humans in the past, as their strongly electric members generate shocks 'producing pain and numbness so violent that it is impossible to describe the nature of the feeling they excite' as described by von Humboldt (von Humboldt and Bonpland, 1853). The investment with powers beyond our own perceptions engaged humans in a profound curiosity. This curiosity was shared throughout antiquity, such as Plato, Aristotle, Pliny, Plutarch, Galen, Lorenzini, Cavendish, Volta, Galvani, and Darwin. Attempts to explain electric organ function led to the invention of batteries. This invention provided sources for electricity used by engineers such as Faraday, Edison, and Tesla, who started the technical revolution we are still enjoying today. The template for Volta's battery (Volta, 1800) was based on the electric stingray organ as described by Lorenzini (1678). The battery was meant to prove that bioelectricity was the same as the electricity generated with

DOI: 10.1201/9781003092810-9

Galvani's metal plates (Moller and Fritzsch, 1993, 1995). After this discovery, many used batteries to elicit sensory stimulations and even make cadavers twitch, much like Galvani had done earlier with his frog leg experiments (Galvani, 1792).

It was over 200 years ago that the first insights of an electric sense were proposed: the existence of a sense that was radically different from any sense humans possess (Carlson and Sisneros, 2019). In particular, it was noticed that the electric eel was able to detect whether an electric circuit was open or closed (Ingenhousz, 1782; Walsh, 1773). However, Emil du Bois-Reymond (Du Bois-Reymond, 1848), the founder of neurophysiology, argued against previous ideas of Walsh, rejecting electroreception (Moller and Fritzsch, 1995). It took another 100 years before electroreception would become a designated sense (Lissmann, 1958). With the development of sensitive amplifiers and easily available recording equipment, Hans Lissmann discovered a whole new world of fishes capable of producing and perceiving weak electric signals (Lissmann, 1958). Today, we may witness a second revolution again driven by electric fish: we may soon use our genetic understanding of electrocytes to generate electric organs in humans to drive electric devices designed to replace dysfunctional organs or implanted tools such as cochlear implants or pacemakers (Amin Karami and Inman, 2012). Moreover, investigations of the bioelectric fields produced by electric fish continuously expand the knowledge on how their nervous system works and how it developed (Moller and Fritzsch, 1993).

This review provides an overview of the electroreceptive sense, focusing extensively on similarities and differences between electroreception in various species. We will specifically discuss electroreceptor sensory neurons as well as their peripheral sensory cells and central nuclei targets to provide an understanding of their interaction. In addition, we will emphasize the comparative aspect of adult stages of these three interlinked parts of a given electroreceptor system.

9.2 ELECTRORECEPTOR SENSORY NEURONS SHARE A COMMON PLACODAL ORIGIN WITH THE LATERAL LINE

Electroreception is only found in a relatively small number of species (Figure 9.1). Most of the approximately 65,000 vertebrate species, such as hagfish, most derived teleosts, all anurans, some salamanders and caecilians, and all amniotes, lack it (Braun and Northcutt, 1997; Fritzsch and Neary, 1998; Kishida et al., 1987; McCormick, 1989; Schlosser, 2002; Wullimann and Grothe, 2013). Only about 5,000 craniates have electroreception, which is in contrast to the more commonly found lateral line system, despite their shared placodal origin (Gillis et al., 2012; Modrell et al., 2011). This shared origin was demonstrated in non-teleosts using lateral line placode ablations and grafting experiments conducted on pigmented and albino axolotl embryos (amphibians). These experiments showed that ampullary organs, neuromasts, and the afferent neurons all originate from an individual elongating lateral line placode (Northcutt et al., 1994, 1995). Furthermore, a lateral line placode origin of ampullary organs has also been demonstrated in a chondrostean ray-finned bony fish, namely the paddlefish (closely related to sturgeons), and a cartilaginous fish (little skate) by focal labeling experiments using

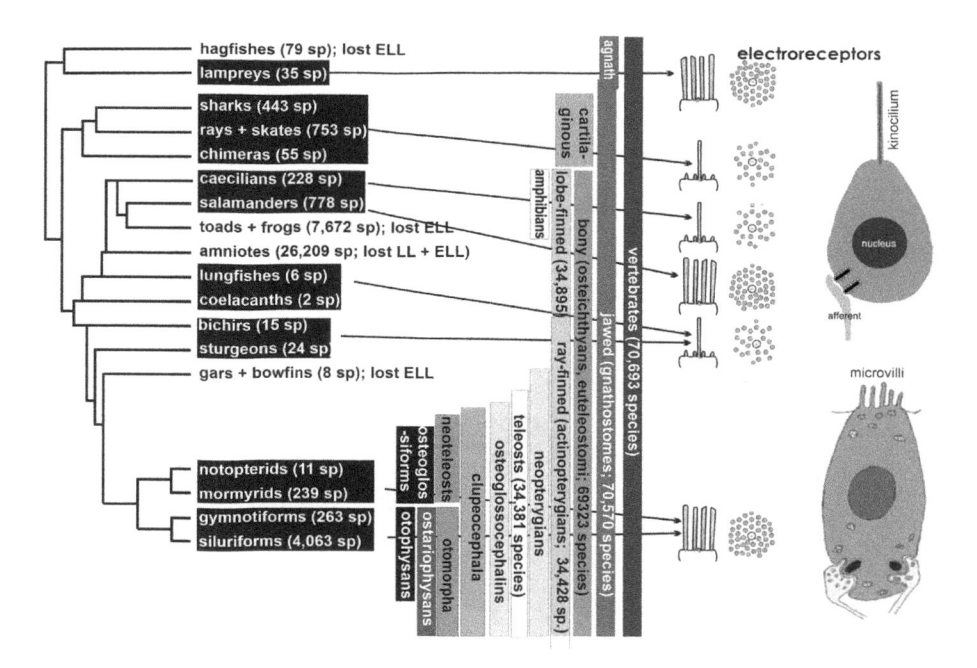

FIGURE 9.1 Electroreception is a general sense that has been lost multiple times. It has been lost in hagfish, neopterygians, frogs, amniotes, and in many adult salamanders and caecilians. Dark boxes indicate species that have electroreception. The sensory receptor can have either a kinocilium or variable numbers of stereovilli/microvilli. Electroreceptors are innervated by at least one afferent neuron. (Modified from Baker, 2019; Elliott and Fritzsch, 2020.)

fluorescent lipophilic dyes (Gillis et al., 2012; Modrell et al., 2011). Experimental evidence for the embryonic origin of lamprey electroreceptors is lacking but there is strong support for the homology to non-teleost jawed vertebrate electroreceptors from the physiology of the receptors as well as from the innervation pattern (see below). Fate-mapping experiments that can confirm the embryonic origin of electroreceptors in teleosts are overdue (Baker, 2019). Even though electroreceptive neurons share a similar organization with lateral line fibers, electroreceptor neurons and lateral line neurons run independently of each other. Electroreceptor afferents branch to innervate single electroreceptors or clusters of electroreceptors at the periphery (Fritzsch, 1981). However, in contrast to inner ear and lateral line hair cells, which receive efferent innervation, electroreceptive hair cells are not innervated by efferent fibers (Fritzsch and Elliott, 2017a; Münz et al., 1984; Roberts and Meredith, 1992).

Innervation of electroreceptive sensory neurons is well described in lampreys (Fritzsch, 1998; Fritzsch et al., 1984; Fritzsch and Northcutt, 1993; Koyama et al., 1993; Pombal and Megías, 2019; Ronan and Northcutt, 1987). Electroreceptive afferent fibers project with the anterior lateral line (aLL) and the posterior lateral line (pLL) via a recurrent anterior lateral line nerve ramus (Figure 9.7). Whereas, in all other electroreceptive vertebrates, electroreceptive afferent fibers project exclusively

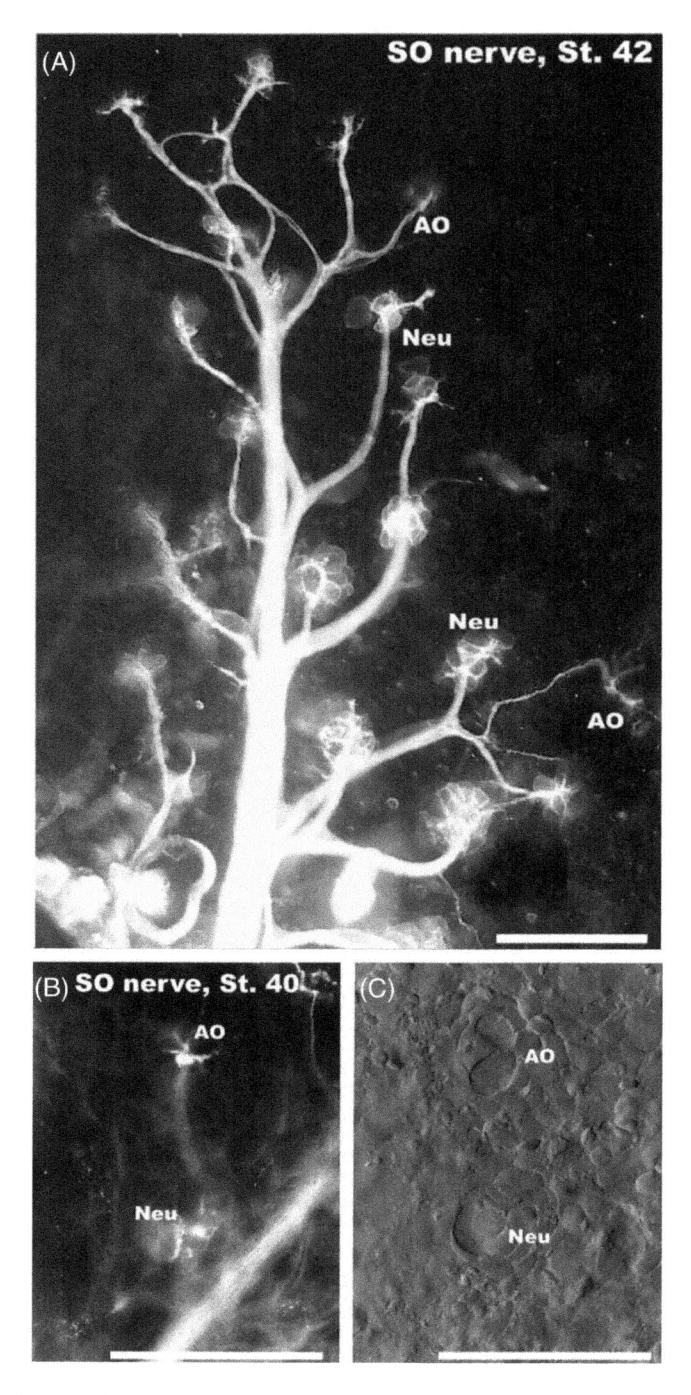

FIGURE 9.2 Labeling of the lateral line and electroreceptive nerves in axolotl. Image showing single afferent fibers to ampullary organs (AO) compared to two nerve bundles to the

FIGURE 9.2 *(Continued)* neuromast (Neu) (A). Detailed analysis of superficial ophthalmic nerve (SO) fibers labeled with lipophilic dyes shows the distinct innervation of an ampullary organ (AO) and neuromasts (Neu) (B, C). Bars indicate 100 µm in all images. (Modified from Fritzsch et al., 2005.)

with the aLL (Fritzsch, 1981; 1993; Ronan and Northcutt, 1987; Wullimann and Grothe, 2013). In lampreys, and similarly in the lateral line of hagfish, all aLL and electroreceptive nerve fibers are devoid of myelin (Gelman, 2007). Innervation has also been well described in the electroreceptive gnathostomes (Baker and Modrell, 2018; Bell et al., 1981; 1983; Bodznick and Boord, 1986; Bullock and Heiligenberg, 1986; Fritzsch, 1981; Jørgensen, 1989; Szabo, 1974). Coelacanths and lungfish have unique dorsal fiber projections (Nieuwenhuys et al., 1998; Roth and Tscharntke, 1976); however, characterizing the distinct projections requires further nerve fiber tracing (Jørgensen, 1989; 2005). In addition to coelacanths and lungfish, there are a few detailed descriptions for salamanders (Fritzsch, 1981; 1989; Northcutt et al., 1994) and caecilians (Fritzsch et al., 1985). While electroreceptive afferents project together with the lateral line, their central targets are distinct, with electroreceptive afferents projecting to the dorsal nucleus and lateral line afferents projecting to the intermediate nucleus (Figures 9.2 and 9.5).

Evidence exists for a parallel evolution of electroreception in certain bony fish (silurids, gymnotids, mormyrids) (Bell et al., 1989; Leitch and Julius, 2019; Zakon, 1986), given that basic holosteans lack electroreception (McCormick, 1989). In silurids, gymnotids, and mormyrids, electroreceptive afferents show a divergent pattern of innervation (Baker, 2019; Bell and Maler, 2005; Wullimann and Grothe, 2013). Among silurids, there are unique projections of electroreceptive fibers with aLL and pLL afferents (Bleckmann et al., 1991; Tong and Finger, 1983) mormyrids have complicated projections with aLL and pLL (Bell et al., 1981; 1983). In gymnotids electroreceptive afferents project with a branch of a recurrent aLL (Figure 9.7; Lannoo et al., 1989; 1990; Wullimann and Grothe, 2013).

In summary, only about 5000 species have electroreception. Furthermore, electroreception likely evolved twice in teleosts (silurids, mormyrids, gymnotids). At the periphery, electroreceptive hair cells synapse on at least one afferent but do not receive efferent innervation. In addition, electroreceptive afferents project with lateral line fibers to reach the hindbrain. Centrally, electroreceptive afferents target the dorsal nucleus in the hindbrain.

9.3 ELECTRORECEPTIVE 'HAIR CELL' STRUCTURE AND THE GENES INVOLVED IN THEIR DEVELOPMENT

First described in cartilaginous fish in 1678, electroreceptive organs had been named the 'ampullae of Lorenzini' in this group (Carlson and Sisneros, 2019; Lorenzini, 1678; Moller and Fritzsch, 1993). Electroreceptive organs have since been described by many individuals (Bell and Maler, 2005; Bullock et al., 2006; Bullock and Heiligenberg, 1986; Hetherington and Wake, 1979; Sarasin and Sarasin, 1890; Szabo, 1974). Non-teleost ampullary organs show common principle (Figure 9.3). Each of these electroreceptive

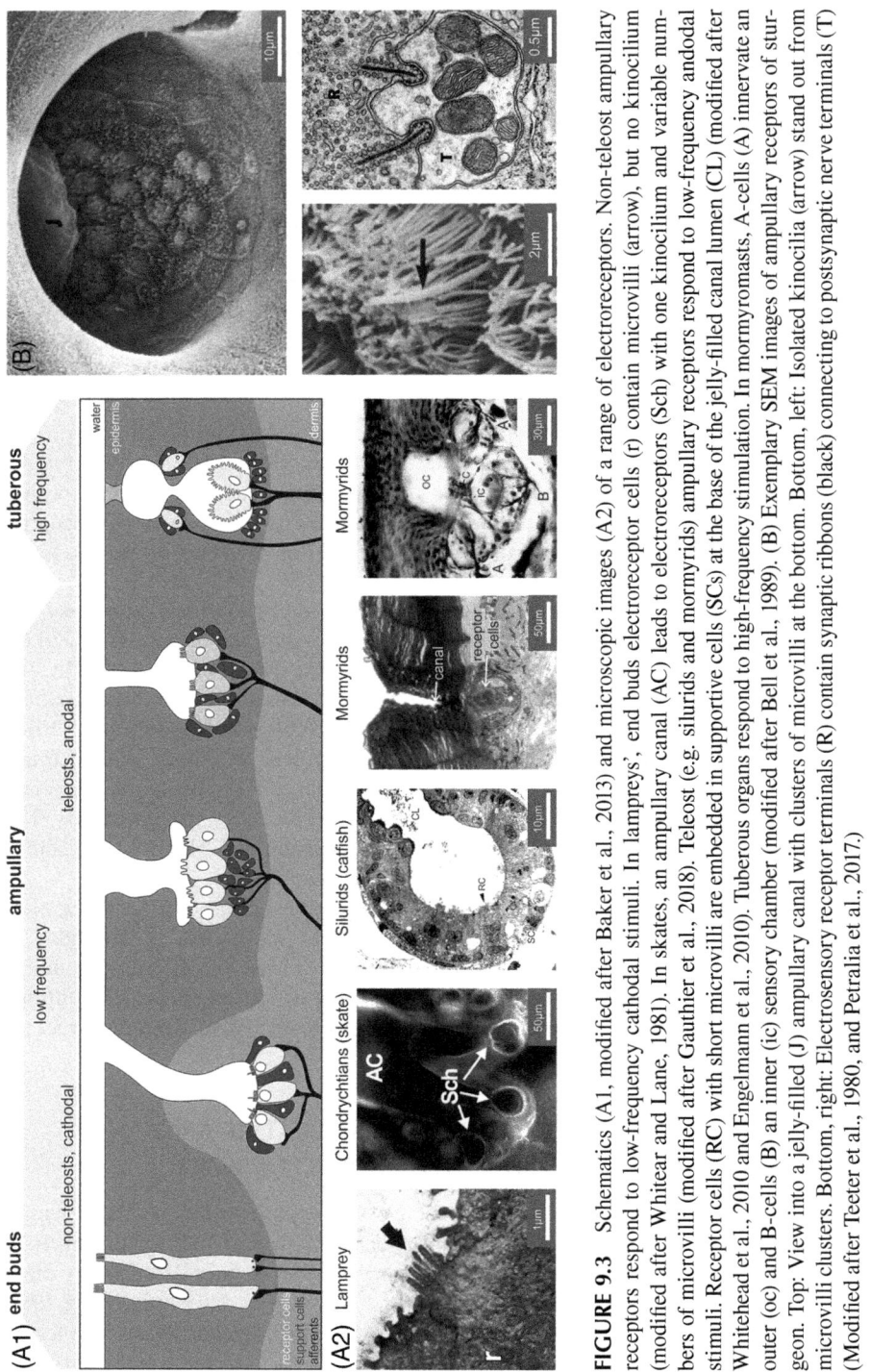

FIGURE 9.3 Schematics (A1, modified after Baker et al., 2013) and microscopic images (A2) of a range of electroreceptors. Non-teleost ampullary receptors respond to low-frequency cathodal stimuli. In lampreys', end buds electroreceptor cells (r) contain microvilli (arrow), but no kinocilium (modified after Whitear and Lane, 1981). In skates, an ampullary canal (AC) leads to electroreceptors (Sch) with one kinocilium and variable numbers of microvilli (modified after Gauthier et al., 2018). Teleost (e.g. silurids and mormyrids) ampullary receptors respond to low-frequency anodal stimuli. Receptor cells (RC) with short microvilli are embedded in supportive cells (SCs) at the base of the jelly-filled canal lumen (CL) (modified after Whitehead et al., 2010 and Engelmann et al., 2010). Tuberous organs respond to high-frequency stimulation. In mormyromasts, A-cells (A) innervate an outer (oc) and B-cells (B) an inner (ic) sensory chamber (modified after Bell et al., 1989). (B) Exemplary SEM images of ampullary receptors of sturgeon. Top: View into a jelly-filled (J) ampullary canal with clusters of microvilli at the bottom. Bottom, left: Isolated kinocilia (arrow) stand out from microvilli clusters. Bottom, right: Electrosensory receptor terminals (R) contain synaptic ribbons (black) connecting to postsynaptic nerve terminals (T) (Modified after Teeter et al., 1980, and Petralia et al., 2017.)

organs has sensory hair cells, separated by supporting cells, located at the base of a bulbous chamber from which a conductive jelly-filled canal leads to a pore at the surface. The channel walls have high impedance properties and are formed by layers of flattened epithelial cells (Jørgensen, 1989; 2005). Hair cells are secondary sensory cells (i.e. lacking an axon). On the apical end of the hair cells are stereovilli, and in some species, the stereovilli are replaced with a single kinocilium (Figure 9.1). On the basal end of the hair cell, their synaptic ribbons synapse onto a single afferent fiber (Fritzsch and Wahnschaffe, 1983; Jørgensen, 2005; Leitch and Julius, 2019). The homology of electroreceptors of lampreys and non-teleost jawed vertebrates is supported by their physiology as they respond both to weak low-frequency cathodal (exterior negative) stimuli with excitation and anodal (exterior positive) stimuli with inhibition (Baker, 2019).

Electroreceptor organs in adult lampreys are often referred to as end buds (Figure 9.3). These organs contain 3–30 hair cells that have multiple stereovilli without a kinocilium at the apical surface and have single round or plesiomorphic ribbons at the basal surface that synapse onto a single, branched afferent fiber (Jørgensen, 2005; Whitear and Lane, 1981). Although the specific pattern of hair cell innervation is not clear in lampreys (Ronan and Bodznick, 1986; Whitear and Lane, 1981), most likely hair cells are innervated by individual nerve terminals. Hair cells form round bars or ribbons approximately 0.4 µm in diameter at these terminals (Whitear and Lane, 1981). Specifically, one type of 'multivillous' hair cell has large ribbons at the base with an enlarged single innervation (Whitear and Lane, 1983) in larvae and adult lampreys. From here one can conclude that electroreceptors function as independent 'multivillous cells' (Whitear and Lane, 1981; 1983), whereas they are aggregated in other vertebrates (Bolz and Fritzsch, 1986; Fritzsch and Wahnschaffe, 1983; Northcutt et al., 1994). In older larvae and adults, electroreceptors often occur in distinct areas that are innervated by a branch of a single nerve fiber (Baker and Modrell, 2018; Fritzsch and Wahnschaffe, 1983; Jørgensen, 2005). Unfortunately, the 'multivillous cell' innervation was incompletely characterized in the original work (Baker, 2019; Jørgensen, 2005). Multiple attempts to demonstrate the identification of 'hair cells' have not been successful (Gelman, 2007; Ronan, 1988; Ronan and Bodznick, 1986).

Hair cells in gnathostomes have two major forms: one with a short kinocilium that is surrounded by many taller stereovilli or one with a taller kinocilium surrounded by shorter stereovilli/microvilli (Jørgensen, 2005; Roberts and Ryan, 1971). In sharks, rays, and ratfish, hair cells within the ampullae of Lorenzini have a short kinocilium that extends from a basal body. Within the bony fish, there is a large diversity of electroreceptors with respect to their shape (Figure 9.3) and to the different lengths of their stereovilli and/or kinocilia (Jørgensen, 1989; 2005; Zakon, 1986). Electroreceptor cells of caecilians have a single kinocilium with a few microvilli (Wahnschaffe et al., 1985), whereas most salamanders have only microvilli (Figures 9.1 and 9.3). These electroreceptive hair cells are typically clustered within sensory ampullae located on the skin, where there may be between 150 and 2000 separate clusters (Bodznick and Boord, 1986). Each electroreceptive hair cell synapses on a single afferent terminal from one of the few afferents innervating each ampullary organ (Bleckmann et al., 1987; Bodznick and Montgomery, 2005). In bichir, sturgeons, lungfish, and caecilians, electroreceptors have a short kinocilium, with or without stereovilli (Jørgensen, 2005; Wahnschaffe et al., 1985). Electroreceptors in bichirs are organized with a single kinocilium surrounded by short stereovilli (Roth and Tscharntke, 1976). These are particularly dense in the snout (Jørgensen, 2005). Similarly, electroreceptors of sturgeons have a kinocilium and short stereovilli. Their electroreceptors are innervated by a branch of a single nerve fiber (Baker and Modrell, 2018; Jørgensen, 2005). Likewise, the electroreceptors of lungfish have a single kinocilium that is surrounded by a few microvilli (Jørgensen, 2005). Electroreceptors in coelacanths appear rather unusual and require a more detailed analysis (Jørgensen, 2005).

Ampullary electroreceptors of teleost are evolutionary distinct from the ampullary organs in non-teleosts (Figure 9.3). Although still highly debated, the most parsimonious hypothesis is that in the osteoglossomorph lineage, ampullary reception has evolved in the stem leading to the common ancestor of mormyrids and notopterids whereas in the ostariophysan lineage ampullary reception has evolved in the stem leading to the common ancestor of Siluriformes and Gymnotiformes (Baker, 2019). The diversity of electroreceptive teleost species is reflected in a rich variety of ampullary organ morphologies. The ampullary canal is mainly short but elongated in saltwater teleosts. Teleost electroreceptors commonly have microvilli without a kinocilium (except *Xenomystus nigri*; Baker, 2019) but differ in the shape of synaptic ribbons (Baker and Modrell, 2018; Jørgensen, 2005; Wullimann and Grothe, 2013). Synaptic ribbons can be exceptionally large (round ribbons that are 2 µm in diameter; Fritzsch and Wahnschaffe, 1983), or they can be rather small (round ribbons that are 0.2 µm in diameter; Whitear and Lane, 1981). Typically, synaptic ribbons are elongated in shape and can be extensive (Baker, 2019; Leitch and Julius, 2019; Zakon, 1986). For example, *X. nigri* has multiple elongated ribbons that are inside a deep depression. Electroreceptors in silurids, gymnotids, and mormyrids have a highly derived pattern of oval or elongated ribbons (Figure 9.3) (Bell and Maler, 2005; Jørgensen, 2005; Lannoo et al., 1990; Zakon, 1986). Teleost ampullary electroreceptors also respond to weak low-frequency signal. However, in contrary to non-teleost electroreceptors they response to anodal (exterior positive) stimuli with excitation and cathodal (exterior negative) stimuli with inhibition (Baker, 2019; Leitch and Julius, 2019).

An additional set of teleost electroreceptors, referred to as tuberous organs, are loosely packed on top of epidermal cells with protruding sensory cells that open into an intraepidermal cavity (Figure 9.3). These tuberous organs are found in silurids, gymnotids, and mormyrids. The tuberous organs are covered by many microvilli that are up to 3 µm in length and 0.1–0.15 µm in diameter. A single nerve fiber divides to form several boutons on the individual sensory cells, which stereotypically have shorter and more elongated ribbons (Bell and Maler, 2005; Jørgensen, 2005; Zakon, 1986). Typically, the ampullary electroreceptors respond to low-frequency stimulation of 1–20 Hz whereas the tuberous organs in teleosts are responsive to a higher frequency of 100 Hz–20 kHz (Bartels et al., 1990; Leitch and Julius, 2019; Münz et al., 1984). The difference in frequency sensitivity is reflected in their different roles in electroperception. Ampullary receptors respond primarily to external electric fields as they are induced, for example, by the bioelectricity of living organism in the fish's close vicinity. The ability to perceive external weak electric fields is known as passive electrolocation. In fish capable of active electrolocation, a self-generated electric field is emitted by a discharge of an electric organ (EOD) in the tail of the fish. Nearby objects with different conductivity than the surrounding water distort the electric field which is sensed by tuberous receptors providing a spatial representation of the electrosensory environment. In addition, tuberous receptors respond to EODs of conspecifics (Carlson and Sisneros, 2019) allowing for electrocommunication.

Electroreceptors of teleost and non-teleost species also differ in the composition of ion channels responsible for synaptic transmission. Whereas silurids, gymnotids, and mormyrids have low resistance voltage-gated channels, sharks, rays, bichir, sturgeons, salamanders, and caecilian have high resistance ones (Leitch and Julius, 2019). In sharks and skates, these electroreceptor ion channels were identified as $Ca_v1.3$ (Bellono et al., 2017, 2018), which may also be present in salamanders (Münz et al., 1984). These voltage-gated Ca^{2+} channels together with K^+ channels mediate electroreceptor membrane oscillations. In the little skates, K^+ channels are identified as large-conductance Ca^{2+} activated K^+ channels (BK channel; Bellono et al., 2017), whereas in sharks the voltage-gated K^+ channel Kv1.3 is crucial (Bellono et al., 2018). Homologous of the Cav1.3, BK channels and Kv channels are also enriched in both the lateral line mechanoreceptive organs and hair cells of the cochlear but not for example in photoreceptors further supporting the evolutionary nearness between electroreceptors and hair cells (Baker, 2019; Leitch and Julius, 2019).

The knowledge about the genes to be involved in electroreceptive hair cell differentiation as well as afferent neuron development is incomplete (Bolz and Fritzsch, 1986; Fritzsch and Wahnschaffe, 1983; Modrell et al., 2014; Northcutt, 1992; Northcutt et al., 1994; Piotrowski and Baker, 2014). Presumably, electroreceptor hair cells will follow the general rule of mechanosensory hair cell development. Thus, it is likely that electroreceptive hair cells will require *Atoh1* expression, which acts downstream of other essential genes in mechanosensory hair cell development, such as *Sox2* (Baker, 2019; Elliott and Fritzsch, 2020) and *Eya1* (Elliott et al., 2021; Xu et al., 2021). Experiments in several organisms, such as axolotls, sturgeons, and sharks demonstrated that transplantation of ectodermal tissue can develop both lateral line hair cells and ampullary organs when transplanted to non-sensory areas (Gilland and Baker, 1993; Modrell et al., 2011; Northcutt et al., 1995). However, none of these transplantations were reversed to investigate whether non-sensory tissue could be induced to form electroreceptors or lateral line hair cells when transplanted to sensory regions.

Recent gene expression studies have suggested that electroreceptor cells and lateral line hair cells evolved as sister cell-types (Baker, 2019; Baker and Modrell, 2018). Lateral line hair cells, electroreceptive hair cells, as well as inner ear hair cells require micro RNAs, in particular *miR-183* (Figure 9.4; Pierce et al., 2008; Soukup et al., 2009; Weston et al., 2011). Loss of the gene *Dicer* affects not only the lateral line but also affects the olfactory system, retina, and much of the telencephalon (Kersigo et al., 2011; Soukup et al., 2009). Studying the development of the mechanosensory and electroreceptive systems could help elucidate the earliest steps of gene expression in lateral line and electroreceptor hair cells (Figure 9.4; Baker, 2019; Pierce et al., 2008). Electroreceptive hair cells and lateral line hair cells can develop near each other and demonstrating their adjacent distribution aided in their identification (Figure 9.4; Bolz and Fritzsch, 1986; Northcutt et al., 1994).

Unlike lateral line hair cells, which are not affected by nerve cutting, electroreceptors will rapidly disappear following transection of the nerve (Fritzsch

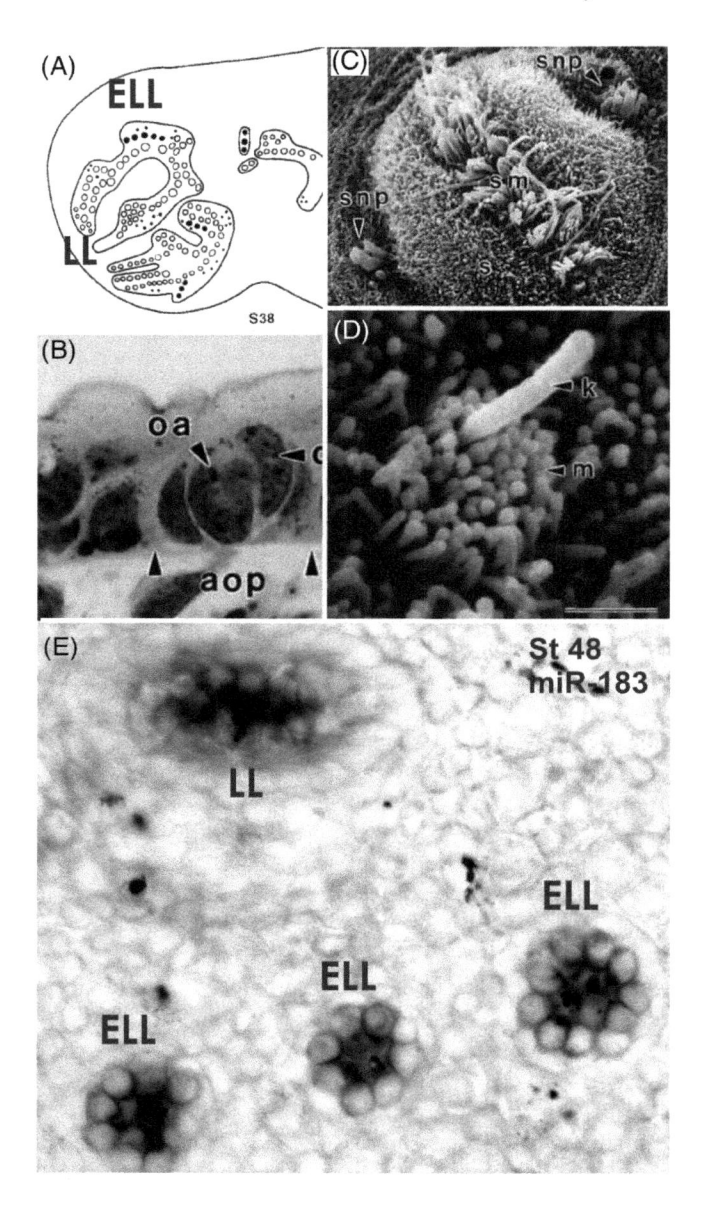

FIGURE 9.4 Overview (A) of lateral line (LL) and electroreceptors (ELL) in the axolotl. The first identification of electroreceptors (ampullary organ primordium, aop) has a single ampullary organ next to the lateral line (B). SEM images of a lateral line receptor (C) and electroreceptor (D). Lateral line neuromasts are shown in an older animal compared to the earliest budding of ampullary organs adjacent to a kinocilium (E). miR-183 is expressed in early development of both lateral line (E, top) and electroreceptors (E, bottom). aop, ampullary organ primordium; oa, oval cell: snp, secondary neurosensory primordium; sensory macula; k, kinocilium; m, multivilli; LL, lateral line; ELL, electroreceptive organ. (Modified from Northcutt et al., 1994; Pierce et al., 2008.)

et al., 1990). Furthermore, it has been shown that, while lateral line hair cells are unaffected, electroreceptors will not develop when the nerve is cut prior to their formation (Roth, 2003). Other types of hair cells also depend on innervation. For instance, neuronal loss following neurotrophin elimination resulted in the loss of cochlear hair cells within several months; however, there is more variation in innervation dependency for vestibular hair cell survival (Kersigo and Fritzsch, 2015). These findings suggest that electroreceptors and cochlear hair cells require innervation, whereas lateral line and vestibular hair cells have a less critical dependence on innervation. Similarly, taste buds, which depend on brain-derived neurotrophic factor (BDNF), are lost following deafferentation (Fritzsch et al., 1998). Early gene duplication in bony fish has resulted in multiple neurotrophins. This makes it more complicated to elucidate which neurotrophic factor(s) is used (Hallböök et al., 2006).

In summary, stereovilli/microvilli of lampreys, mormyrids, gymnotids, silurids, and most salamanders seem to be all similar in their overall length, whereas sharks, rays, bichir, sturgeon, lungfish, and caecilians have a long kinocilium and may lack stereovilli. Electroreceptors have a single afferent innervating the round or elongated ribbon synapses. Lampreys, chondrichthyes, lungfish, caecilians and salamanders have a ribbon synapse as compared to the elongated ribbons found in mormyrids, gymnotids and silurids.

9.4 ELECTRORECEPTIVE SENSORY NEURONS PROJECT CENTRALLY TO THE DORSAL NUCLEI

Primitive and derived electroreceptors are innervated by homologous lateral line nerves but their primary targets in the medulla are not homologous (Bullock et al., 2006; Bullock and Heiligenberg, 1986; McCormick, 1999; Northcutt, 2005; Northcutt et al., 1995). Central projections are split into two groups. The primary medullary target of primitive electroreceptors (lampreys, chondrichthyes, bichir, sturgeon, coelacanths, lungfish, caecilians, and salamanders) is the dorsal nucleus (DN; Bell and Maler, 2005; Bodznick and Boord, 1986; Münz et al., 1984; Ronan, 1988), whereas the primary medullary target of derived electroreceptors (teleost: silurids, gymnotids, notopterids, mormyrids) is the electrosensory lateral line lobe (ELL; Bell et al., 1989; Carr and Maler, 1985; Finger and Tong, 1984).

Despite not being homologous, DN and ELL do share common neuroarchitectural principles and are thus referred to as a *cerebellum-like* structure (Figure 9.5; Bell and Maler, 2005), as both receive input from the periphery in their deep layers and parallel fiber input in their molecular layers (Bell et al., 2008). The primary afferent fibers terminate in the deep layer where they innervate the basilar dendrites of principal cells either directly or indirectly via interneurons. The parallel fibers contact the apical dendrites of the principal cells in the molecular layer. These fibers arise from a mass of granule cells located in the eminentia granularis posterior (EGp) that receive input from a variety of sources, including inputs from other sensory modalities, higher-level input of the same modality and, in mormyrids, even corollary discharge signals. The cerebellum-like structures have been intensely investigated in

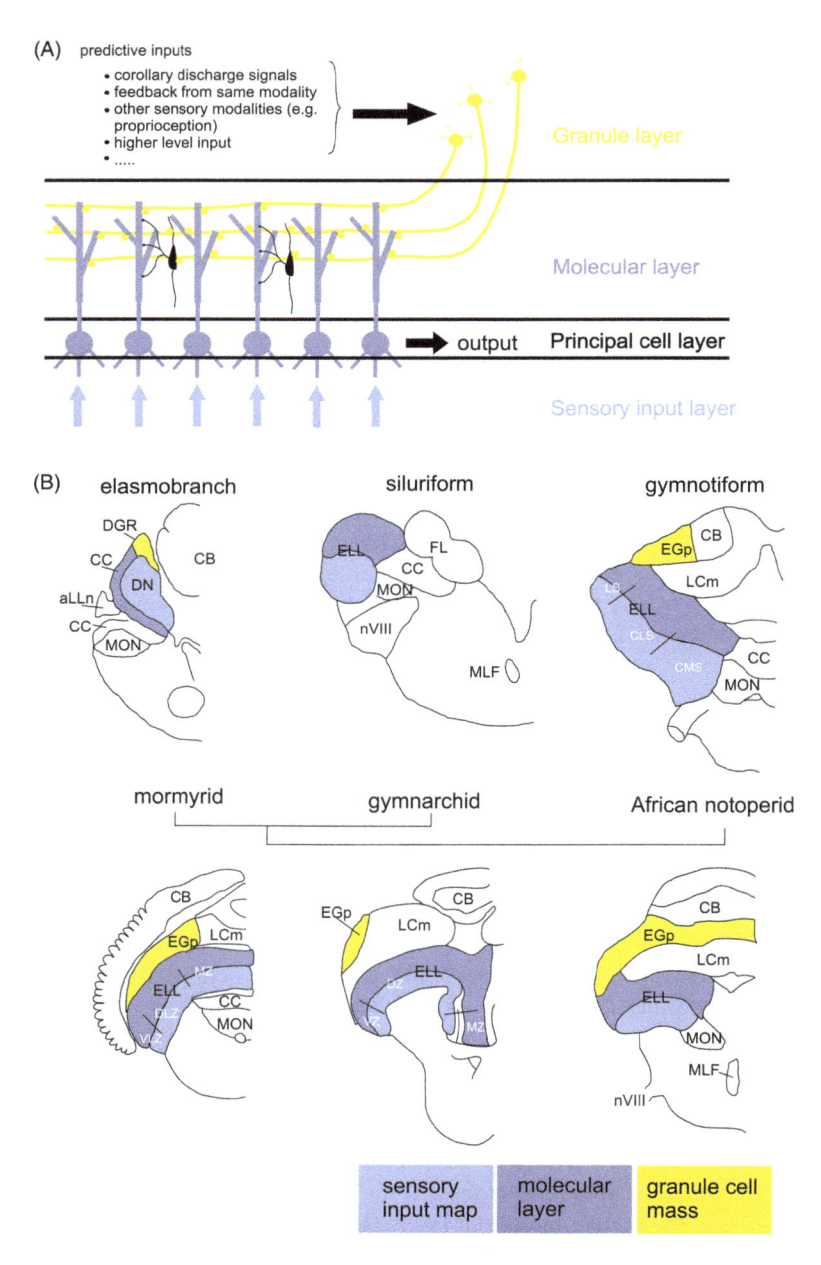

FIGURE 9.5 (A) Scheme of major features of cerebellum-like structures. Primary afferent fibers terminate in the deep layer where they innervate the basilar dendrites of principal cells either directly or indirectly via interneurons (not shown). Parallel fibers contact the apical dendrites of the principal cells in the molecular layer. These fibers arise from a mass of granule cells located in the eminentia granularis posterior (EGp) that receive input from a variety of sources. (B) Overview of cerebellum-like structures in different vertebrate groups. The molecular layer, granule cell mass, and sensory input map are shown in different

FIGURE 9.5 *(Continued)* colors, as indicated at the lower right. aLLN, anterior lateral line nerve; CB, cerebellum; CC, cerebellar crest; CLS, centrolateral segment; CMS, centromedial segment; DN, dorsal nucleus; DZ, dorsal zone, EGp, eminentia granularis posterior; ELL, electrosensory lobe; gran, granular layer; LS, lateral segment; IN, intermediate nucleus (corresponding to MON, medial octavolateral nucleus); MZ, medial zone; nVIII, eighth nerve; VZ, ventral zone. ([a] Modified from Bell et al., 2008; [b] Modified from Bell and Maler, 2005; Bell et al., 1997.)

three groups of fish (e.g. elasmobranch, gymnotids, and mormyrid). One of the key features of both the cerebellum and cerebellum-like structures is likely generating and subtracting predictions of the sensory consequence of motor commands (Warren and Sawtell, 2016). The feedback signals are conveyed by parallel fibers to the principal cells of the cerebellum-like structures. Adaptive spike-timing plasticity at these synapses is at the base of the cancelation of predictable sensory input, including ego-motion induced self-stimulation (reafference). There is evidence that multiple behavior-specific cancellation signals are contributing to the suppression of predictable sensory reafference (Kennedy et al., 2014; Lai et al., 2021). As an example, this is important for the detection of behavioral relevant stimuli, e.g. prey items, on a background of behaviorally irrelevant stimuli, e.g. tail movement.

The DN receives input from electroreceptor afferents in all major divisions of craniates that have electroreceptors. All lateral line projections are entering near the 'rhombomeric' region r4 and project rostral to r2 and caudal to r6 (Elliott et al., 2021). Electroreceptive fibers of non-neopterygians enter the medulla via the dorsal branch of the anterior lateral line nerve (Bell and Maler, 2005). The posterior lateral line nerve ganglion has only mechanosensory ganglion cells innervating the trunk neuromasts. In lampreys, however, electrosensory fibers join the posterior lateral line nerve via a recurrent anterior lateral line nerve ramus to innervate trunk electroreceptors (Figure 9.7a; Fritzsch et al., 1984; Koyama et al., 1993; Ronan and Northcutt, 1987; Wullimann and Grothe, 2013).

The termination of electroreceptor afferents in the DN is somatotopically organized in some, but not all, non-neopterygian fish (Bell and Maler, 2005). In lamprey for example, afferent fibers from electroreceptors do not terminate somatotopically (Ronan and Bodznick, 1986). Electrosensory afferents and principal cells of the DN in paddlefish (Chondrostei) are also not topographically organized (Hofmann et al., 2005). Primary afferent projections have been well described in cartilaginous fish (sharks, skates, rays, and chimaeras). Here, afferents terminate in the ipsilateral portion of the DN. Projections are somatotopic with the anterior electroreceptor afferents projecting to the ventral portion of the DN, whereas those of the posterior receptors project to the dorsal DN (Bodznick and Boord, 1986; Newton et al., 2019). No data are available regarding projection patterns in non-neopterygian bony fish (bichir, lungfish, and coelacanth).

As pointed out before, electroreception reappeared independently at least twice if not four times during teleost (neopterygii) radiation (Bullock and Heiligenberg, 1986): once in gymnotids and silurids and a second time in notopterids and mormyrids. Whether silurids and notopterids acquired electroreception independently of gymnotids and mormyrids, respectively, is debatable. Therefore, we discuss each group separately. Central electrosensory projections are quite diverse. Members of these groups are so distinct that we touch on them only briefly as they have been presented exquisitely by others (Bell and Maler, 2005; Wullimann and Grothe, 2013).

Electrosensory fibers in **silurids** provide input to the ELL via both the anterior and the posterior lateral line nerve (Figures 9.5b and 9.7c; Bell and Maler, 2005). Here, they form a single map with the anterior body mapped more medially and the posterior body mapped more laterally. Hence, it seems that there is at least a coarse somatotopy (Bleckmann et al., 1991; Finger and Tong, 1984; McCormick and Braford Jr, 1993; Tong and Finger, 1983).

In **gymnotids**, the electrosensory fibers of both ampullary and tuberous electroreceptors only provide input to the ELL of the medulla via the anterior lateral line nerve (Figures 9.5b and 9.8a). Tuberous electroreceptors are divided into two types, phase (time) coders (T-units) and probability (amplitude) coders (P-units) (Bell and Maler, 2005). The former only respond to each EOD with a single action potential and can therefore encode the frequency of an EOD whereas in the latter the discharge probability is modulated by the amplitude of the EOD. Primary afferent fibers of T-units project to spherical cells (Maler, 1979). Anatomical and physiological studies have shown that the time-coding pathways preserve fine temporal information and neurons receive no descending feedback from central nervous system (CNS) circuits (Fortune and Chacron, 2011). P-units project either directly to basilar pyramidal cells or indirectly to non-basilar pyramidal cells via inhibitory interneurons. Hence, basilar pyramidal cells (ON-cells) response to an increase in EOD amplitude with an increasing firing rate and non-basilar pyramidal cells (OFF-cells) respond to a decrease in EOD amplitude with an increasing firing rate and vice versa. Pyramidal cells have a center-surround structure. Having distinct populations of neurons responding respectively to increased or decreased intensity within their receptive field centers has long been thought to be important for contrast enhancement, but recently their role in detecting changes in motion direction has been investigated (Clarke et al., 2014). The ELL is divided into a medial (MS), centromedial (CMS), centrolateral (CLS), and a lateral (LS) segment (Bell and Maler, 2005; Lannoo et al., 1989). The MS only receives input from the ampullary receptors, whereas the other three maps get tuberous electroreceptor inputs. The head and tail are represented in the same orientation in each map, whereas the dorsoventral axis is inverted in the CMS and LS compared to the other two segments (Figure 9.8; Krahe and Maler, 2014). The maps differ in their receptive field size, with the CMS having the smallest receptive fields and the LS having the largest receptive fields. Furthermore, each map is composed of columns, consisting of six pyramidal cell classes (three superficial, intermediate, or deep and two basilar or non-basilar pyramidal cells). They differ particularly in the amount of feedback projections they receive from higher brain areas, with the superficial pyramidal cells receiving the most, and deep pyramidal cells receiving the least feedback. The interaction of receptive field size and feedback shape the spatiotemporal tuning of the output neurons of these maps. The differences of both results in functional differences between the segments: the CMS pyramidal neurons predominantly response to local, low-frequency prey-like signals, whereas the LS pyramidal neurons respond to global, high-frequency communication signals (Clarke and Maler, 2017; Krahe and Maler, 2014). Interestingly, first experiments on motion processing show surprisingly little differences across maps as all pyramidal cells seems to have high-pass velocity tuning curves (Khosravi-Hashemi and Chacron, 2014). In the CLS, it has further been shown that feedback synthesizes a neural code for motion while at the same time reduces the response to distracting stimuli. This has been postulated to play a critical role in forming spatial attention (Clarke and Maler, 2017).

Notopterids (*African knifefish*) have only the ampullary type of electroreceptors. Afferent fibers enter the medulla via the anterior and posterior lateral line nerve where they terminate in the ELL. There is at least a coarse somatotopy as electrosensory afferents from the anterior body terminate medially and fibers from the posterior body terminate laterally. The ELL itself is a three-layered structure consisting of a deep input layer and a molecular layer that frame a layer of large cells (Braford Jr, 1986).

Mormyrids. In *Gymnarchus niloticus*, all electroreceptive projections enter the medulla via both the anterior and posterior lateral line nerve (Figure 9.5b and 9.8b). Fibers from the anterior part of the body terminate anteriorly and those that innervate the posterior body terminate posteriorly in the ELL (Bell and Maler, 2005). The ELL consists of bilateral lobes, each comprising three distinctive zones: dorsal (DZ), ventral (VZ), and medial (MZ) (Bass and Hopkins, 1983). *Gymnarchus* has three types of electroreceptors: ampullary and two types of tuberous electroreceptors. These are the S- and O-type receptors. *Gymnarchus* are wave-type mormyrids and interestingly their tuberous receptors are functionally comparable to the tuberous receptors of the evolutionary distinct group of gymnotiformes. In particular, O-type afferents show functional similarities to P-units of gymnotids as they encode the EOD amplitude modulation whereas S-type afferents have functional similarities to T-units as they encode differences in EOD phase (Fortune and Chacron, 2011). Detailed anatomical studies in this species are rare, but there are anatomical and physiological evidences that ampullary receptors occur in the same areas in the DZ and MZ in which also tuberous-type neurons are found (Kawasaki and Guo, 1998). In this study, all neurons recorded and labeled in the ELL had pyramidal morphology with large and extensive apical dendrites and less extensive basal dendrites. Neurons in the DZ of the ELL respond selectively to amplitude modulation. Neurons in the MZ are categorized to be sensitive to the amplitude, the differential phase, or to both (Kawasaki and Guo, 1998).

Fish within the mormyrid family all have pulse-type EODs and have three types of electroreceptors: ampullary, mormyromast, and knollenorgan receptors. Of these, the mormyromasts have a further sub-division with differentially tuned A- and B-cells (Figure 9.3a). Both A and B fibers respond to an amplitude increase of the EOD with a decrease in spike latencies and an increase in spike number but only B fibers change their first spike latency response if the EOD is phase-shifted. These differences in response properties likely enable the distinction of resistant and capacitive objects (von der Emde and Bleckmann, 1992). Electrosensory fibers of the head project via the aLL and those of the trunk via the pLL and terminate in distinct regions of the ELL (Bell et al., 1981; Bell and Maler, 2005). Similar to the ELL of *G. niloticus*, there is a somatotopy in the ELL of mormyrids with fibers from the anterior part of the body terminating anteriorly and those of the posterior body terminating posteriorly. The cortex of the ELL is divided into three zones: the ventrolateral zone which exclusively receives input of the ampullary receptors afferents, the dorsolateral zone where afferents of the B-cells of the mormyromasts terminate, and the medial zone which exclusively receives A-cell mormyromast afferent input. These three zones are layered and contain a great variety of cell types. In addition to the three-layered zones, the ELL also contains a nucleus that receives knollenorgan afferents. The somatopy of the nucleus is rather coarse. Knollenorgans are exquisitely tuned to the EODs of conspecifics and play an important role in electrocommunication (Baker et al., 2013; Bell et al., 1989). Unique to this group is that all mechanosensory lateral line fibers and electroreceptors project together to the ELL, but with limited overlap in the ELL and the crista cerebellaris (eminentia granularis). Both lateral line and electroreceptors project bilaterally (Bell and Maler, 2005).

In the following, we will briefly summarize the role of different genes in the development of the formation of central nuclei (for a more detailed review, see Elliott et al., 2021; Fritzsch and Elliott, 2017b). The formation of central nuclei is mediated in part by basic helix-loop-helix (bHLH) transcription factors (TFs; Figure 9.6) that also are important in auditory and/or vestibular nuclei development (*Atoh1*, *Neurog1*, *Ptf1a*), and their expression is specified by the dorsoventral patterning morphogen gradients in the hindbrain (bone morphogenetic proteins (*BMPs*), *Wnts*, and sonic

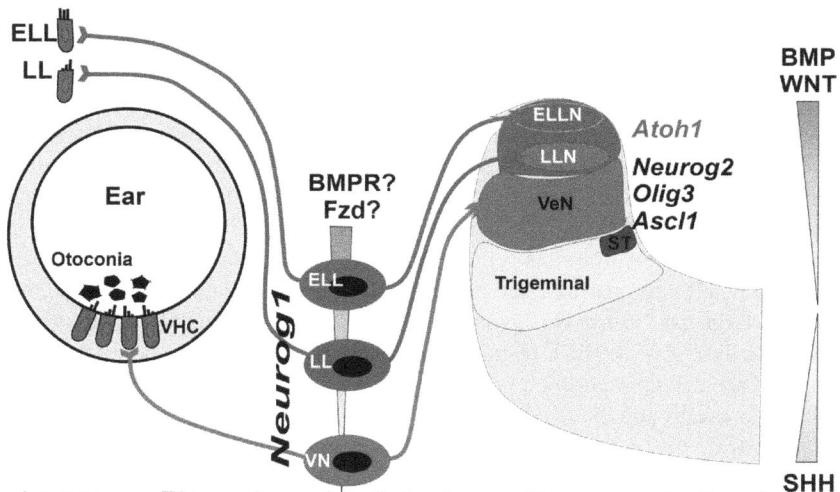

Lampreys, Elasmobranchs, Polypterus, Sturgeons, Latimeria, Lungfish, many Salamander, Caecilians, some Teleosts

FIGURE 9.6   Scheme of differences in vestibular, auditory, mechanoreception and electroreception organ projections in the hindbrain as revealed by tracing experiments. The various peripheral sensory organs (vestibular [VN], auditory [AN], mechanosensory lateral line [LL], or electrosensory ampullary organs [ELL]) each have unique central projections that terminate in distinct, non-overlapping regions in the hindbrain. Formation of central nuclei is mediated in part by bHLH TFs that also are important in ear neurosensory development (*Atoh1, Neurog1*), and their expression is specified by the dorsoventral patterning morphogen gradients in the hindbrain (*BMPs*, *Wnts*, and *Shh*). (Modified from Elliott and Fritzsch, 2020.)

hedgehog (*Shh*). *Neurog1* is important in sensory neuron differentiation, likely including all lateral line and electroreceptors (Elliott et al., 2021; Kageyama et al., 2018). *Atoh1* is expressed dorsally along the spinal cord and hindbrain to the cerebellum in mice (Bermingham et al., 2001). Similarly, BMPs and WNTs are dominant in the roof plate. They comprise the main signaling pathways of TFs that set up dorsal cell type identity (Briscoe and Ragsdale, 2018; Hernandez-Miranda et al., 2017; Lai et al., 2016). *Shh*, on the other hand, is produced at the floor plate and is instrumental for the formation of ventral cell type identities by activating or repressing the expression of TFs in a concentration-dependent manner. Recent work showed that the deletion of both *Lmx1a* and *Lmx1b* leads to the loss of the auditory nuclei in mice (Chizhikov et al., 2021). This loss in *Lmx1a/b* null mice is likely due to the loss of the downstream genes, *Atoh1* and certain Wnts. In line with the data from *Lmx1a/b* null mice, it can be postulated that neither lamprey nor hagfish would develop any electroreceptor or lateral line nuclei following loss of the *Lmx1a/b* gene.

Among bony fish, only zebrafish data have provided insights to the role of different genes in development. A genome duplication of *Atoh1* into *Atoh1a* and *Atoh1b* is believed to have occurred early in the teleost lineage. Overlapping but yet distinct functions of gene duplicates likely reflect evolutionary 'sub-functionalization' (Force et al., 1999). These genes cross-regulate each other but are differentially required during distinct developmental periods, first in the preotic placode and later in the otic vesicle. In zebrafish, only *Atoh1b* but not *Atoh1a* is required for development of tether cells, which are analogous to primary neurons. This probably reflects an ancestral *Atoh1* function. *Atoh1a* on the other hand is essential for later hair cell development, which continues to form well beyond embryonic development, which is probably also an ancestral *Atoh1* function (Millimaki et al., 2007; 2010; Wullimann et al., 2011). In addition, the ubiquitous early gene expression of *Sox2* necessary for early proneural bHLH gene expression (Kageyama et al., 2018) show interesting interactions of *Sox2* and *Sox3* with downstream genes of *Atoh1, Neurod1*, and *Neurog1* (Gou et al., 2018; Undurraga et al., 2019). Knockdown of *Sox2* does not prevent hair cell production; however, *Sox2* is required for hair cell survival, as well as for transdifferentiation of support cells into hair cells during regeneration (Millimaki et al., 2010).

In summary, the DN projection is common among non-teleost vertebrates. In contrast, electroreceptive projections in teleosts are unique and show an astonishing complexity compared to other vertebrates (Figures 9.5). We are beginning to gain novel insight about the development of these systems driven by newly characterized genes that are essential for all dorsal parts of the hindbrain (Figure 9.6). The common theme might be that these genes are driving development of the dorsal parts of the alar plate, including dorsal, intermediate, and auditory nuclei (Elliott and Fritzsch, 2020).

9.5 HIGHER-ORDER PROJECTIONS OF ELECTROSENSORY NUCLEI TO THE TORUS AND INTERACTION WITH CORTICAL INPUT

Lampreys possess an octavolateralis area that is segregated into three areas receiving inputs from different modalities – the intermediate nucleus (IN, corresponding to MON) for lateral line inputs, the ventral octavolateralis column (VN) for octaval

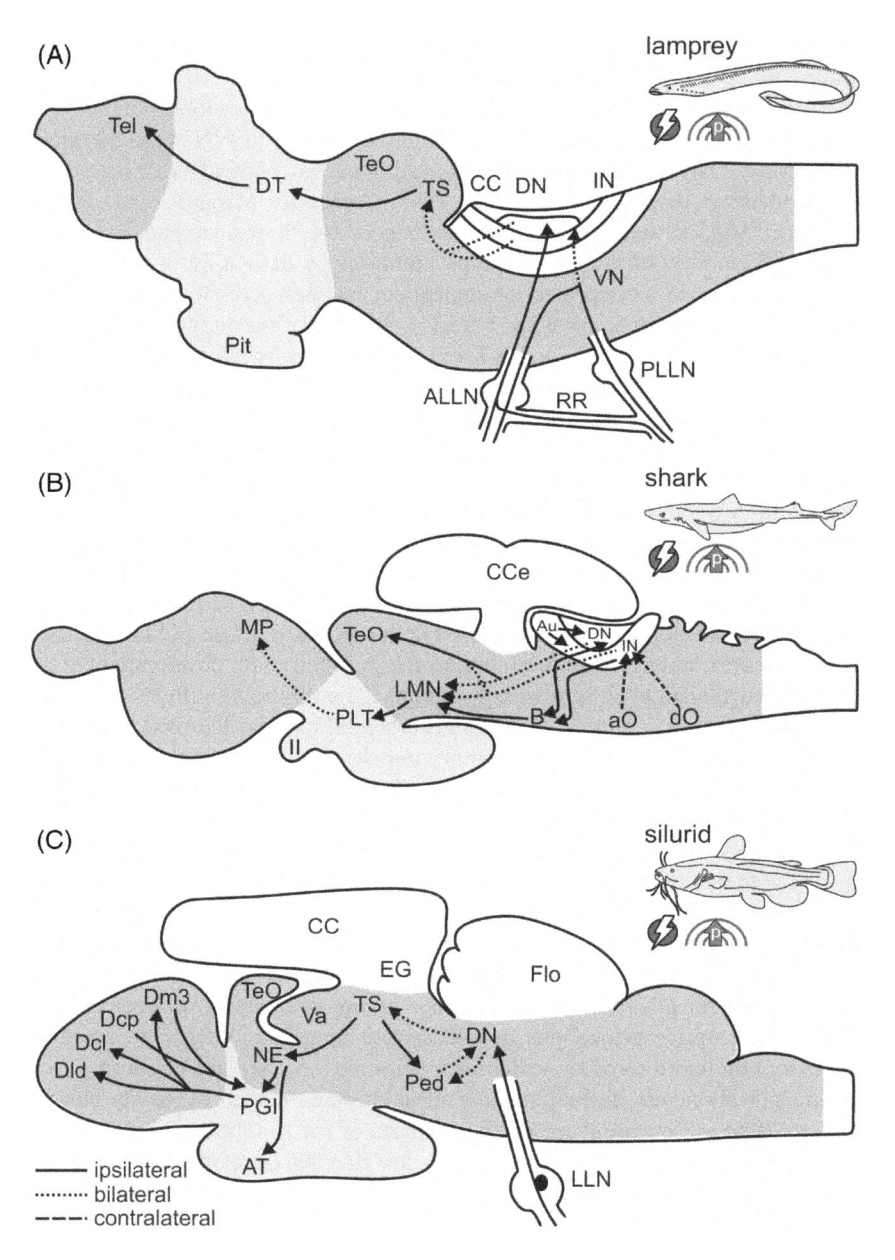

FIGURE 9.7   Electroreceptive pathways in passive electric fish. (A) In lampreys, electro-receptors are innervated by lateral line nerves that reach the dorsal nucleus via a recurrent ramus. Fibers project bilaterally to the torus semicircularis, continue to the dorsal thala-mus, and terminate in the telencephalon. (B) Comparable central projections terminate in the torus (or lateral mesencephalic nucleus) in sharks (B) and silurids (C). Projections extend via the posterior lateral thalamic nucleus to the medial pallium in sharks (B). In silurids, the torus innervates a pretectum nucleus electrosensorius to provide feedback to the DN

FIGURE 9.7 *(Continued)* via preeminential nuclei and contacts the lateral preglomerular nucleus where various branches reach to the dorsal telencephalon. aO, anterior octaval nucleus; Au, auricle; ALLN, anterior lateral line nerves; B, nucleus B; CC, cerebellar crest; CCe, corpus cerebelli; Dcl, Dcp, and Dld Dm3, centrolateral, centroposterior, and laterodorsal, medial zone 3 of pallial area dorsalis telencephalic; DN, dorsal nucleus; dO, descending octaval nucleus; DN, dorsal (octavolateralis) nucleus; DT, dorsal thalamus; EG, eminentia granularis; Flo, facial lobe; II optic nerve; IN, intermediate nucleus; LLN, lateral line nerves; LMN, lateral mesencephalic nucleus; MP, medial pallium; NE, nucleus electrosensorius (pretectum); Ped, dorsal preeminential nucleus; Pit, pituitary; PLLN, posterior lateral line nerve; PLT, posterior lateral thalamic nucleus; RR, recurrent ramus; Tel, telencephalon (primordium hippocampi); TeO, optic tectum; TS, torus semicircularis; Va, valvula cerebelli; VN, ventral octavolateralis column. (Modified from Wullimann and Grothe, 2013.)

inputs and the dorsal nucleus (DN; corresponding to DON) for electrosensory inputs (Elliott and Fritzsch, 2020). Both mechanosensory and electrosensory information is transferred via bilateral projections to the torus semicircularis (TSl), acting as a potential center for multimodal integration (González et al., 1999b; Pombal and Megías, 2019). Electrosensory fibers can be followed from the torus (Figure 9.7a) to the tectum (Bodznick and Northcutt, 1981) and reach the ipsilateral pallium (the so-called primordium hippocampi) via the ventral and dorsal thalamus. A few fibers project back to the torus from the telencephalon, specifically only from the medial pallium, but not from the lateral pallium (González et al., 1999a; Northcutt and Wicht, 1997; Polenova and Vesselkin, 1993).

In **cartilaginous fish** (comprising *elasmobranchs* including sharks, skates, rays, and *holocephalans* with only one recent group, the Chimaerans), electrosensory pathways coming from the DN project bilaterally to the optic tectum (TeO) and to the lateral mesencephalic nucleus (LMN). This nucleus is considered to be the chondrichthyan equivalent of the torus semicircularis (Bleckmann, 2020; Wullimann and Grothe, 2013). While the mechanosensory fibers terminate in the dorsolateral LMN, the electrosensory projections target the ventromedial LMN (Elliott and Fritzsch, 2020). From here, ascending input reaches a thalamic nucleus which provides input to mainly contralateral regions of the dorsal telencephalon (Figure 9.7b; Boord and Montgomery, 1989; Smeets and Northcutt, 1987). In skates, DN ascending fibers innervate the torus and the optic tectum (Bell and Maler, 2005). The DN also has a major somatotopic projection to the LMN which feeds back to the DN via a descending connection to the paralemniscal nucleus. In addition, it has a recurrent connection to the lateral posterior nucleus of the diencephalon which ascends to the medial pallium of the telencephalon (Bleckmann, 2020; Bodznick and Boord, 1986; Wullimann and Grothe, 2013).

Similar to lampreys and chondrichthyans, the electrosensory afferents from lateral line nerve fibers in **ray-finned fishes** like bichir, sturgeon, and paddlefish terminate in the dorsal nucleus (McCormick, 1989; Piotrowski and Northcutt, 1996). It is unclear whether or not the DN is topographically organized (Hofmann et al., 2005; Montgomery et al., 2012). Electrosensory pathways arising from the DN reach the optic tectum, the torus semicircularis and the LMN (Bleckmann, 2020; Hofmann et al., 2005). Differences in receptive fields and direction-selectivity of paddlefish'

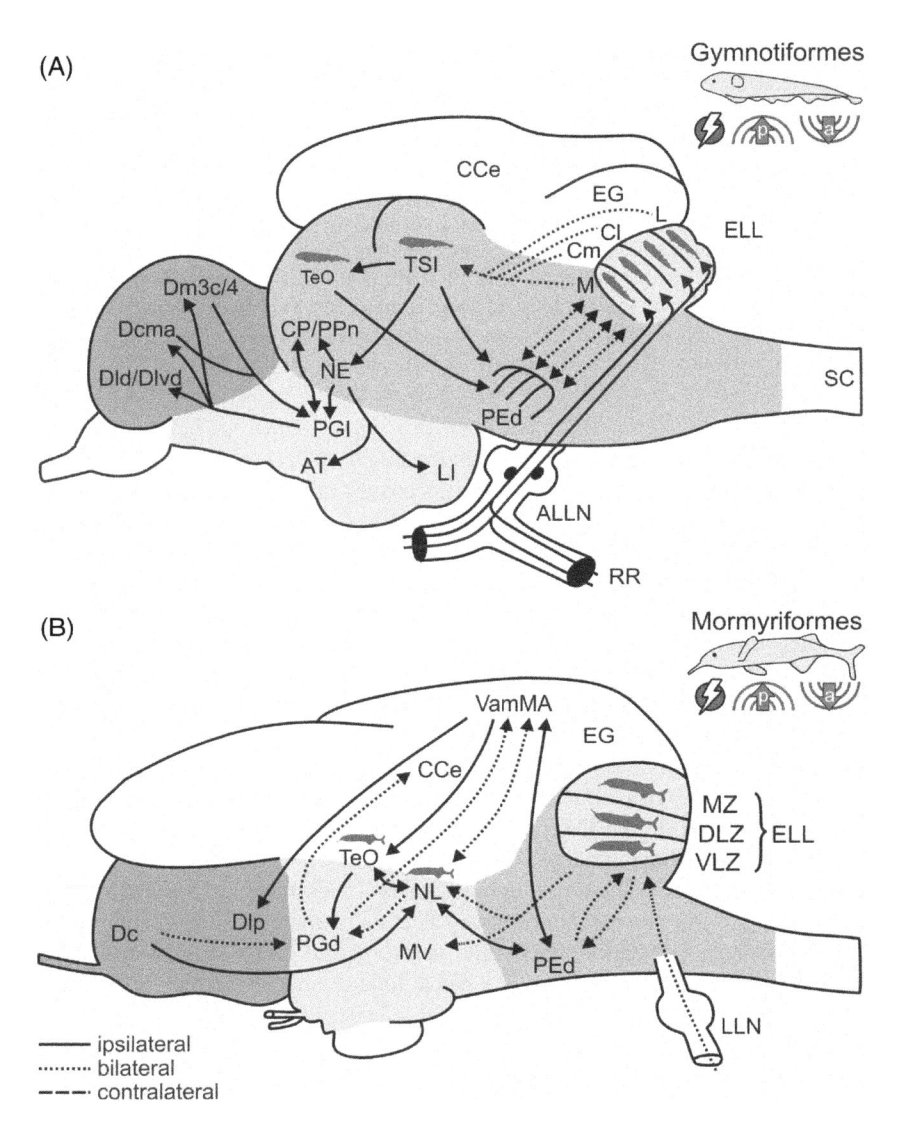

FIGURE 9.8 Electroreceptive pathways in active electric fish. (A) The gymnotids' electronsensory lateral line lobe (ELL) consists of four topographic segments that receive recurrent feedback via the preeminential nucleus (fish icons indicate the topographic representation of electrosensory surface). Torus and tectum feed back to the PEd that also gets direct ELL input and topographically projects back to the ELL. This loop has been found to selectively enhance salient electrosensory information (see Clarke and Maler, 2017). Torus connects to the nucleus electrosensorius that plays a role in the jamming avoidance response. The preglomerual nucleus distributes the electrosensory input into multiple pallial layers, including strong input to DL (see Giassi et al., 2012). (B) The ELL of mormyrids is separated into three zones (ventral, dorsal, and medial zone), each with a somatotopic map of the ELL (fish icons indicate the topographic representation of electrosensory surface). The torus (lateral nucleus) is part of the feedback loop from the PEd and disperses information into pallial layers via the PGd. ALLN, anterior lateral line nerve; AT, anterior tuberal nucleus; CCe, corpus

FIGURE 9.8 *(Continued)* cerebelli; Cl/CM, centrolateral/centromedial segment of ELL; Cp/PPn, central posterior thalamic/prepacemaker nucleus; Dc/Dlp, central/lateroposterior zone of pallial area dorsalis telencephali; Dcma, anterior part of centromedial zone of pallial area dorsalis telencephali; Dld/Dlvd, laterodorsal/dorsal part of lateroventral zone of pallial area dorsalis telencephli; DLZ, dorsolateral zone of ELL; DM, medial zone of pallial area dorsalis telencephali; ELL, electrosensory lateral line lobe; EG, eminentia granularis; L, lateral; ELL, segment; LLN, lateral line nerves; M, medial; ELL, segment (ampullary organs); MV, medioventral nucleus of torus semicircularis; LI, lobus inferior; LLN, lateral line nerves; MZ, medial zone of ELL; NE, nucleus electrosensorius (pretectum); NL, lateral nucleus of torus semicircularis; PEd, dorsal preeminential nucleus; PGd, dorsal part of preglomerular nucleus; PGl, lateral preglomerular nucleus; RR, recurrent ramus of ALLN; SC, spinal cord; TeO, optic tectum; TSl, lateral ventral nucleus of torus semicircularis; VamMA, medial leaf of valvula cerebelli (mormyromast/ampullary); VLZ, ventrolateral zone of ELL. ([A] Modified from Carr et al., 1982; Wullimann and Grothe, 2013; [B] Modified from Wullimann and Grothe, 2013; Zeymer et al., 2018.)

LMN and optic tectum in response to electrosensory stimuli suggest separate roles of these regions in electrosensory processing (LMM for longer-range orienting responses and tectum for prey capture; Chagnaud et al., 2008; 2017; Hofmann et al., 2008; Wullimann and Grothe, 2013).

Central reciprocal projections from the ventral torus nucleus have been described in detail for silurids (Figure 9.7c), gymnotids, and mormyrids (Figure 9.8; Wullimann and Grothe, 2013).

In **silurids**, electroreceptors from both anterior and posterior fibers reach the DN in a topological fashion selectively separated by the cerebellar crest from the lateral line input (Figure 9.7c; Bleckmann et al., 1991; Tong and Finger, 1983). It is unclear whether the ascending lateral line and electrosensory information is processed in parallel or is integrated (Wullimann and Grothe, 2013). The DN sends electrosensory information to the torus as well as to the preeminential nucleus. The torus provides input to the preeminential nucleus which provides feedback back to the DN. A separate recurrent system is provided through parallel fibers that terminate in the molecular zone of the DN, originating from the eminentia granularis (EG; Finger, 1986; Finger and Tong, 1984). Electrosensory information is transferred to the pretectal nucleus electrosensorius and the lateral preglomerular nucleus to extend to the pallium in various areas (Wullimann and Grothe, 2013).

In **gymnotids**, electrosensory information reaches the ELL via the anterior lateral line nerve where it disperses into four zones (three receiving tuberous and one receiving ampullary afferents) which all have a topographic representation of the receptor input (Carr et al., 1986; Metzner, 1999). Second-order projection from here passes the torus (TSl) bilaterally to the optic tectum (TeO). Both TeO and TSl send topographically ordered electrosensory feedback to the preeminent nucleus (PEd) through which this feedback reaches the ELL (Bell and Maler, 2005; Lannoo et al., 1990). The TeO might provide the pallium via the dorsal subdivision of the dorsolateral telencephalon (DL) with electrolocation signals that could be used for learning associations regarding the electrosensory environment. A descending input from the central division of the dorsal telencephalon (DM) to TeO, TSl and the nucleus preementialis might enable adjusted behavioral responses (Giassi et al., 2012).

In addition, the TSl sends non-topographic information to the nucleus electrosensorius (NE) which plays a role in the jamming avoidance responses (JAR) contacting the prepacemaker nucleus (pPN) that triggers electric organ discharges (Bell and Maler, 2005) as well as to the lateral preglomerular nucleus (PGI; Figure 9.8a). The PGI serves as the main hub that provides electrosensory input to telencephalic DL (Wullimann and Grothe, 2013). A descending connection of the DC to the NE might allow for memory-related adaptation of electrocommunication signals (Giassi et al., 2012).

Mormyrids and their monotypic sister group Gymnarchidae, with *Gymnarchus niloticus* as the only representative, share an enlarged electrosensory lateral line lobe (Szabo, 1974). Electrosensory and mechanosensory input is transmitted in parallel by both anterior and posterior nerve ganglia to end in three zones, each with a somatotopic map of the electrosensory surface (Meek et al., 1999). Ampullary receptors for passive electrolocation terminate in the ventrolateral ELL, tuberous receptors for active electrolocation target the medial and dorsolateral zones (Figure 9.8b; Bell and Szabo, 1986). Fibers from knollenorgans, the organs dedicated to electrocommunication, project to the nucleus of the ELL from where they largely ascend along the other lemniscal projections of the electroreceptive pathways to the torus (Bell et al., 1989).

The torus is parcellated into five nuclei in mormyrids with medioventral (MV) and lateral toral nuclei (NL) receiving parallel input from ELL. Both process different aspects of the electrosensory input, i.e. the lateral torus somatotopically represents dorsoventral and rostro-caudal ELL maps, maintaining phase and amplitude information from mormyromasts (Hollmann et al., 2016) while the medioventral nucleus is one of the few sites converging information from mormyromast, ampullary, and knollenorgan receptors (Bell and Maler, 2005). MV and NL provide feedback to the ELL through the eminentia granularis and the preeminential nucleus (PEd), where the three ELL cortical maps converge into one (Figure 9.8b). This is different in Gymnotiformes where the segregation of the ELL into four segments is preserved in PEd (Bell et al., 1989; Bell and Maler, 2005). From the NL electrosensory information topographically reaches the optic tectum and the preglomerular nucleus (PGd) that projects to the dorsal pallium (Bell and Maler, 2005; Wullimann and Grothe, 2013). A unique pathway for electrosensory input to the pallium arises from a direct cerebello-thalamic projection (Wullimann and Rooney, 1990). The input to the valvula originates from reciprocal connections with the NL and PGd (Bell et al., 1981; Wullimann and Grothe, 2013).

The higher-order projections in electroreceptive fish species are comparable to central projections of mechano- and electroreceptors in **amphibians** (Fritzsch and Neary, 1998; Münz et al., 1984; Northcutt et al., 1994). In axolotls, higher-order neurons in the striatum receive electroreceptive information and, besides visual and mechanosensory maps, an electroreceptive topography of the sensory periphery is present in the optic tectum (Bartels et al., 1990). This projection is similar to the lateral line system in non-electric *Xenopus*, which gets mechanosensory input via ascending lateral lemniscal fibers to the torus semicircularis and the optic tectum (Dean and Claas, 2020; Mohr and Görner, 1996; Wullimann and Grothe, 2013).

In summary, ascending electrosensory pathways are comparable in all electroreceptive vertebrates. Central electroreceptive contralateral projections extend from the first medullary nucleus to reach the midbrain (i.e. torus semicircularis). From here, multiple branches send recurrent feedback to the ELL either directly to the first-order structures themselves and/or indirectly via the eminentia granularis (Bell and Maler, 2005). Consistently the torus further targets thalamic regions and the optic tectum. Except for the cerebello-telencephalic input in mormyrids, the thalamic nuclei provide the major electroreceptive pathways to the dorsal telencephalon, for which – except for gymnotids, details are as of yet scarce.

9.6   SUMMARY AND CONCLUSION

Electroreception is a major sensory modality found in many aquatic vertebrates which has apparently been reinvented several times independently. With its finely tuned interplay of distinct receptors and central nuclei in the medulla, electroreception constitutes as a perfect paradigm for studying evolutionary questions such as convergence. Non-teleost ampullary receptors share a common placodal origin with the lateral line, whereas the embryonic origin of both lamprey and teleost electroreceptors is largely unknown. Electroreceptor 'hair cells' have either a short kinocilium surrounded by multiple larger stereovilli or a single large kinocilium surrounded by shorter stereovilli/microvilli. Ampullary receptors of non-teleosts respond to cathodal low-frequency signals, whereas teleost ampullary receptors respond to anodal low-frequency signals. The projection patterns of the ampullary system to the primary hindbrain region are relatively conserved in all vertebrates. Although there is still a lot to learn about the development of these hindbrain regions, it is strongly evident that similar gene sets are involved in the formation of different hindbrain nuclei (e.g. electrosensory, auditory, and/or vestibular nuclei). In particular, bHLH TFs and their expression patterns along the dorsoventral axis of the hindbrain seem to be crucial.

In addition to the low-frequency passive electroreception, several species of teleost fish are capable of active electroreception, i.e. measuring the distortion of a self-generated electric field produced by an electric organ in their tail. These high-frequency signals require a new receptor type, namely tuberous receptors that appear evolutionarily independent at least two times within teleost fish. Electroreceptive projections in teleosts are very complex as projections from different receptors innervate various areas of the hindbrain creating numerous somatotopic maps which allow for parallel processing of electrosensory information. In derived teleosts, additional feedback loops enable an adaptive filtering of reafference signals already in the hindbrain and might furthermore provide mechanisms for more complex computational tasks such as motion processing and spatial attention. Ascending electrosensory pathways project to midbrain areas (i.e. torus semicircularis) and further to thalamic regions as well as the optic tectum, and to higher-order brain areas such as the telencephalon, allowing for the generation of behavior adapted to the electrosensory scene.

REFERENCES

Amin Karami, M., Inman, D.J., 2012. Powering pacemakers from heartbeat vibrations using linear and nonlinear energy harvesters. Applied Physics Letters 100, 042901.

Baker, C.A., Kohashi, T., Lyons-Warren, A.M., Ma, X, Carlson, B.A., 2013. Multiplexed temporal coding of electric communication signals in mormyrid fishes. Journal of Experimental Biology 216, 2365–2379.

Baker, C.V., 2019. The development and evolution of lateral line electroreceptors: Insights from comparative molecular approaches, in: Electroreception: Fundamental Insights from Comparative Approaches. New York: Springer Nature Switzerland AG, pp. 25–62.

Baker, C.V., Modrell, M.S., 2018. Insights into electroreceptor development and evolution from molecular comparisons with hair cells. Integrative and Comparative Biology 58, 329–340.

Baker, C.V.H., Modrell, M.S., Gillis, J.A., Krahe, R., Fortune, E., 2013. The evolution and development of vertebrate lateral line electroreceptors. The Journal of Experimental Biology 216 (13), 2515–2522. doi: 10.1242/jeb.082362.

Bartels, M., Münz, H., Claas, B., 1990. Representation of lateral line and electrosensory systems in the midbrain of the axolotl, *Ambystoma mexicanum*. Journal of Comparative Physiology A 167, 347–356.

Bass, A.H., Hopkins, C.D., 1983. Hormonal control of sexual differentiation: Changes in electric organ discharge waveform. Science 220, 971–974.

Bell, C.C., Bodznick, D., Montgomery, J., Bastian, J., 1997. The generation and subtraction of sensory expectations within cerebellum-like structures. Brain Behavior Evolution 50, 19–31.

Bell, C.C., Finger, T., Russell, C., 1981. Central connections of the posterior lateral line lobe in mormyrid fish. Experimental Brain Research 42, 9–22.

Bell, C.C., Han, V., Sawtell, N.B., 2008. Cerebellum-like structures and their implications for cerebellar function. Annual Review of Neuroscience 31, 1–24.

Bell, C.C., Libouban, S., Szabo, T., 1983. Pathways of the electric organ discharge command and its corollary discharges in mormyrid fish. Journal of Comparative Neurology 216, 327–338.

Bell, C.C., Maler, L., 2005. Central neuroanatomy of electrosensory systems in fish, in: Electroreception. New York: Springer, pp. 68–111.

Bell, C.C., Szabo, T., 1986. Electroreception in mormyrid fish: Central anatomy. Electroreception 375, 421.

Bell, C.C., Zakon, H., Finger, T., 1989. Mormyromast electroreceptor organs and their afferent fibers in mormyrid fish: I. Morphology. Journal of Comparative Neurology 286, 391–407. doi: 10.1002/cne.902860309.

Bellono, N.W., Leitch, D.B., Julius, D., 2017. Molecular basis of ancestral vertebrate electroreception. Nature 543, 391.

Bellono, N.W., Leitch, D.B., Julius, D., 2018. Molecular tuning of electroreception in sharks and skates. Nature 558, 122.

Bermingham, N.A., Hassan, B.A., Wang, V.Y., Fernandez, M., Banfi, S., Bellen, H.J., Fritzsch, B., Zoghbi, H.Y., 2001. Proprioceptor pathway development is dependent on Math1. Neuron 30, 411–422.

Bleckmann, H., 2020. Central lateral line pathways and central integration of lateral line information., in: Fritzsch, B. (Ed.), The Senses. Amsterdam and London: Elsevier, pp. 163–184.

Bleckmann, H., Bullock, T., Jørgensen, J., 1987. The lateral line mechanoreceptive mesencephalic, diencephalic, and telencephalic regions in the thornback ray, *Platyrhinoidis triseriata* (Elasmobranchii). Journal of Comparative Physiology A 161, 67–84.

Bleckmann, H., Niemann, U., Fritzsch, B., 1991. Peripheral and central aspects of the acoustic and lateral line system of a bottom dwelling catfish, *Ancistrus* sp. Journal of Comparative Neurology 314, 452–466.

Bodznick, D., Boord, R., 1986. Electroreception in *Chondrichthye*s: Central anatomy and physiology. Electroreception 8, 225–256.

Bodznick, D., Montgomery, J.C., 2005. The physiology of low-frequency electrosensory systems, in: Electroreception. New York: Springer, pp. 132–153.

Bodznick, D., Northcutt, R.G., 1981. Electroreception in lampreys: Evidence that the earliest vertebrates were electroreceptive. Science 212, 465–467.

Bolz, D., Fritzsch, B., 1986. On the development of electroreceptive ampullary organs of Triturus alpestris (Amphibia: Urodela). Amphibia-Reptilia 7, 1–9.

Boord, R.L., Montgomery, J.C., 1989. Central mechanosensory lateral line centers and pathways among the elasmobranchs, in: The Mechanosensory Lateral Line. New York: Springer, pp. 323–339.

Braford Jr, M., 1986. African Knifefishes. The Xenomystines. Electroreception. New York: John Wiley & Sons, 453–464.

Braun, C.B., Northcutt, R.G., 1997. The lateral line system of hagfishes (Craniata: Myxinoidea). Acta Zoologica 78, 247–268.

Briscoe, S.D., Ragsdale, C.W., 2018. Homology, neocortex, and the evolution of developmental mechanisms. Science 362, 190–193.

Bullock, T.H., Heiligenberg, W., 1986. Electroreception. New York: Wiley and Sons.

Bullock, T.H., Hopkins, C.D., Fay, R.R., 2006. Electroreception. New York: Springer Science & Business Media.

Carlson, B.A., Sisneros, J.A., 2019. A Brief History of Electrogenesis and Electroreception in Fishes, Electroreception: Fundamental Insights from Comparative Approaches. Switzerland: Springer Nature Switzerland AG, pp. 1–23.

Carr, C.E., Maler, L., 1985. A Golgi study of the cell types of the dorsal torus semicircularis of the electric fish Eigenmannia: Functional and morphological diversity in the midbrain. Journal of Comparative Neurology 235, 207–240.

Carr, C.E., Maler, L., Sas, E., 1982. Peripheral organization and central projections of the electrosensory nerves in gymnotiform fish. Journal of Comparative Neurology 211, 139–153.

Carr, C.E., Maler, L., Taylor, B., 1986. A time-comparison circuit in the electric fish midbrain. II. Functional morphology. Journal of Neuroscience 6, 1372–1383.

Chagnaud, B.P., Engelmann, J., Fritzsch, B., Glover, J.C., Straka, H., 2017. Sensing external and self-motion with hair cells: A comparison of the lateral line and vestibular systems from a developmental and evolutionary perspective. Brain, Behaviour and Evolution 90, 98–116.

Chagnaud, B.P., Wilkens, L.A., Hofmann, M.H., 2008. Receptive field organization of electrosensory neurons in the paddlefish (*Polyodon spathula*). Journal of Physiology-Paris 102, 246–255.

Chizhikov, V.V., Iskusnykh, I.Y., Fattakhov, N., Fritzsch, B., 2021. Lmx1a and Lmx1b are Redundantly required for the development of multiple components of the mammalian auditory system. Neuroscience 452, 247–264.

Clarke, S.E., Longtin, A., Maler, L., 2014. A neural code for looming and receding motion is distributed over a population of electrosensory ON and OFF contrast cells. Journal of Neuroscience 34, 5583–5594.

Clarke, S.E., Maler, L., 2017. Feedback synthesizes neural codes for motion. Current Biology 27, 1356–1361.

Dean, J., Claas, B., 2020. Hydrodynamic sensing by the African clawed frog, *Xenopus laevis*, in: Fritzsch, B. (Ed.), The Senses. Amsterdam and London: Elsevier, pp. 185–214.

Di Bonito, M., Studer, M., Puelles, L., 2017. Nuclear derivatives and axonal projections originating from rhombomere 4 in the mouse hindbrain. Brain Structure and Function 222, 3509–3542.

Du Bois-Reymond, E., 1848. Untersuchungen ueber thierische Elektricitie. Berlin: Reimer.

Elliott, K.L., Fritzsch, B., 2020. Evolution and development of lateral line and electroreception: An integrated perception of neurons, hair cells and brainstem nuclei, in: Fritzsch, B. (Ed.), The Senses. Elsevier, pp. 95–115.

Elliott, K.L., Pavlínková, G., Chizhikov, V.V., Yamoah, E.N., Fritzsch, B., 2021. Development in the mammalian auditory system depends on transcription factors. International Journal of Molecular Sciences 22, 4189.

Engelmann, J., Gertz, S., Goulet, J., Schuh, A., von der Emde, G., 2010. Coding of stimuli by ampullary afferents in *Gnathonemus petersii*. Journal of Neurophysiology 104 (4), 1955–1968.

Finger, T., 1986. Electroreception in catfish: Behavior, anatomy, and electrophysiology. Electroreception, 287–317.

Finger, T.E., Tong, S.L., 1984. Central organization of eighth nerve and mechanosensory lateral line systems in the brainstem of *Ictalurid* catfish. Journal of Comparative Neurology 229, 129–151.

Force, A., Lynch, M., Pickett, F.B., Amores, A., Yan, Y.-l., Postlethwait, J., 1999. Preservation of duplicate genes by complementary, degenerative mutations. Genetics 151, 1531–1545.

Fortune, E., Chacron, M., 2011. Physiology of tuberous electrosensory systems. Encyclopedia of Fish Physiology: From Genome to Environment 1, 366–374.

Fritzsch, B., 1981. The pattern of lateral-line afferents in urodeles: A horseradish-peroxidase study. Cell Tissue Research 218, 581–594.

Fritzsch, B., 1989. Diversity and regression in the amphibian lateral line and electrosensory system, in: The Mechanosensory Lateral Line. New York: Springer, pp. 99–114.

Fritzsch, B., 1993. Evolutionary gain and loss of non-teleostean electroreceptors. Journal of Comparative Physiology 173, 710–712.

Fritzsch, B., 1998. Evolution of the vestibulo-ocular system. Otolaryngology-Head and Neck Surgery 119, 182–192.

Fritzsch, B., Barbacid, M., Silos-Santiago, I., 1998. Nerve dependency of developing and mature sensory receptor cells a. Annals of the New York Academy of Sciences 855, 14–27.

Fritzsch, B., de Caprona, M.-D.C., Wächtler, K., Körtje, K.-H., 1984. Neuroanatomical evidence for electroreception in lampreys. Zeitschrift für Naturforschung C 39, 856–858.

Fritzsch, B., Elliott, K.L., 2017a. Evolution and development of the inner ear efferent system: Transforming a motor neuron population to connect to the most unusual motor protein via ancient nicotinic receptors. Frontiers in Cellular Neuroscience 11, 114.

Fritzsch, B., Elliott, K.L., 2017b. Gene, cell, and organ multiplication drives inner ear evolution. Developmental Biology 431, 3–15.

Fritzsch, B., Gregory, D., Rosa-Molinar, E., 2005. The development of the hindbrain afferent projections in the axolotl: Evidence for timing as a specific mechanism of afferent fiber sorting. Zoology 108, 297–306.

Fritzsch, B., Neary, T., 1998. The octavolateralis system of mechanosensory and electrosensory organs. Amphibian Biology 3, 878–922.

Fritzsch, B., Northcutt, R.G., 1993. Cranial and spinal nerve organization in amphioxus and lampreys: Evidence for an ancestral craniate pattern. Acta Anatomica 148, 96–109.

Fritzsch, B., Wahnschaffe, U., 1983. The electroreceptive ampullary organs of urodeles. Cell and Tissue Research 229, 483–503.

Fritzsch, B., Wahnschaffe, U., Caprona, M.-D.C., Himstedt, W., 1985. Anatomical evidence for electroreception in larval Ichthyophis kohtaoensis. Naturwissenschaften 72, 102–104.

Fritzsch, B., Zakon, H.H., Sanchez, D.Y., 1990. Time course of structural changes in regenerating electroreceptors of a weakly electric fish. Journal of Comparative Neurology 300, 386–404.

Galvani, L., 1792. De Viribus Electricitatis in Motu Musculari. Bologna: Universita di Bologna.

Gauthier, A.R.G., Whitehead, D.L., Tibbetts, I.R., Cribb, B.W., Bennett, M.B., 2018. Morphological comparison of the ampullae of Lorenzini of three sympatric benthic rays. Journal of Fish Biology 92(2), 504–514. doi: 10.1111/jfb.13531.

Gelman, S., 2007. The mechanosensory lateral line system: Morphological, physiological, and behavioral study in pre-and post-metamorphic lampreys.

Giassi, A.C., Duarte, T.T., Ellis, W., Maler, L., 2012. Organization of the gymnotiform fish pallium in relation to learning and memory: II. Extrinsic connections. Journal of Comparative Neurology 520, 3338–3368.

Gilland, E., Baker, R., 1993. Conservation of neuroepithelial and mesodermal segments in the embryonic vertebrate head. Cells Tissues Organs 148, 110–123.

Gillis, J.A., Modrell, M.S., Northcutt, R.G., Catania, K.C., Luer, C.A., Baker, C.V.J.D., 2012. Electrosensory ampullary organs are derived from lateral line placodes in cartilaginous fishes. 139, 3142–3146.

González, A., Smeets, W.J., Marín, O., 1999a. Evidences for shared features in the organization of the basal ganglia in tetrapods: Studies in amphibians. European Journal of Morphology 37, 151–154.

González, M.J., Yáñez, J., Anadón, R., 1999b. Afferent and efferent connections of the torus semicircularis in the sea lamprey: An experimental study. Brain Research 826, 83–94.

Gou, Y., Vemaraju, S., Sweet, E.M., Kwon, H.-J., Riley, B.B., 2018. sox2 and sox3 play unique roles in development of hair cells and neurons in the zebrafish inner ear. Developmental Biology 435, 73–83.

Hallböök, F., Wilson, K., Thorndyke, M., Olinski, R.P., 2006. Formation and evolution of the chordate neurotrophin and Trk receptor genes. Brain, Behavior and Evolution 68, 133–144.

Hernandez-Miranda, L.R., Müller, T., Birchmeier, C., 2017. The dorsal spinal cord and hindbrain: From developmental mechanisms to functional circuits. Developmental Biology 432, 34–42.

Hetherington, T.E., Wake, M.H., 1979. The lateral line system in larval *Ichthyophis* (Amphibia: Gymnophiona). Zoomorphologie 93, 209–225.

Hofmann, M.H., Chagnaud, B., Wilkens, L.A., 2005. Response properties of electrosensory afferent fibers and secondary brain stem neurons in the paddlefish. Journal of Experimental Biology 208, 4213–4222.

Hofmann, M.H., Jung, S., Siebenaller, U., Preissner, M., Chagnaud, B., Wilkens, L., 2008. Response properties of electrosensory units in the midbrain tectum of the paddlefish (*Polyodon spathula* Walbaum). Journal of Experimental Biology 211, 773–779.

Hollmann, V., Hofmann, V., Engelmann, J., 2016. Somatotopic map of the active electrosensory sense in the midbrain of the mormyrid *Gnathonemus petersii*. Journal of Comparative Neurology 524, 2479–2491.

Ingenhousz, J., 1782. Vermischte Schriften, Molitor, N.K. (Ed. and trans.). Wien: Krauss, p. 276.

Jørgensen, J.M., 1989. Evolution of octavolateralis sensory cells, in: Coombs, S., Goerner, P., Muenz, H. (Eds.), The Mechanosensory Lateral Line: Neurobiology and Evolution. New York: Springer Verlag, pp. 99–115.

Jørgensen, J.M., 2005. Morphology of electroreceptive sensory organs, in: Electroreception. New York: Springer, pp. 47–67.

Jørgensen, J.M., Bullock, T.H., 1987. Organization of the ampullary organs of the African knife fish *Xenomystus nigri* (Teleostei: Notopteridae). Journal of Neurocytology 16, 311–315.

Kageyama, R., Shimojo, H., Ohtsuka, T., 2018. Dynamic control of neural stem cells by bHLH factors. Neuroscience Research.

Kawasaki, M., Guo, Y.-X., 1998. Parallel projection of amplitude and phase information from the hindbrain to the midbrain of the African electric fish *Gymnarchus niloticus*. Journal of Neuroscience 18, 7599–7611.

Kennedy, A., Wayne, G., Kaifosh, P., Alvina, K., Abbott, L.F., Sawtell, N.B, 2014. A temporal basis for predicting the sensory consequences of motor commands in an electric fish. Nature Neuroscience 17, 416–422.

Kersigo, J., D'Angelo, A., Gray, B.D., Soukup, G.A., Fritzsch, B., 2011. The role of sensory organs and the forebrain for the development of the craniofacial shape as revealed by Foxg1-cre-mediated microRNA loss. Genesis 49, 326–341.

Kersigo, J., Fritzsch, B., 2015. Inner ear hair cells deteriorate in mice engineered to have no or diminished innervation. Frontiers in Aging Neuroscience 7, 33.

Khosravi-Hashemi, N., Chacron M.J., 2014. Motion processing across multiple topographic maps in the electrosensory system. Physiological Reports 2(3), 1–13.

Kishida, R., Goris, R., Nishizawa, H., Koyama, H., Kadota, T., Amemiya, F., 1987. Primary neurons of the lateral line nerves and their central projections in hagfishes. Journal of Comparative Neurology 264, 303–310.

Koyama, H., Kishida, R., Goris, R., Kusunoki, T., 1993. Giant terminals in the dorsal octavolateralis nucleus of lampreys. Journal of Comparative Neurology 335, 245–251.

Krahe, R., Maler, L., 2014. Neural maps in the electrosensory system of weakly electric fish. Current Opinion in Neurobiology 24, 13–21.

Lai, H.C., Seal, R.P., Johnson, J.E., 2016. Making sense out of spinal cord somatosensory development. Development 143, 3434–3448.

Lai, N.Y., Bell, J.M., Bodznick, D., 2021. Multiple behavior-specific cancellation signals contribute to suppressing predictable sensory reafference in a cerebellum-like structure. Journal of Experimental Biology 224, jeb240143.

Lannoo, M.J., Maler, L., Tinner, B., 1989. Ganglion cell arrangement and axonal trajectories in the anterior lateral line nerve of the weakly electric fish *Apteronotus leptorhynchus* (Gymnotiformes). Journal of Comparative Neurology 280, 331–342.

Lannoo, M.J., Maler, L., Vischer, H.A., 1990. Development of the electrosensory nervous system of Eigenmannia (gymnotiformes): II. The electrosensory lateral line lobe, midbrain, and cerebellum. Journal of Comparative Neurology 294, 37–58.

Leitch, D.B., Julius, D., 2019. Electrosensory transduction: Comparisons across structure, afferent response properties, and cellular physiology, in: Electroreception: Fundamental Insights from Comparative Approaches. Switzerland: Springer Nature Switzerland AG, pp. 63–90.

Lissmann, H., 1958. On the function and evolution of electric organs in fish. Journal of Experimental Biology 35, 156–191.

Lorenzini, S., 1678. Osservazioni Intorno Alle Torpedini L'Onofri, Florence.

Maler, L., 1979. The posterior lateral line lobe of certain gymnotiform fish. Quantitative light microscopy. Journal of Comparative Neurology 183, 323–363.

McCormick, C.A., 1989. Central lateral line mechanosensory pathways in bony fish, The Mechanosensory Lateral Line. Springer, pp. 341–364.

McCormick, C.A., 1999. Anatomy of the central auditory pathways of fish and amphibians, in: Fay, R.R., Popper, A.N. (Eds.), Comparative Hearing: Fish and Amphibians. New York: Springer-Verlag, pp. 155–217.

McCormick, C.A., Braford Jr, M.R., 1993. The primary octaval nuclei and inner ear afferent projections in the otophysan *Ictalurus punctatus*. Brain, Behavior and Evolution 42, 48–68.

Meek, J., Grant, K., Bell, C., 1999. Structural organization of the mormyrid electrosensory lateral line lobe. Journal of Experimental Biology 202, 1291–1300.

Metzner, W., 1999. Why are there so many sensory brain maps? Cellular and Molecular Life Sciences 56, 1–4.

Millimaki, B.B., Sweet, E.M., Dhason, M.S., Riley, B.B., 2007. Zebrafish atoh1 genes: Classic proneural activity in the inner ear and regulation by Fgf and Notch. Development 134, 295–305.

Millimaki, B.B., Sweet, E.M., Riley, B.B., 2010. Sox2 is required for maintenance and regeneration, but not initial development, of hair cells in the zebrafish inner ear. Developmental Biology 338, 262–269.

Modrell, M.S., Bemis, W.E., Northcutt, R.G., Davis, M.C., Baker, C.V., 2011. Electrosensory ampullary organs are derived from lateral line placodes in bony fishes. Nature Communications 2, 496.

Modrell, M.S., Hockman, D., Uy, B., Buckley, D., Sauka-Spengler, T., Bronner, M.E., Baker, C.V., 2014. A fate-map for cranial sensory ganglia in the sea lamprey. Developmental Biology 385, 405–416.

Mohr, C., Görner, P., 1996. Innervation patterns of the lateral line stitches of the clawed frog, Xenopus laevis, and their reorganization during metamorphosis. Brain, Behavior and Evolution 48, 55–69.

Moller, P., Fritzsch, B., 1993. From electrodetection to electroreception: The problem of understanding a non-human sense. Journal of Comparative Physiology A 173, 734–737.

Moller, P., Fritzsch, B., 1995. History of electroreception, in: Moller, P. (Ed.), Electric Fishes: History and Behavior. London: Chapman & Hall, pp. 238–315.

Montgomery, J.C., Bodznick, D., Yopak, K.E., 2012. The cerebellum and cerebellum-like structures of cartilaginous fishes. Brain, Behavior and Evolution 80, 152–165.

Münz, H., Claas, B., Fritzsch, B., 1984. Electroreceptive and mechanoreceptive units in the lateral line of the axolotl *Ambystoma mexicanum*. Journal of Comparative Physiology A 154, 33–44.

Newton, K.C., Gill, A.B., Kajiura, S.M., 2019. Electroreception in marine fishes: Chondrichthyans. Journal of Fish Biology 95, 135–154.

Nieuwenhuys, R., Puelles, L., 2016. The fundamental morphological units (FMUs) of the CNS. Towards a New Neuromorphology. Berlin Springer, pp. 143–196.

Nieuwenhuys, R., ten Donkelaar, H., Nicholson, C., 1998. The Central Nervous System of Vertebrates. Berlin: Springer.

Northcutt, R.G., 1992. Distribution and innervation of lateral line organs in the axolotl. Journal of Comparative Neurology 325, 95–123.

Northcutt, R.G., 2005. Ontogeny of electroreceptors and their neural circuitry, in: Electroreception. New York: Springer, pp. 112–131.

Northcutt, R.G., Brändle, K., 1995. Development of branchiomeric and lateral line nerves in the axolotl. Journal of Comparative Neurology 355, 427–454.

Northcutt, R.G., Brändle, K., Fritzsch, B., 1995. Electroreceptors and mechanosensory lateral line organs arise from single placodes in axolotls. Developmental Biology 168, 358–373.

Northcutt, R.G., Catania, K.C., Criley, B.B., 1994. Development of lateral line organs in the axolotl. Journal of Comparative Neurology 340, 480–514.

Northcutt, R.G., Wicht, H., 1997. Afferent and efferent connections of the lateral and medial pallia of the silver lamprey. Brain, Behaviour and Evolution 49, 1–19.

Petralia, R.S., Wang, Y.-X., Mattson, M.P., Yao, P.J., 2017. Invaginating presynaptic terminals in neuromuscular junctions, photoreceptor terminals, and other synapses of animals. NeuroMolecular Medicine 19, 193–240.

Pierce, M.L., Weston, M.D., Fritzsch, B., Gabel, H.W., Ruvkun, G., Soukup, G.A., 2008. MicroRNA-183 family conservation and ciliated neurosensory organ expression. Evolution & Development 10, 106–113.

Piotrowski, T., Baker, C.V., 2014. The development of lateral line placodes: Taking a broader view. Developmental Biology 389, 68–81.

Piotrowski, T., Northcutt, R.G., 1996. The cranial nerves of the Senegal Bichir, *Polypterus senegalus* [Osteichthyes: Actinopterygii: Cladistia]. Brain, Behavior and Evolution 47, 55–66.

Polenova, O.A., Vesselkin, N.P., 1993. Olfactory and nonolfactory projections in the river lamprey (*Lampetra fluviatilis*) telencephalon. Journal für Hirnforschung 34, 261–279.

Pombal, M.A., Megías, M., 2019. Development and functional organization of the cranial nerves in lampreys. The Anatomical Record 302, 512–539.

Riddiford, N., Schlosser, G., 2016. Dissecting the pre-placodal transcriptome to reveal presumptive direct targets of Six1 and Eya1 in cranial placodes. eLife 5, e17666.

Roberts, B.L., Meredith, G.E., 1992. The efferent innervation of the ear: Variations on an enigma, in: The Evolutionary Biology of Hearing. New York: Springer, pp. 185–210.

Roberts, B.L., Ryan, K., 1971. The fine structure of the lateral-line sense organs of dogfish. Proceedings of the Royal Society of London. Series B. Biological Sciences 179, 157–169.

Ronan, M., 1988. Anatomical and physiological evidence for electroreception in larval lampreys. Brain Research 448, 173–177.

Ronan, M., Bodznick, D., 1986. End buds: Non-ampullary electroreceptors in adult lampreys. Journal of Comparative Physiology A 158, 9–15.

Ronan, M., Northcutt, R.G., 1987. Primary projections of the lateral line nerves in adult lampreys. Brain, Behavior and Evolution 30, 62–81.

Roth, A., 2003. Development of catfish lateral line organs: Electroreceptors require innervation, although mechanoreceptors do not. Naturwissenschaften 90, 251–255.

Roth, A., Tscharntke, H., 1976. Ultrastructure of the ampullary electroreceptors in lungfish and Brachiopterygii. Cell and Tissue Research 173, 95–108.

Sarasin, P., Sarasin, F., 1890. Ergebnisse naturwissenschaftlicher Forschungen auf Ceylon. II. Heft 4. Zur Entwicklungsgeschichte und Anatomie derceylonischen Blindwühle, Ichthyophis glutinosus. Wiesbaden 4, 153–263.

Schlosser, G., 2002. Development and evolution of lateral line placodes in amphibians—II. Evolutionary diversification. Zoology 105, 177–193.

Smeets, W.J., Northcutt, R.G., 1987. At least one thalamotelencephalic pathway in cartilaginous fishes projects to the medial pallium. Neuroscience Letters 78, 277–282.

Soukup, G.A., Fritzsch, B., Pierce, M.L., Weston, M.D., Jahan, I., McManus, M.T., Harfe, B.D., 2009. Residual microRNA expression dictates the extent of inner ear development in conditional Dicer knockout mice. Developmental Biology 328, 328–341.

Szabo, T., 1974. Anatomy of the specialized lateral line organs of electroreception, in: Electroreceptors and Other Specialized Receptors in Lower Vertebrates. Berlin, Heidelberg: Springer, pp. 13–58.

Teeter, J., Szamier, R., Bennett, M., 1980. Ampullary electroreceptors in the sturgeon *Scaphirhynchus platorynchus* (rafinesque). Journal of Comparative Physiology 138, 213–223.

Tong, S.L., Finger, T.E., 1983. Central organization of the electrosensory lateral line system in bullhead catfish *Ictalurus nebulosus*. Journal of Comparative Neurology 217, 1–16.

Undurraga, C.A., Gou, Y., Sandoval, P.C., Nuñez, V.A., Allende, M.L., Riley, B.B., Hernández, P.P., Sarrazin, A.F., 2019. Sox2 and Sox3 are essential for development and regeneration of the zebrafish lateral line. bioRxiv, 856088.

Volta, A., 1800. XVII. On the electricity excited by the mere contact of conducting substances of different kinds. In a letter from Mr. Alexander Volta, FRS Professor of Natural Philosophy in the University of Pavia, to the Rt. Hon. Sir Joseph Banks, Bart. KBPR S. Philosophical Transactions of the Royal Society of London, 403–431.

von der Emde, G., Bleckmann, H., 1992. Differential responses of two types of electroreceptive afferents to signal distortions may permit capacitance measurement in a weakly electric fish, *Gnathonemus petersii*. Journal of Comparative Physiology A 171, 683–694.

Von Humboldt, A., Bonpland, A., 1853. *Personal Narrative of Travels to the Equinoctial Regions of America, During the Years 1799–1804*. HG Bohn.

Wahnschaffe, U., Fritzsch, B., Himstedt, W., 1985. The fine structure of the lateral-line organs of larval Ichthyophis (Amphibia: Gymnophiona). Journal of Morphology 186, 369–377.

Walsh, J., 1773. Of the electric property of the torpedo. Philosophical Transcription of the Royal Society 63, 478–489.

Warren, R., Sawtell, N.B., 2016. A comparative approach to cerebellar function: Insights from electrosensory systems. Current Opinion in Neurobiology 41, 31–37.

Weston, M.D., Pierce, M.L., Jensen-Smith, H.C., Fritzsch, B., Rocha-Sanchez, S., Beisel, K.W., Soukup, G.A., 2011. MicroRNA-183 family expression in hair cell development and requirement of microRNAs for hair cell maintenance and survival. Developmental Dynamics 240, 808–819.

Whitear, M., Lane, E.B., 1981. Bar synapses in the end buds of lamprey skin. Cell and Tissue Research 216, 445–448.

Whitear, M., Lane, E.B., 1983. Multivillous cells: Epidermal sensory cells of unknown function in lamprey skin. Journal of Zoology 201, 259–272.

Whitehead, D.L., Tibbetts, I.R., Daddow, L.Y., 2000. Ampullary organ morphology of freshwater salmontail catfish, *Arius graeffei*. Journal of Morphology 246 (2), 142–149. doi: 10.1002/1097-4687(200011)246:2<142::AID-JMOR8>3.0.CO;2-D. PMID: 11074581.

Wullimann, M.F., Grothe, B., 2013. The central nervous organization of the lateral line system, in: The Lateral Line System. New York: Springer, pp. 195–251.

Wullimann, M.F., Mueller, T., Distel, M., Babaryka, A., Grothe, B., Köster, R.W., 2011. The long adventurous journey of rhombic lip cells in jawed vertebrates: A comparative developmental analysis. Frontiers in Neuroanatomy 5, 27.

Wullimann, M.F., Rooney, D.J., 1990. A direct cerebello-telencephalic projection in an electrosensory mormyrid fish. Brain Research 520, 354–357.

Xu, J., Li, J., Zhang, T., Jiang, H., Ramakrishnan, A., Fritzsch, B., Shen, L., Xu, P.-X., 2021. Chromatin remodelers and lineage-specific factors interact to target enhancers to establish proneurosensory fate within otic ectoderm. Proceedings of the National Academy of Sciences 118.

Zakon, H.H., 1986. The electroreceptive periphery, in: Electroreception. New York: Wiley, pp. 103–156.

Zeymer, M., von der Emde, G., Wullimann, M.F., 2018. The mormyrid optic tectum is a topographic interface for active electrolocation and visual sensing. Frontiers in Neuroanatomy 12, 79.

10 An Integrated Perspective of Commonalities and Differences across Sensory Receptors and Their Distinct Central Inputs

Karen L. Elliott, Bernd Sokolowski,
Ebenezer N. Yamoah, Bernd Fritzsch

CONTENTS

10.1 INTRODUCTION

Finding food, avoiding predators, and finding a mate are vital for the survival and propagation of an individual. Sensory organs detect information from the outside world for each of these tasks and relay it to the brain to elicit proper motor responses. Within the brain, neurons from each sensory modality form a primary sensory 'map', often representing the topographical organization of the peripheral sensory receptors. This is especially true for the visual, somatosensory, and auditory systems. Other sensory systems have less defined 'maps', such as the vestibular system, where there is an incomplete segregation of movement detection since angular movements always produce accompanying linear acceleration. Many of these primary sensory 'maps' become integrated in higher-order processing centers in the brain, for example, the combination of taste and olfaction or of vestibular and vision.

The chemo-affinity and activity-mediated synaptic plasticity theories for the formation of sensory maps have dominated the field during the last century. These theories have provided insights into the various molecular cues that are used to guide sensory neurons to their proper location within specific brain nuclei and in addition, insights into the plasticity mediated by activity to sharpen precise connections. Each of the eight primary sensory maps of vertebrates have unique features and seemingly use distinct molecular cues, cell cycle exit mechanisms, and activity combinations during development, regeneration, and plasticity. Here, we will introduce the evolution of deuterostome neurosensory development, comparing the organization of the dorsal spinal cord and brainstem in vertebrates with precursor structures observed in two chordates, for which we have limited information. These structures are the neural crest and placodes, which are found in 31 species of lancelets and 3,100 species of ascidians.

10.1.1 Lancelet (Amphioxus)

Among chordates, lancelets share a notochord, a dorsal spinal cord, and an associated 'brainstem', but they differ from most chordates concerning the distribution of neural crest cells and placodes (Figure 10.1). For example, while the *Neurog* gene is expressed across all chordates, as it defines early peripheral sensory neuron formation during development, lancelets have no well-defined neural crest or placodes (Holland, 2020). Moreover, the lancelet, which is among the most 'primitive' of the chordates, did not undergo the gene duplication found in vertebrates that were quadrupled to the 46 chromosomes found in humans (Holland and Daza, 2018), and followed by a third whole-genome duplication in teleosts.

Unlike 'olfactorans' (i.e., urochordates, vertebrates), amphioxus has no distinctive olfactory organ, but does have cells expressing olfactory genes along their lateral wall (Schlosser, 2018). Moreover, the 'frontal eye' is identified in lancelets as having only a few sensory cells reaching out of the neuropore (Holland, 2020; Lacalli et al., 1994), while not showing the multilayers that characterize vision in vertebrates (Lamb, 2013; Suzuki and Grillner, 2018). Twenty-one opsin genes and melanopsin are partially characterized in the lancelet (Pergner and Kozmik, 2017), but further details are needed to understand the evolution of the vertebrate eye from the small origin of the lancelet's 'frontal eye.' In addition, lancelets have Joseph cells and dorsal ocelli, which lie separately and posterior to the 'frontal' eye and have an excitation of 470 nm (del Pilar Gomez et al., 2009). Moreover, the lancelet does not express the same order of genes as expressed in the basic optic ganglion of vertebrates, which depends on *Atoh7* expression (Miesfeld et al., 2020; Wu et al., 2021), which is downstream of *Eya1/Six1*, *Pax6*, *Sox2*, and *Notch* (Chapter 3). Finally, trigeminal fibers may be a continuation of other spinal projections (Holland, 2020),

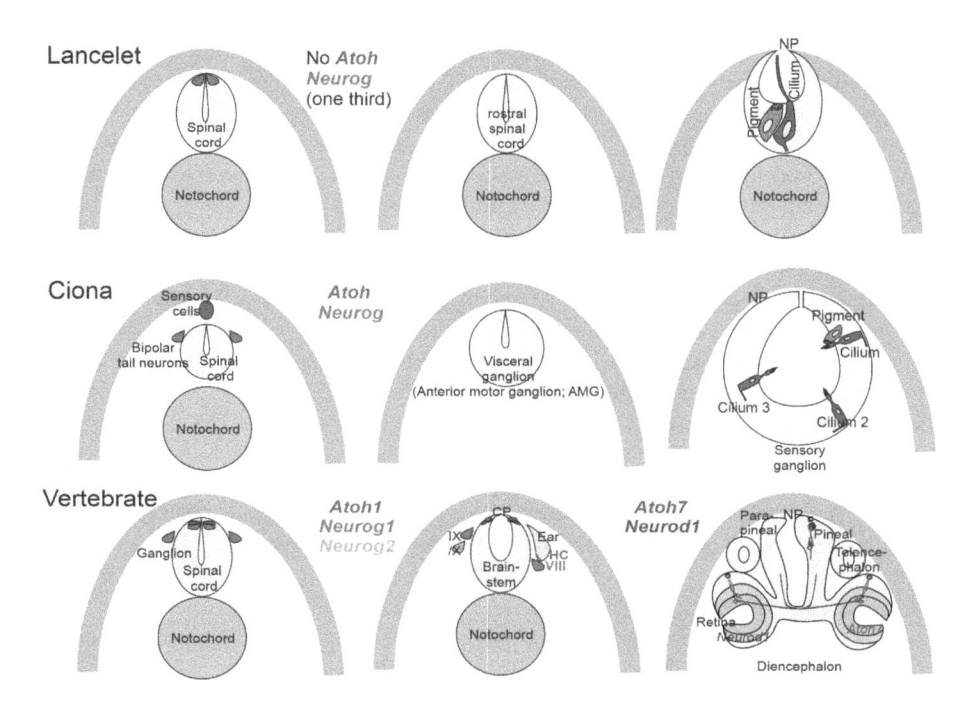

FIGURE 10.1 Comparison of gene expression and critical brain and sensory formations in chordates. Lancelet has *Neurog* expression in the caudal one-third of the spinal cord that is not positive for *Neurog* more rostrally and has no expression of *Atoh*. The central lumen of the terminal sac extends cilia into the neuropore (NP), carrying neurosensory cells that project to the 'hypothalamus.' In ascidians, *Atoh1* positive cells are located in sensory cells dorsal to the spinal cord. The visceral and rostral spinal cord connects the 'bipolar tail neurons' to reach the anterior motor neurons (AMG). The 'sensory ganglion' forms one cilium associated with pigment and two ciliary sensory cells that are not associated with pigments, comparable to the split into three pigment cells in certain ascidians. In contrast to the other chordates, vertebrates have *Atoh1* expressed in the dorsal-most part of the spinal cord and the brainstem, followed by *Neurog1/2*. *Neurog1/2* is expressed in the geniculate of the spinal cord and is expressed in developing placodes except for the retina. Mechanosensory cells depend on Atoh1 and retinal ganglion neurons depend on *Atoh7*. In addition, Neurod1 is needed for retina sensory receptors in vertebrates. Note that the pineal is asymmetric in lampreys and some other vertebrates that start as an expansion to form the bilateral telencephalon that extends beyond the neuropore. (Modified from Alsina, 2020; Dennis et al., 2019; Holland and Somorjai, 2020; Stolfi et al., 2015.)

continuing the equivalent of Rohan-Beard cells in lancelets (Fritzsch and Northcutt, 1993; Chapter 4), which are not found in taste buds (Chapter 5).

A central population of cells expands from the neural plate defined by *Sox1/2/3* expression and a short period of *Neurog* expression (Figure 10.1) in the posterior third of the neural tube of the lancelet (Holland, 2020). However, lancelet lacks the expression of *Neurog1/2* that defines the neural crest and placodes in vertebrates (Ma et al., 1998). Interestingly, there is no homolog for *Atoh* in lancelets suggesting that this gene is not required for single sensory cells in this chordate, as it is in the 'hair cells' of vertebrates (Bermingham et al., 1999; Fritzsch et al., 2010). Detailed analysis of *Wnt* signals shows dissimilarities among *Wnt* genes of lancelets as well (Somorjai et al., 2018), particularly in the caudal expression of *Wnt1* and *Wnt3*. Both are important for the dorsal brainstem and spinal cord of vertebrates, where they are needed to induce the roof plate and the choroid plexus (Glover et al., 2018). *Lmx1a/b* are needed for *Gfp7* and other BMPs to define the roof plate and initiate the dorsal expression of *Atoh1* (Chizhikov et al., 2021; Wang et al., 2005). The lack of *Atoh* expression and *Neurog* expression in the partial, posterior region of the caudal spinal cord results in the absence of crucial gene expression in the lancelet (Holland, 2020). However, many genes are expressed, and those in the adjacent future neural plate among lancelets, such as *BMP2/4, Pax, SoxE, Chordin, Sox1/2/3, Snail, Wnt8, Eya, Six1/2*, and *Fgf8/17/19,* are known but do not generate a placode or neural crest beyond glial-like cells. Lancelets lack the duplication of critical *SoxE* genes (*Foxd, Sox9/10, Snail*) to define the glial cells as 'Schwann cells' that are now found in astrocytes (Bozzo et al., 2021). Data derived from serial sections (Holland and Somorjai, 2020), showing the formation of glial cells along the peripheral nerves, fit with details from tract-tracing methods in the hindbrain of the lancelet (Fritzsch, 1996). Likewise, lancelets form sensory cells derived from local ectoderm cells (Holland and Somorjai, 2020), including the Quadrefages cells that are the closest multicellular cells of vertebrates and tunicates (Fritzsch and Elliott, 2017; Manni et al., 2018). Clear identification of taste buds (Chapter 5) or hair cell types of sensory cells (Chapters 6–9) are unknown in the lancelet that lack the *Tmc1/2* expression needed to provide hair cell function (Erives and Fritzsch, 2020; Pan et al., 2017).

Detailed gene expression that resembles certain aspects of other chordates is known among lancelets (Holland, 2020). For example, *Foxg1, Emx, Otx,* and *Gbx* are common among lancelets and vertebrates. However, the expression patterns of *Pax4/6* and *Pax2/5/8* genes are different than in the lancelet (Figure 10.2). Studies show lancelet follows the hourglass of the phylotypic period of vertebrates (Marlétaz et al., 2018). *Foxg1* has an early interaction with *Sox2* and shows the lack of olfactory formation (Dvorakova et al., 2020) downstream of *Eya1/Six1* (Moody and LaMantia, 2015). In contrast to vertebrates (Dvorakova et al., 2020; Zhang et al., 2021), olfactory expression of *Foxg1* and *Eya/Six* in lancelets does not occur outside the CNS (Holland, 2020). The 'sensory vesicle' (Lacalli et al., 1994) may be distinct from the rhombencephalon/spinal cord (Figure 10.2). *Hox* genes are essential for the transition of the spinal cord to 'rhombencephalon' (Holland, 2020). Importantly, *Fgf8/17/18* expression throughout the cerebral vesicle is in stark contrast to the expression in the midbrain/hindbrain isthmus that is important for the cerebellum (Watson et al., 2017) and the prosencephalon. Notable is the absence of motoneurons that provide the ocular motoneurons in vertebrates (Fritzsch, 1996; Jahan et al., 2021).

In summary, the 'frontal eye' and central sensory projections are common to trigeminal/spinal cords and are present in lancelets. However, this chordate does not have defined placodes or neural crest-derived neurons. Thus, it lacks the formation of sensory cells representing the 'hair cells', 'olfactory receptors', and 'taste buds' found in vertebrates.

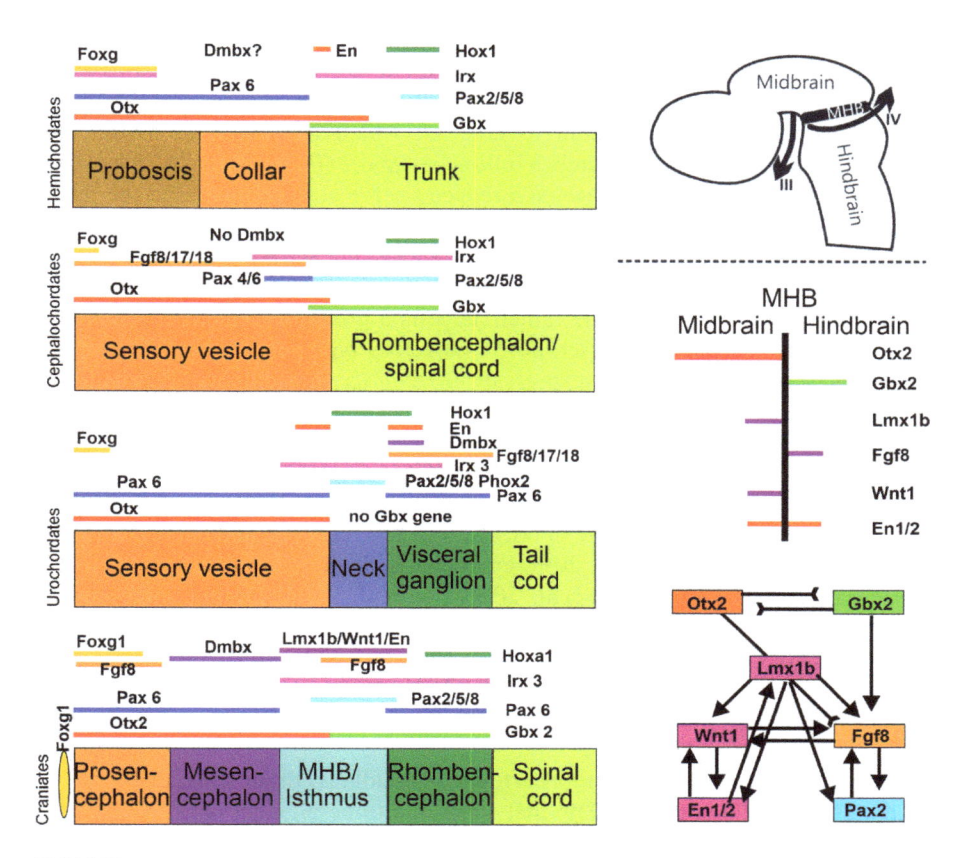

FIGURE 10.2 The evolution of gene expression at the midbrain-hindbrain boundary (MHB) in deuterostomes. The MHB of vertebrates exhibits abutting domains of *Otx2* and *Gbx2* expression. This stabilizes the expression of *Fgf8* (lower right), which in turn stabilizes the expression of *Wnt1* and engrailed (*En1*). Mutation of *Otx2, Gbx2, Fgf8,* or *Wnt1* eliminates the MHB. *Pax2/5/8* are also expressed at the MHB, whereas the expression of Dmbx occurs immediately rostral to the MHB in the midbrain to later expand into the hindbrain and spinal cord (right). Note the partial overlap of *Pax2/5/8* with the caudal expression of *Otx2* and the rostral expression of *Gbx2*. Hemichordates, top left, have an overlapping *Gbx, Otx, Irx,* and *En* in the rostral trunk. *Pax6* abuts *Gbx2*, whereas *Pax2/5/8* overlaps with the caudal expression of *Gbx2*. Outgroup data suggest that coelenterates have a *Dmbx* ortholog, thus raising the possibility that hemichordates also have a *Dmbx* gene. Cephalochordates have no *Dmbx* expression in the 'brain.' The *Otx* expression domain abuts the Gbx expression domain, as in craniates. However, *Gbx* overlaps with Pax2/5/8 and most of *Irx3*. Urochordates have no *Gbx* gene but have a *Pax2/5/8* and *Pax6* configuration comparable to vertebrates. *Dmbx* coincides with the caudal end of the *Irx3* expression, whereas *Dmbx* expression is rostral to *Irx3* in vertebrates. Together, these data show that certain gene expression domains are topographically conserved (*Foxg1, Hox, Otx*), whereas others show varying degrees of overlap. It is conceivable that the evolution of nested expression domains of transcription factors is causally related to the evolution of specific neuronal features such as the evolution of oculomotor and trochlear motoneurons (top right) around the MHB. Experimental work has demonstrated that the development of these motor centers depends on the formation of the MHB. (Modified from Glover et al., 2018.)

10.1.2 Ascidians

Tunicates have a reduced genome of 70–170 million base pairs (Mbp) that is decreased secondarily compared to lancelet, which has about 520 Mbp, and humans, which have approximately 3 billion base pairs (Fodor et al., 2020; Holland, 2020). Two of five tunicates develop a reduced nerve cord compared to the lancelet, with approximately 20,000 neurons in the lancelet compared to only about 89 neurons in the tunicate, *Ciona* (Holland, 2020; Ryan et al., 2018). Recent evidence in tunicates suggests the presence of an 'olfactory' system that lacks the expression of 'olfactory receptors' and rather suggests a group of 'olfactores' (Bassham and Postlethwait, 2005; Schlosser, 2015). Olfactory type gene expression supports this idea as suggested by the expression of *Eya/Six* in ascidians (Bassham and Postlethwait, 2005). Further evidence for this idea comes from *Eya1/Six1* null mice, which lack olfactory formation (Riddiford and Schlosser, 2016; Xu et al., 1997) as well as all placodes. Consistent with the idea of gene homology across olfactory development is the presence of genes downstream to *Eya1/Six1*, known to be dependent on *Sox2* and *Neurog1* in olfactory structures, eyes, and ears (Dvorakova et al., 2020; Panaliappan et al., 2018) and that also underlies a significant defect in vertebrate ears when absent (Li et al., 2020; Zou et al., 2004). Despite some similarity in gene expression between the tunicates and vertebrates for specific genes present in olfactory receptors (Touhara et al., 2016), data indicate no specific receptor placode or epithelium in tunicates (Chapter 2). However, detailed descriptions show the formation of a neuropore that opens directly to the mouth (Veeman et al., 2010; Veeman and Reeves, 2015).

A complex set of vision receptors is found in ascidians (Braun et al., 2020; Konno et al., 2010; Winkley et al., 2020) that splits into three pigment cups in *Thalia*. A unique opsin-1 gene is expressed in ascidians (Pisani et al., 2020) that responds to retinoic acid, comparable to all chordates (Campo-Paysaa et al., 2008). Two additional ciliated receptors are known that also lack associated pigment cells (Figure 10.1; Konno et al., 2010). The three complex receptors conceivably evolved into the bipartite retina and the pineal/parapineal in lampreys and certain gnathostomes (Figure 10.1; Chapter 3). *Atoh* is present in a small population of ectoderm cells (Tang et al., 2013). Current evidence suggests the expression of *Atoh8* (Negrón-Piñeiro et al., 2020) could induce these cells to evolve into ganglion neurons (Figure 10.1; Wu et al., 2021). The trigeminal ganglion starts as pre-placodal ectoderm (PPE) adjacent to the neural crest (NC; Buzzi et al., 2019; Moody and LaMantia, 2015) that lies adjacent to the neural plate (NP) in vertebrates, both of which are absent in ascidians (Chapter 4). Moreover, there is no clear identification of taste buds that depend on soluble chemical compounds as characterized for vertebrates (Chapter 5).

In contrast to *Neurog* expression in the lancelet, which is limited to the caudal third; (Holland, 2020), ascidians express both *Neurog* and *Atoh,* as also found in vertebrates (Kim et al., 2020; Negrón-Piñeiro et al., 2020; Stolfi et al., 2015; Tang et al., 2013; Figure 10.1). Moreover, *Neurog* is found near the dorsal 'spinal cord' of ascidians, which is similar to that in vertebrates, where *Neurog1/2* is differently distributed in all neural crest and placode neurons (Ma et al., 1998). In contrast, there is limited *Neurog* expression in lancelets, where it lies adjacent to the caudal spinal cord (Stolfi et al., 2015). At the same time, *Atoh* is expressed in ventral and dorsal cells along the ectoderm (Tang et al., 2013; Figure 10.1), suggesting that expression is differentially distributed compared to the vertebrate's spinal cord ganglia.

Ascending projections are described in the four 'bipolar tail neurons' that connect with the anterior motor ganglion in ascidians (Ryan et al., 2018), which is reminiscent of the terminal ganglion neurons that end at the rhombencephalon to express second projections from the gracile and cuneate nuclei of vertebrates (Nieuwenhuys and Puelles, 2015). In vertebrates, a dorsal expression of *Atoh1* exists from the spinal cord to the brainstem, including the cerebellum (Bermingham et al., 2001). In addition, a reciprocal interaction of *Atoh1* with *Neurog1* exists in vertebrates (Gowan et al., 2001; Lai et al., 2016). In contrast, neither *Atoh* nor *Neurog* are expressed in the 'spinal cord' of ascidians (Negrón-Piñeiro et al., 2020; Ryan et al., 2018). This suggests a limited cross-interaction with *Neurog1/2* and *Atoh1* in the rhombencephalon (Fritzsch et al., 2006) that is certainly pertinent to ear development (Fritzsch and Elliott, 2017). While there are differences in *Neurog* and *Atoh* expression in ascidians as compared to vertebrates, many other genes are common among both chordates, such as *Sox1/2/3, Pax3, Zic, Pax2/5/8, Gata2/3, Eya/Six, Hes, ID, Foxg1, Otx, Emx* (Glover et al., 2018; Holland, 2020; Liu and Satou, 2020; Schlosser, 2015). In addition, *Wnt* signaling defines A-P positioning in neural induction (Figure 10.2; Meinhardt, 2015). Chordate complexity reflects the shuffling of genes, ranging from gene duplication in vertebrates as well as the loss of some bHLH genes (Ryan et al., 2018). The roof plate is likely positive for *Lmx* genes in ascidians (Glover et al., 2018). An additional gene is needed for a related BMP-like gene, *Gfp1/2* (Liu and Satou, 2020), uncharacterized in ascidians. A comparison of a large number of different ascidians show that the formation of secondary sensory cells or 'hair cells' resemble those found in vertebrates (Caicci et al., 2013; Fritzsch et al., 2007; Fritzsch and Elliott, 2017; Manni et al., 2018), including the expression of the recently defined *Tmc* genes involved in mechanotransduction (Erives and Fritzsch, 2020). The evolution of *Tmc1/2* happened after the ascidians split from the vertebrates and gave rise to the three bifurcations of *Tmc1/2/3, Tmc4/8/7*, and *Tmc5/6*, which occurred before the formation of choanoflagellate cell communication (Göhde et al., 2021).

The gene regulatory network that controls gene expression in ascidian embryonic development and leads to the tadpole larval stage has revealed evolutionarily conserved gene circuits between ascidians and vertebrates (Liu and Satou, 2020; Ryan et al., 2018). The previous description showed that the neurogenic plate depends on *Foxg1, Six1/2, Emx*, and *Zic* expression, which is alike in vertebrates and lancelets (Liu and Satou, 2019). The earliest manifestation of *Foxg* plays a role in the anterior neural plate (Figure 10.2) that is expressed in vertebrates in the olfactory, retina, and ear placodes (Dvorakova et al., 2020; Kersigo et al., 2011), suggesting a different pattern of early gene regulation exists, particularly in the telencephalon. Interestingly, feedback exists between *Foxg1* and *Fgf8* in vertebrates (Liu and Satou, 2020) but does not show similar expression in ascidians (Figure 10.2). *Otx/Gbx* provides vital feedback in vertebrates that partially overlaps in lancelet and hemichordates, whereas there is no *Gbx* gene in ascidians (Figure 10.2). In contrast to the absence of *Dmbx* in the lancelet, the expression of *Dmbx* in ascidians does not fit the role of the midbrain for vertebrates (Figure 10.2; Glover et al., 2018). Ascidian gene regulatory network expression suggests vertebrate and ascidian similarities across these two chordates. Notable is the expression of *Atoh* and *Neurog,* which closely resembles that in tunicate dorsal root and sensory cells (Figure 10.1), respectively. In contrast, *Eya/Six* expression is among the earliest in the pre-placodal region of vertebrates, but is absent outside the neural plate in ascidians (Moody and LaMantia, 2015; Thiery et al., 2020).

In summary, while ascidians express the *Neurog* and *Atoh* homologs found in vertebrates, they are found in cells with a morphology similar to 'hair cells' with separate neuronal innervation (Manni et al., 2018). An early olfactory organ that forms the basis for olfaction in vertebrates, seems not to develop in ascidians, while the organization of an olfactory system is unclear in amphioxus (Holland, 2020; Moody and LaMantia, 2015; Schlosser, 2018; Touhara et al., 2016). While the beginnings of sensory cells that respond to vision are found among ascidians, there is no multilayer eye as found in vertebrates (Figure 10.1). Moreover, ascidians lack the expression of *Atoh7* to make retinal ganglion neurons and *Neurod1/Otx2* to drive the development of rods and cones (Wu et al., 2021). The absence of an early and broad expression of *Foxg* in ascidians lies in contrast to vertebrates, where *Foxg1* is essential for the development of olfaction, eyes, and ears (Dvorakova et al., 2020; Liu and Satou, 2019).

10.1.3 VERTEBRATES

Vertebrates have eight major sensory systems overall but show some differences in the number of senses present per species. For instance, not all vertebrates have an auditory (found in gnathostomes), lateral line (absent in amniotes), or electroreceptive (absent in hagfish, most teleosts, anurans, and amniotes) system. In addition to variations in number, there are many variations within a given sensory system across vertebrates. For example sensory cells, such as in the olfactory system, could become nearly absent as in dolphins or quite prominent as in elephants (Niimura et al., 2020), there could be limited color vision, as in some teleosts, versus multicolor vision, as in birds (Cuthill et al., 2000), there could be an increase in the trigeminal nerve of different sensory reception (Erzurumlu et al., 2010), or even have an expansion of taste buds that cover the skin in addition to the tongue (Finger, 2008). All sensory neurons depend on *Neurog1/2*, except for the retinal ganglion cells of the visual system, which depends on *Atoh7*, a divergent bHLH gene (Wu et al., 2021). A comparison of sensory cells shows that there is a different organization in the olfactory system, where first-order neurons also function as sensory cells, while in other sensory systems, the sensory cells synapse with first-order neurons (Imai et al., 2009). *Atoh1* is expressed in the major sensory receptors of five sensory cells, including hair cells of the vestibular, auditory, lateral line, and electroreceptive systems and the Merkel cells of the somatosensory system. In contrast, taste bud receptors depend on *Sox2/Shh* (Okubo et al., 2006). In contrast, vision receptor cells depend on *Neurod1/Otx2* (Ochocinska et al., 2012). A common theme is the use of G protein-coupled receptors (*GPCR*), which are found in three sensors (olfaction, vision, taste) and that also use three common receptors (vestibular, auditory, lateral line). A different set of channels is used for trigeminal sensors (*Piezo*; Coste et al., 2010), and has recently been identified in certain electroreceptors.

All vertebrates have a primary central organization that extends from the spinal cord to the telencephalon (Figure 10.2) but differ in their distribution among the eight sensory inputs. Olfaction is unique in cyclostomes and gnathostomes olfactory receptors that is absent in other chordates. Visual projections spread from the chiasm to reach the superior colliculus that expands secondary connections to the telencephalon. Trigeminal (NV), taste buds (NVII, IX, X), vestibular (NVIII), cochlear (NVIII), lateral line (NVII + IX), and electroreception (NVII + IX) projections end up in different layers in the brainstem, progressing from the lowest rhombencephalon (trigeminal) to the highest layer in electroreception and cochlea (Fritzsch et al., 2019). The segregation of olfaction and vision are separated by a discrete expression of *Otx2/Gbx2*, to define the remaining six sensors of the brainstem (Figure 10.2). Notable interactions define the midbrain-hindbrain boundary (MHB) that regulates the unique development of oculomotor/trochlear motoneurons and allows secondary projections to reach the midbrain and telencephalon from the olfactory and visual systems. The telencephalon is a novel expansion that forms the bilateral forebrain of vertebrates but is absent in lancelets and tunicates that include the opening of a neuropore.

In summary, all vertebrates have a unique set of distinct sensory receptors but can share certain commonplace features. Central discrete projections of nearly all sensory inputs are now known for lampreys and all other gnathostomes to the brainstem, midbrain and telencephalon.

10.2   bHLH GENES ACT DOWNSTREAM OF *Sox2* AND *Eya1*

Preplacodal ectoderm will develop into distinct sensory placodes (olfactory, eye, trigeminal, otic and, epibranchial placode; Moody and LaMantia, 2015; Thiery et al., 2020) that are adjacent to the neural crest (Simões-Costa and Bronner, 2015) and that require a set of genes that lead, eventually to the upregulation of *Eya1/Six1* (Riddiford and Schlosser, 2016; Xu et al., 1999). Recently, *Eya2* expression was described that suggests a redundancy with *Eya2*, with limited additional null defects (Zhang et al., 2021). All sensory neurons of olfactory, visual, trigeminal, epibranchial, taste, vestibular, and auditory systems express *Eya1/2* early in development. In addition, there is *Eya1/2* expression in the dorsal spinal cord and brainstem that will later be expressed in the midbrain and in certain motoneurons (Figure 10.3). Consistent with the earliest description of *Eya1*, null mutants show an additional dependence on this gene outside the brain and placodes, also referred to as branchio-oto-renal spectrum disorder (BOR) (Moody and LaMantia, 2015; Muthusamy et al., 2021; Xu et al., 1999)]. *Eya1/Six1*, *Sox2*, and various *Pax* genes (*Pax6, Pax3, Pax2/8*) play a major role in all sensory neuron development in the vertebrate forebrain and brainstem (Dvorakova et al., 2020; Moody and LaMantia, 2015). In addition, they interact with *Brg1* (SWI/SNF) to regulate all the earliest steps in neuronal precursor development (Xu et al., 2021).

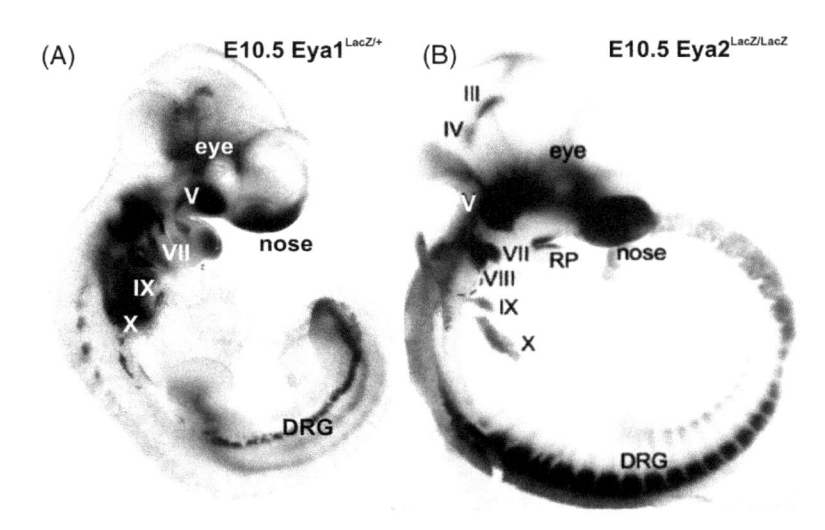

FIGURE 10.3 This image shows the different sensory neurons using *Eya1*-LacZ (A) and *Eya2*-LacZ/LacZ (B). From the nose through the eyes, we show the trigeminal (V), facial (VII), otic (VIII), glossopharyngeal (IX), and vagus (X). Note that specific areas of the brainstem are *Eya1/Eya2* positive. (Image provided by Pin-Xian Xu [A] and Zhang et al., 2021].

A common feature of vertebrates is the combined expression of conserved and unique genes that evolve specific interactions in the various sensory systems. Among these unique representations is the dependence on *Foxi3* (Birol et al., 2016) and *Fgf3/10* (Urness et al., 2011; Wright and Mansour, 2003) for the development of the otic placode. Induction of an otic placode does not occur without either *Foxi3* or *Fgf3/10*. Variations are prominent with the presence or the absence of specific genes such as *Fgf8* (olfaction, trigeminal), *Tbx1* (trigeminal, vestibular, auditory), and *Foxg1* (olfactory, anterior half of the retina, taste, vestibular, auditory; Table 10.1). In addition, taste buds depend on *Phox2b* and *Foxi2* as well as *Pax2, Foxg1, Isl1*, and *BMP4* (Alsina, 2020; Buzzi et al., 2019). A unique use of *Pax3* is in the separation of a trigeminal ganglia that follows the segregation of *Otx1/ Gbx2* (Moody and LaMantia, 2015; Steventon et al., 2012) to express a variation of *Irx, Tbx, Fox1, Runx*, and *BMP4*. Olfactory and eye development depends on *Pax6* expression and will not form in its absence in chordates (Sokpor et al., 2018; Thiery et al., 2020). Otocyst development depends on *Pax2/8* and will not form in the absence of both of these genes (Bouchard et al., 2010). Moreover, in the absence of *Eya1*, the ear otocyst forms, but without any other sensory neuron formation (Xu et al., 2021; Zou et al., 2004). *Pax2* is also needed for epibranchial neuron formation (Alsina, 2020).

Importantly, every peripheral nervous system (PNS), as well as every central nervous system (CNS), depends on *Sox2* (Dvorakova et al., 2020; Imayoshi and Kageyama, 2014; Kageyama et al., 2019; Panaliappan et al., 2018). All *Sox2* genes are involved in stem cell replacement, multipotency, and cell fate decisions that induce different gene expression (Imayoshi and Kageyama, 2014). Downstream of

TABLE 10.1

Summary of the Genes Involved in Neurosensory Development of the Various Sensory Systems

	Pre-neurons	Neurons	Central Projections	Sensory Cells	Sensory Transduction	Distinct Sensory Receptions
Olfaction	*Eya1/Six1, Sox2, Otx2, Foxg1, Pax6, Emx2, Ebf2. RA. Fgf8, Shh, BMP4*	**N I** (*Ascl1*) *Neurog1/2, Neurod1, Runx, Isl1*	Olfactory sensory neurons (OSN) *Nrp1, Nrp2, Robo/ Slit, Ac3, EphA/ephrinA, Kirrel2/3*		Odorant receptors (OR) G protein-coupled receptors	Main olfactory epithelium (MOE; *OR, TAAR, MS4A*) Vomeronasal organ (VNO; *V1R, V2R, FPR*)
Vision	*Eya1/Six1, Sox2, Otx2, Pax6, Mitf, RA, BMP4/7*	**N II** *Atoh7, Pou4f2, Isl1 Shh*	Retinal ganglion cells (RGC); SCN, BOT, LGN, Sup. Collic. (SC), *ephrinA/B- EphA/B, Ryk, Wnt3, En-2, dak/box, Robo2*	***Cones, Rods*** *Neurod1 Otx2*	Opsins G protein- coupled receptors	Cones> *Neurod1, Thrb2* Rods > *Neurod1, Rorb*
Trigeminal	*Eya1/Six1, Sox2, Pax3, Fgf8, Irx, Tbx1/3, Isl1, Fox1, Runx, BMP4*	**N V** *Neurog1, Neurod1, Pou4f1*	Trigeminal ganglion (TG) *Neurog1/2, Olig2, Ptf1a, Ascl1, Pou4f1, Lmx1b, Nrp2, Oc1/2, Hmx1*	***Merkel cells*** *Atoh1, Pou4f3*	Trigeminal neurons, Merkel cell, *Piezo1.2, ASIC1*	Aβ, Aδ and C fibers, Merkel cells>*Atoh1*, Ruffini, Meissner, Pacinian, lanceolate, free nerve ending
Taste buds	*Eya1/Six1, Sox2, Pax2, Foxi2, Phox2b, Foxg1, Isl1, BMP4*	**N VII, IX, X** *Neurog2, Neurod1, Pou4f1*	Solitary tract (ST), *Ascl1, Olig2. Neurog2, Tlx3, Isl1, Pou4f1, Phox2b, Lmx1b*	***Taste buds*** *Sox2, Prox1, Shh, Pou2f3*	Taste receptors G protein-coupled receptor Type II (sweet/ bitter/ umami) (*Tas1r1-3, Tas2r*) Type III (sour) *Otop1*, Salty (?)	Fungiform, Foliate, (Circum) vallate taste buds Geniculate, Petrosal, Nodose

(Continued)

TABLE 10.1 (Continued)
Summary of the Genes Involved in Neurosensory Development of the Various Sensory Systems

	Pre-neurons	Neurons	Central Projections	Sensory Cells	Sensory Transduction	Distinct Sensory Receptions
Vestibular	Foxi3, Fgf3/10, Eya1/Six1, Sox2, Dlx5/6; Pax2/8, Gbx2, Hmx3, Otx1/2, Foxg1, Fgfr2b, Tbx1, RA, BMP4	**N VIII** Neurog1, Neurod1, Isl1, Pou4f1	Vestibular nuclei (VN) superior, medial, lateral, inferior VN, Fgf8 (r0), Hoxb1 (LVN) Lbx1, Phox2b, Pou3f1, Lhx1, Ascl1, Ptf1a, Neurog1, Olig2	**Hair cells** Atoh1, Pou4f3, Gfi1, Srrm3/4, miR-183	Sensory channel Tmc1/2 (K+) CDH23, PCDH15 TMHS, TMIE Planar cell polarity (PCP) Vangl, Frz, Dsh1, Emx2	Gravistatic macula receptor. (saccule, utricle, lagena) canal cristae (anterior, posterior, horizontal CC) Type I (calyx) Type II (boutons) Atoh1
Auditory Gnathostomes	Eya1/Six1; Brg1, Sox2, Pax2, Gata3, Lmx1a/b, Foxg1, Tbx1, RA, Shh, BMP4	**N VIII** Neurog1, Neurod1, Pou4f1, Isl1, Wnt3a, Prox1	Cochlear nuclei (CN) dorsal, anteroventral, posteroventral CN (AVCN, PVCN, DCN) Atoh1, Olig3, BMPs, Lmx1a/b, Gdf7, Wnt3a, Npr2, Neurod1, Frzd3	**Hair cells** Atoh1, Pou4f3, Gfi1, Barhl1, Srrm3/4, miR-183	Sensory channel Tmc1/2 (K+) CDH23, PCDH15 TMHS, TMIE Planar cell polarity (PCP) Vangl2, Dsh1, Celsr1, Gal2	Inner hair cells (IHC) Atoh1, Srrm3/4, Fgf8 Outer hair cell (OHC) Atoh1, Emx2, Jag1, Fgf20, MANF, Vangl2, Dsh1, Celsr1
Lateral line Vertebrates no Amniotes	Eya1/Six1; Sox2/3; Neurog1, Neurod1	**aLL, pLL** Neurog1, Neurod1, Pou4f1	Latera line nuclei Atoh1, Olig3, BMPs, Lmx1a/b, Gdf7, Wnt3a	**Hair cells** Atoh1 (a/b), miR-183	Sensory channel Tmc1/2 (K+) CDH23, PCDH15 Planar cell polarity (PCP) Vangl2, Dsh1, Celsr2, Gal3	Lateral line hair cells (LL) Atoh1, Emx2, Notch1, Fgf3/10, Wnt, Dkk1b Vangl2, Dsh1, Celsr2
Electroreception Vertebrates lost in hagfish, teleosts, anurans, amniotes	Eya1/Six1?; Sox2/3; Neurog1	**aLL** Neurog1, Neurod1	Anterior lateral line (silurids, gymnotids, mormyrids with aLL+pLL) Atoh1, Ptf1a	**Hair cells** Atoh1(a/b) miR-183	Sensory channel $Ca_v1.3$ (sharks, skates, salamanders) different polarity in bony fish **NO PCP**	Electroreceptors (ELL) ampullary organs tuberous or knollen organs

Sox2 and other *Sox* genes (Reiprich and Wegner, 2015) are the bHLH genes *Neurog1*, *Neurog2*, and *Ascl1* (Ma et al., 1998). bHLH genes interfere with the expression of *Sox2* that will eventually overcome *Sox2*, to drive neurons instead of pluripotent cell development (Reiprich and Wegner, 2015). Murine neural-specific bHLH genes with proneural activity are partially characterized, including *Neurog1, Neurog2, Ascl1, Neurod4, Atoh1*, and *Atoh7* (Dennis et al., 2019). Specifically, without *Neurog1* expression, there is no ear and trigeminal neuron formation (Ma et al., 1998). *Neurog2*, meanwhile, is needed for the epibranchial neurons that form taste bud neurons (Fode et al., 1998). An interaction of *Neurog1* and *Neurog2* is needed to drive olfactory neurons (Shaker et al., 2012). Upstream of *Neurog1/2* is the expression of *Ascl1*, and in its absence, very few olfactory neurons develop (Krolewski et al., 2012). Feedback between *Ascl1* and *Gli3* regulates the development of vomeronasal neurons (Taroc et al., 2020). Furthermore, *Ascl1* and *Gli3* expression is regulated by *Hes1* and *Hes5* that define supporting cells and neurons (Bertrand et al., 2002; Sokpor et al., 2018). In the spinal cord, sequential expression of *Neurog1* and *Neurog2* can compensate for the initial expression of *Neurog1* (Ma et al., 1999). A unique expression of *Atoh7* occurs in retinal ganglion neurons (Miesfeld et al., 2020; Wu et al., 2021). *Atoh7* and *Neurog2* are differentially regulated by *Hes* genes in the eye: *Hes1, Hes3*, and *Hes5* supresses *Atoh7*, but not *Neurog2* expression (Maurer et al., 2014).

Members of the large *NeuroD* family (*Neurod1, Neurod2, Ndrf, Neurod6/Math2*), Nscl (*Nhlh1/Nscl1, Nhlh2/Nscl2*), and Olig (*Olig1, Olig2, Olig3, Bhlhe22*) families interact to regulate neuron and glial cell differentiation (Dennis et al., 2019). All sensory neurons express and depend on *Neurod1* (Bertrand et al., 2002; Kim et al., 2001), a distinct bHLH gene that evolved early (Fritzsch et al., 2010). Levels of *Neurod1* deletion vary between near-complete loss of vestibular/spiral ganglion neurons (Jahan et al., 2010; Macova et al., 2019) and an incomplete loss of *Neurod1* that is upstream of *Ascl1* in the olfactory system (Krolewski et al., 2012; Taroc et al., 2020). Retinal sensory cells, the rods and cones, depend on *Neurod1* (Dennis et al., 2019). Downstream of *Neurod1* is *Pou4f1*, which is needed for the differentiation of brainstem neurons that diverge from *Pou4f2* in the retina, where *Pou4f2* interacts with *Isl1* to develop retinal ganglion neurons (Li et al., 2014; Wu et al., 2021). Another gene, *Nhlh1*, interacts with *Neurod1* in the ear, olfactory, and cerebellum (Krüger et al., 2004; 2006; Pan et al., 2009).

In summary, *Eya1* and *Sox2* interact with *Pax2/3/6/8* to define the earliest steps of neuronal precursor development. The next step is the interaction with several bHLH genes (*Ascl1, Atoh7, Neurog1, Neurog2, Neurod1*) that alone or in combination downregulate *Sox2* and upregulate several or a single bHLH gene(s). Downstream of bHLH genes, we know that *Pou4f1* and *Pou4f2* are needed for neuronal differentiation.

10.3 SENSORY CELLS DISPLAY DISTINCT SIMILARITIES

A common theme in olfaction, vision, and taste sensory cell development is using distinct gene expression to regulate G protein-coupled receptors (GPCR) development (Buck and Axel, 1991; Lamb et al., 2016; Roper and Chaudhari, 2017). Another common receptor is *Tmc1/2*, which regulates the opening or closure of mechanoreceptors found in the vestibular, auditory, and lateral line systems (Erives and Fritzsch, 2020; Pan et al., 2017). These *Tmc* proteins may interact with two stereocilia connector proteins (CDH23, PCDH15) to confer function (Qiu and Müller, 2018). Mechanoreceptors of the trigeminal system requires *Piezo* (Coste et al., 2010) and was only recently identified with $Ca_v1.3$ in certain ampullary electroreceptors (Baker, 2019; Bellono et al., 2018). In essence, GPCR receptors see chemical/ionic/photic stimuli, whereas mechanoreceptors are utilized in touch, vestibular, auditory, lateral line, and electroreception. The varieties of GPCRs reflect an enormous level of diversity, especially for the various olfactory receptors (OR, *TAAR, V1R, V2R*; Chapter 2]. This diversity stands in contrast to the limited differences in mechanoreceptors driven by *Tmc1/2* or *Piezo* expression. Variations are also limited among taste buds (Chapter 5). There are two different types of GPCR taste receptors (Roper and Chaudhari, 2017) that contrast with the distinct yet similar types in the olfactory system. Opsins (Chapter 3) underlie vision but show diversification of receptors of rods and cones in most vertebrates (Arendt et al., 2016; Lamb, 2013).

The transduction systems develop from a set of bHLH genes that are dependent on *Eya1/Six1* expression. A prominent bHLH, *Atoh1*, is needed in vestibular, auditory, lateral line, electroreception, and Merkel cells development (Bermingham et al., 1999; Elliott et al., 2021; Fritzsch and Elliott, 2017; Maricich et al., 2009). The various trigeminal sensors show unique neuronal interactions that function as transducers (Abraira and Ginty, 2013). The unique expression of a different bHLH, *Neurod1*, defines cones and rods in the visual system, collaborating with *Otx2* and functions downstream of *Thrb2* and *Rorb* (Brzezinski and Reh, 2015; Wu et al., 2021). Olfactory transduction receptors are found at the peripheral end of the ganglia, which project directly to the olfactory bulb, sidestepping the formation of independent sensory cells (Fritzsch et al., 2019).

Downstream of *Atoh1* is a large set of genes needed in hair cell and Merkel cell differentiation. *Pou4f3* interacts with *Gfi1* to differentiate mechanosensory cells (Hertzano et al., 2004; Xiang et al., 2003). These genes provide fine-tuning of distinct hair cell types, in particular outer and inner hair cells of the auditory system. *Neurog1, Neurod1*, and *Isl1* show various reductions in the number of hair cells that also change the type of hair cells and their distribution when they are deleted (Elliott et al., 2021; Jahan et al., 2010; Macova et al., 2019; Matei et al., 2005). A transformation of inner hair cells by *Insm1* (Lorenzen et al., 2015) instead of outer hair cells contrasts with *Srrm3/4* that causes the loss of all inner hair cells (Nakano et al., 2020).

Planar cell polarity (PCP) in mechanoreceptors requires shifting the central apical position of the kinocilium to a lateral position and extending the length of stereocilia to develop the staircase form of tip links (Ghimire and Deans, 2019; Montcouquiol et al., 2003; Tarchini and Lu, 2019). Once polarity is set, the next step involves developing the tip links to connect *CDH23* and *PCDH15* to open the channel (Qiu and Müller, 2018; Shibata et al., 2016). Functional details of *Tmc1/2* reveal a complex interaction with additional channel proteins (TMHS, TMIE). Unique among the olfactory and taste bud system is their continued proliferation of precursors (Dennis et al., 2019; Schwob et al., 2017). This ability to proliferate and differentiate permanently is in contrast to the auditory and visual systems, where there is no new formation of lost sensory cells through the induction of proliferation in older mammals (Yamoah et al., 2020). However, vertebrates such as chick, teleosts, etc., show regeneration of hair cells and add later visual sensory cells.

In summary, sensory cells are highly diversified in olfaction, showing great variation without spreading into discrete sensory cell types that add sensory cells. Taste buds, vision, and mechanosensory sensors have limited morphological variations that show replacement in some (taste) but not other sensory cells (retina, ear).

10.4 CENTRAL PROJECTIONS ARE DEFINED BY COMMON THEMES

Projections from each of the eight sensory neuron populations reach separate terminals in the brain (Figure 10.4). Olfactory projections reach the olfactory bulb through a complex guidance system that leads a single neuron to interact with a small set of mitral, tufted, and granule cells within a given glomerulus (Mombaerts et al., 1996; Niimura et al., 2020; Chapter 2). Sorting out the central projections requires topologies driven by *Kirrel2/3, ephrinA/EphA, Nrp1/2, Robo*, and *Ac3* (Imai, 2020). Retinal ganglion neurons reach several central targets that are best characterized by that of the superior colliculus, which shows topological projections driven by *ephrinA/B, EphA/B, Ryk, and Wnt3* (Triplett and Feldheim, 2012). Comparable central projections are known in the auditory system that follows the *Wnt/Fzd* combination and adds ephrinA/EphA. Clear segregation of base/apex spiral ganglion neurons follows dorsoventral projections driven by cell cycle exits. Vestibular (Chapter 6) and taste buds (Chapter 5) have incomplete segregation, while lateral line and electroreceptors fibers segregate (Fritzsch et al., 2019; Chapter 8.9). A clear topology is known in trigeminal projections that are innervated by the three branches, the ophthalmic, maxillary, and mandibular fibers and projections mostly distinct from nearby projections (Erzurumlu et al., 2010; Chapter 4).

The central target nuclei located within the central nervous system depend on the expression of various genes. The bHLH genes, in interaction with *Sox2*, regulate the development of the neural tube (Bertrand et al., 2002; Dennis et al., 2019). Among the bHLH genes, only seven genes show proneural activities (*Neurog1, Neurog2, Ascl1, Neurod4, Atoh1, Atoh7*). Another set of bHLH genes, needed for later development, is downstream from the earliest bHLH genes and includes the *Neurod1, Nscl, Ptf1*, and *Olig* families, which show widespread expression and overlap (Hernandez-Miranda et al., 2017; Lai et al., 2016; Lowenstein et al., 2021). Within the hindbrain, the expression of *Lmx1a/b* is required to regulate choroid plexus development and depends on *Gfp7, Atoh1, Wnt3a*, and others for auditory neurons (Chizhikov et al., 2021; Wang et al., 2005). Without auditory neurons, central projections disappear, including second-order neurons of the MOC complex (Di Bonito et al., 2017) and the migration of the pontine area (Nichols and Bruce, 2006). Downstream is the expression of *Neurog1* (caudal and spinal cord), *Neurog2*, and *Olig3* that depend on vestibular neurons and probably includes more ventral nuclei beyond dA2 (Lunde et al., 2019). Interestingly, central neurons depend on

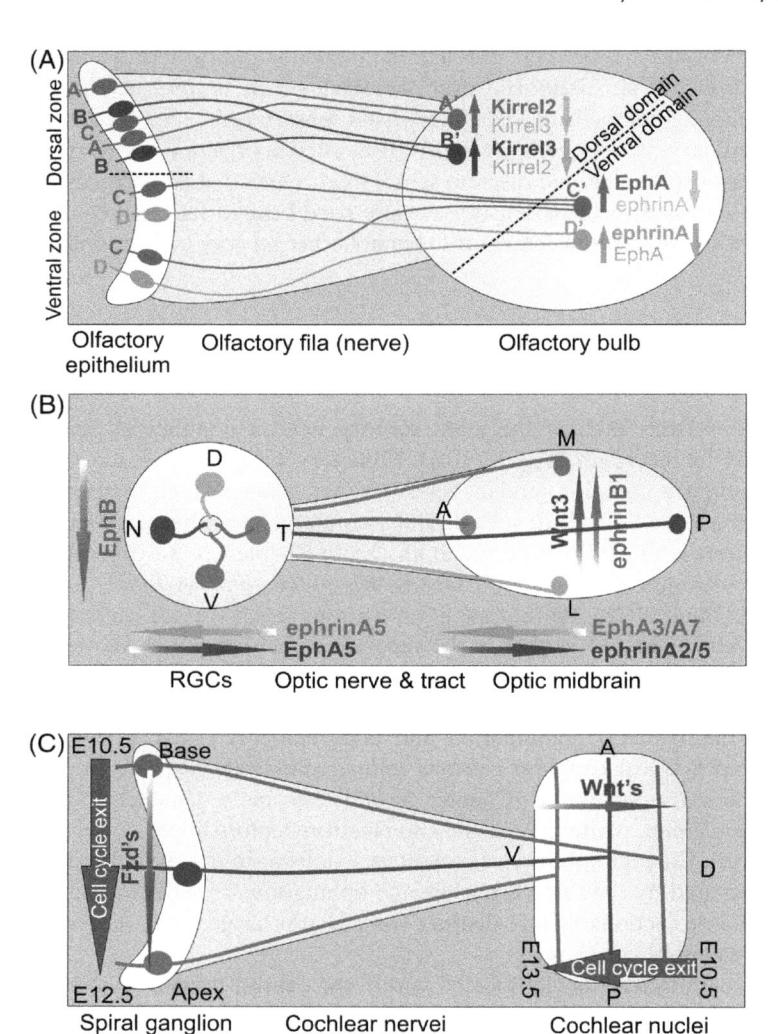

FIGURE 10.4 Development of three distinct mammalian sensory maps. Molecular cues (A, B), and spatio-temporal cues (C) are shown for the early non-spatial olfactory map (A), the two-dimensional (2D) retino-tectal map (B), and the unidimensional auditory map (C). (A) The olfactory map defines different olfactory receptor molecules in the dorsal and ventral zone of the olfactory epithelium. Receptor cells displaying distinct olfactory receptors project their axons to the dorsal and ventral domain of the olfactory bulb, where they converge and initiate olfactory glomeruli formation. Note that olfactory fibers sort before they reach the olfactory bulb and that some ventral zone receptors are expressed in the dorsal zone, but afferents sort to the ventral domain. Different opposing gradients of receptors facilitate further the sorting of olfactory afferents. Within this limited topology, the distribution of specific olfactory receptor-expressing receptor cells is pretty random. (B) The retinotectal system maps a 2D surface (the retina ganglion cells) onto another 2D surface (the midbrain roof or tectum opticum) via highly ordered optic nerve/tract fiber pathways. The presorted fibers are further guided by molecular gradients matching retinal gradients of ligand/receptor

FIGURE 10.4 *(Continued)* distributions within the midbrain. (C) The auditory map is uni-dimensional, projecting a species-specific frequency range from the mammalian hearing organ, the organ of Corti, via orderly distributed spiral ganglion neurons (SGNs), and their fibers in the auditory (cochlear) nerve onto the ventral cochlear nucleus complex. Both SGNs and cochlear nucleus neurons show a matching temporal progression of cell cycle exit followed by matching differentiation that could be assisted by spatio-temporal expression changes of receptors and ligands (shown here are the putative Wnt/Fzd combinations) that further support the fiber sorting. Note that this map projects a single frequency of an inner hair cell of the organ of Corti via a set of SGNs onto longitudinal columns of cochlear nucleus neurons in a cell-to-band projection and thus is not a point-to-point as the olfactory and visual map. Moreover, afferents innervating multiple outer hair cells (OHCs) generate a band-to-band projection centrally. A, anterior; D, dorsal; L, lateral; M, medial; N, nasal; P, posterior; T, temporal; V, ventral. (Modified from Fritzsch et al., 2019.)

Ptf1a derived neurons that migrate to interact with vestibular and cochlear neurons (Fujiyama et al., 2009; Iskusnykh et al., 2016). Further ventral is a complex of four distinct populations (aA3-dB3) positive for *Ascl1* and separated by *Atoh1* (Figure 10.5). How different genes segregate to reach distinct trigeminal neurons is

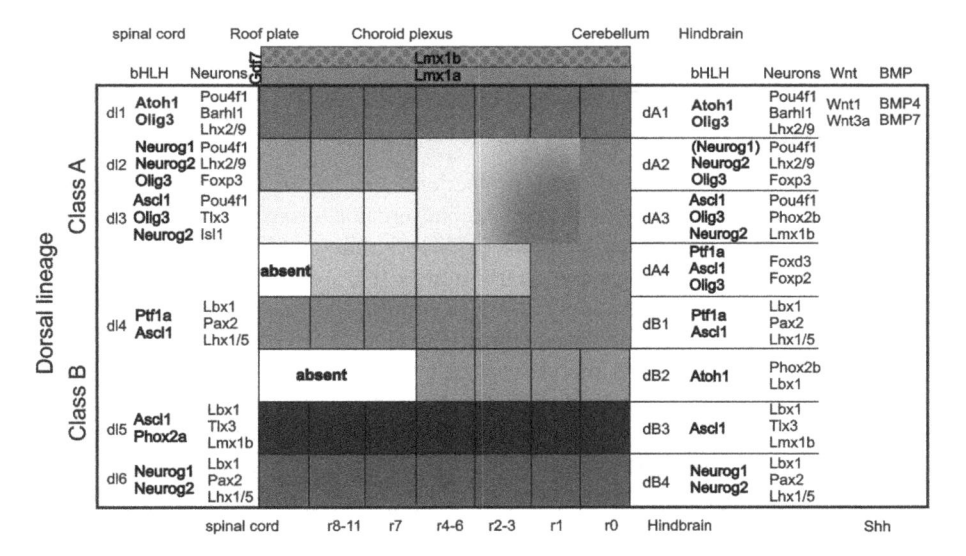

FIGURE 10.5 An overview of gene interactions in the early spinal cord and brainstem (r2–7) displays distinct gains and loss among rhombomeres and the spinal cord. Two classes (A, B) of neuronal domains, identified with different numbers, exist between the brainstem and the spinal cord; however, dA2, dA3, dA4, and dB2 are not continuous. Dorsal most expressions are *Lmx1a/b* and *Gdf7* that define the dorsal parts of the choroid plexus (brainstem) and roof plate (spinal cord). The dorsal to ventral regions are characterized by distinct bHLH genes (dlA1–4, dB1–4) and other neuronal differentiation genes. The two unique domains are dA4 (*Foxd3, Foxp2*) and dB2 (*Phox2b, Lbx1*). These are unique among the brainstem and have progressive reduction in more rostral rhombomeres of dA2 and dA3. Note that the BMP and Wnt genes are reduced progressively in the more ventral nuclei. Note that expression of *Ptf1a* is more rostral. (Compiled from Hernandez-Miranda et al., 2017; Iskusnykh et al., 2016; Lai et al., 2016; van der Heijden and Zoghbi, 2020.)

unclear (Hernandez-Miranda et al., 2017). Of interest is the interaction of *Ascl1* with *Phox2b, Tlx3,* and *Lmx1b* to define taste bud input (Qian et al., 2001). Proneural bHLH genes interact with *Hes* genes to interfere with proneural transcription factors to bind and sequester proneural bHLH genes from dimerization (Fritzsch et al., 2010; Kageyama et al., 2019).

The midbrain and diencephalon reach the central projections of the visual cortex from the geniculate ganglion and the superior colliculi. Genes have been characterized, such as *Neurog1, Neurog2, Neurod1,* and *Ascl1,* by their expression in mammals, frogs, and zebrafish (Osório et al., 2010), but not in detail relative to their null effects, to understand the molecular regions identifying their expression (Arimura et al., 2019). The olfactory bulb depends on *Neurog1, Neurog2, Neurod1,* and *Ascl1,* parallel to the olfactory organ (Shaker et al., 2012). The sequence of *Sox2/Pax2* expression will lead to the upregulation of *Ascl1* followed by *Neurog1, Neurog2,* and *Neurod1,* which initiate the olfactory sensory neurons (Chon et al., 2020; Dennis et al., 2019; Schwob et al., 2017).

In summary, central nuclei can be defined as topological and non-topological organizations of sensory projections. All CNS nuclei depend on *Sox2* expression for bHLH genes that are upregulated for a large set of second-order neurons.

10.5 SECOND-ORDER PROJECTIONS DO NOT FOLLOW A COMMON PLAN

A common theme of second-order projections can be identified among central projections that show many additional second-order neurons that reach the telencephalon. Short second-order neurons from the olfactory bulb reach directly to the telencephalon in lampreys and gnathostomes (Figure 10.6A; Nieuwenhuys and Puelles, 2015; Suryanarayana et al., 2021). In contrast, certain inputs require at least one and up to three intermediate connections (Figure 10.6B–D). In vision, only one related connection links directly to the cortex from the LGB (Figure 10.6B; Dhande and Huberman, 2014). Two connections define intermediate connections: the trigeminal and spinal cord connecting contralaterally, with exceptions, to the cortex after forming synapses in the VPM/VPL (Figure 10.6C; Liao et al., 2020). A second connection is provided from taste buds that reach partial direct and indirect connections from solitary nucleus inputs (Lundy Jr and Norgren, 2015). In contrast, some three levels of the vestibular, auditory, lateral line, and electroreceptors show bilateral interactions from all these inputs (Chagnaud et al., 2017; Grothe et al., 2004; Wullimann and Grothe, 2013).

Olfactory bulb projections have been studied in all representative vertebrates that undergo a restriction of olfactory recipient areas. Olfaction is unique among mammalian sensory systems in that second-order neurons project directly to the primary sensory cortex without an intervening thalamic relay (Striedter and Northcutt, 2019; Wilson and East, 2020). The olfactory cortex consists of a collection of laminar structures arranged along the ventrolateral surface of the mammalian brain that receives direct input from the olfactory bulb, mitral and tufted cells. The olfactory

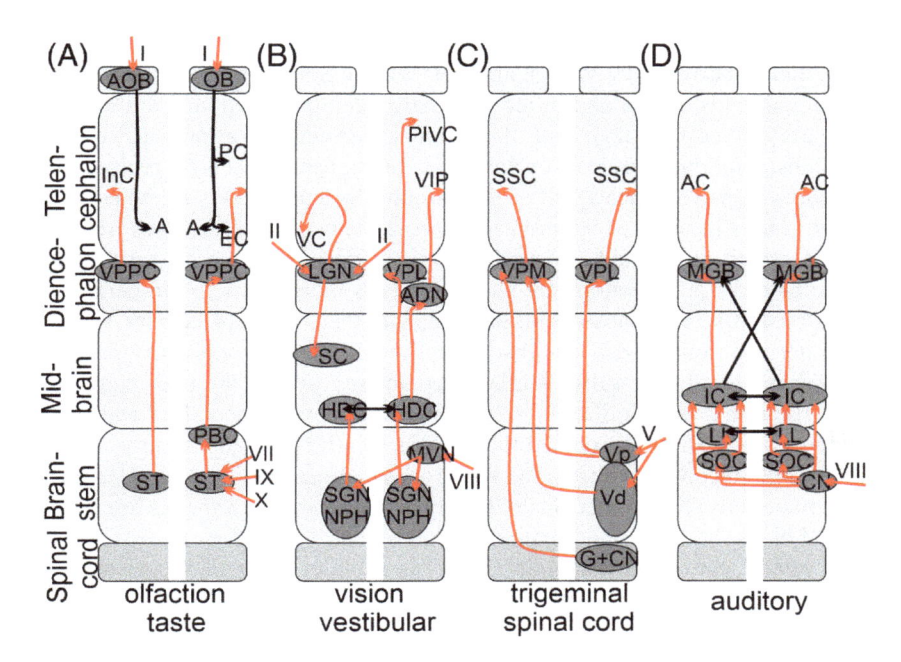

FIGURE 10.6 Details of six sensory system central projections. A direct connection of the olfactory system (A) projects ipsilaterally. A mostly ipsilateral solitary tract (ST) receives the input from three different fibers that, after one or two connections of the diencephalon, reach its target (InC). Visual input (B) from two optic nerves reaches the VC via the LGN. Vestibular output is complicated and reaches at least two cortical areas that reach bilaterally (truncated) connections. Some specific outputs are ipsilateral, while others are bilateral or contralateral. The trigeminal and spinal cord projections (C) combine in the central thalamus to project out to the somatosensory cortex. Auditory projections (D) have at least 3–4 connections before reaching the auditory cortex (AC). In addition, several connections are shown in black to highlight the reciprocal interactions. A, amygdala; AC, auditory cortex; ADN, anterodorsal thalamus; CN, cochlear nucleus; EC, entorhinal cortex; G+CN, gracile and cuneate nucleus, HDC, lateral mamillary and dorsal tegmental nucleus; IC, inferior colliculus; InC, insula cortex; LGN, lateral geniculate nucleus; LL, lateral lemniscus; MGB, medial geniculate body; MVN, medial vestibular nucleus; NPH, nucleus prepositus hypoglossi; PC, piriform cortex; PBC, parabrachial nucleus; PIVC, parieto-insula vestibular cortex; SC, superior colliculus; SGN, supragenual nucleus; SOC, superior olivary complex; SSC, somatosensory cortex; ST, solitary tract; VC, visual cortex; VIP, ventral intraparietal area; VPM, ventral posterior medial; VPL, ventral posterior lateral; VPPC, ventral posterior parvicellular nucleus; Vd, descending trigeminal; Vp, principal trigeminal; I, II, V, VII, VIII, IX, X, sensory projections.

size varies among mammals and is relatively large among basic mammals (Striedter and Northcutt, 2019). Central targets are the piriform cortex, olfactory tubercle, cortical nucleus of the amygdala, and lateral entorhinal cortex (Cleland and Linster, 2003; Haberly, 2001). Except for the entorhinal cortex, which receives lateral olfactory tract terminals in superficial layer 1, a pyramidal cell layer II and layer composed of associated fiber axons forms layer III. The cortex does not show columnar organizations that project back to the olfactory bulb (Haberly and Price, 1978; Neville and Haberly, 2004; Wilson and East, 2020). Similar to the brain's olfactory input in mammals, mitral and tufted cells have been identified also in the lamprey (Suryanarayana et al., 2021). Still, they show a distinct central projection between tufted cells, a unique feature in cyclostomes. In hagfish, the central projections are even more extensive to nearly all of the pallium (Striedter and Northcutt, 2019). In contrast, sharks have a vast ventrolateral part of the lateral pallium free from olfactory projections, designated as dorsal and medial pallium. A progressive reduction in the olfactory spread within ray-finned fishes is more restricted in derived teleosts than fish, like *Polypterus*. This suggests that there is a replacement of a dorsal area in derived ray-finned fishes that do not receive primary olfactory projections (Striedter and Northcutt, 2019). While tufted and mitral cells are known among tetrapods, there is another olfactory relay central projection, the ruffed cells of teleosts (Korsching, 2020; Wilson and East, 2020). Central projections of the accessory olfactory bulb receive input from the vomeronasal olfactory receptors, a feature also seen in mammals (Niimura et al., 2020). The vomeronasal system was lost in crocodiles and birds.

The lateral geniculate ganglion (LGN) is a major relay of the visual cortex (Hubel and Wiesel, 1962; Karten, 2013; Lean et al., 2019; McLaughlin et al., 2000; Stellwagen and Shatz, 2002). For years, no LGB-like connections were considered a novel expansion from gnathostomes that recently expanded in the pallium of lampreys (Suryanarayana et al., 2020). Interestingly, the topology of the LGN projects distinctive uncrossed and crossed fibers (Kalish et al., 2018; Tassinari et al., 1997). Developmental data demonstrate that specific eye domains in the dorsal LGN (dLGN) segregate through spontaneous electrical activity (left and right) prior to the onset of vision (Guido, 2018; Huberman et al., 2008). Initially, ipsilateral and contralateral LGN fibers overlap broadly but are segregated by activity (Dhande and Huberman, 2014). The earliest insights suggested that LGN fibers are a novel feature unique to mammals (Striedter and Northcutt, 2019). More recent studies changed our perspective by demonstrating a connection to the visual retinotopic input, which expanded previous suggestions on the evolution of vertebrates (Briscoe and Ragsdale, 2018; Karten, 2013; Suryanarayana et al., 2020; Tosches and Laurent, 2019). While overall similarities are now established across all vertebrates, this does not mean that there are more similarities in visual input when examining the details. The six-layered cytoarchitecture of mammals is reduced to fewer layers among non-mammals, showing instead three layers or less (Tosches and Laurent, 2019).

The cortical somatosensory cortex was discovered by Penfield, showing distorted human-like figures, referred to as a 'homunculus' (Penfield and Boldrey, 1937). Projections from trigeminal principal sensory (oralis; Vp) and spinal (interpolaris, caudalis) nuclei, and dorsal column spinal cord nuclei (gracile and cuneate) are organized in an inverted position relative to the ventral posterior (VPM) and adjacent ventral posterior lateral (VPL) thalamus. The ascending branches of second-order trigeminal neurons give rise to projections to the thalamus from the Vp, Vd, and spinal cord to form the lemniscal set of fibers that reach the thalamus on the contralateral trigeminal neurons, except for ipsilateral Vp projections (Erzurumlu et al., 2010). Somatosensory inputs are common in all vertebrates (Briscoe and Ragsdale, 2019; Pani et al., 2012), including lampreys (Suryanarayana et al., 2020). The same organization is followed in the topography of neocortical areas with the visual cortex occipitally, the auditory cortex ventrally, and two major somatosensory areas rostrally (Catania et al., 2000). Unique to rodents, the YP fibers are organized into 'barrelettes' to mimic the well-known barrel fields of the cortex (Erzurumlu and Gaspar, 2020). Investigations of the barrel field are driven by their unique organization in rodents (Liao et al., 2020) and have become significant in studies of plasticity (Erzurumlu and Gaspar, 2020; Qi et al., 2020).

From the solitary tract (ST), innervation reaches the parabrachial nuclei (PB) in the dorsal pons to innervate mostly ipsilateral fibers, where they likely end in the thalamus (Gasparini et al., 2021; Herbert et al., 1990; Lundy Jr and Norgren, 2015; Saper and Loewy, 1980). In contrast, many neurons of the ST expand to reach the thalamic gustatory relay directly in primates (Beckstead et al., 1980). The thalamic gustatory or taste relay is adjacent and medial to the VPM and is referred to as the ventral posterior thalamic nucleus parvocellularis (VPPC) (Lundy Jr and Norgren, 2015). This unique ipsilateral connection breaks down into a complex interaction from the VPPC that receives fibers from the ipsilateral and, to a lesser extent, the contralateral PB, to reach the insula cortex. The PB reach the taste cortex without any relay within the thalamus and have reciprocal connections with gustatory and visceral afferents. Fibers expand to reach the medial septum and olfactory tubercle (Lundy Jr and Norgren, 2015). Unfortunately, very little is known about the thalamic connections in non-mammalian vertebrates.

Vestibular projections reach the ventral posterolateral nucleus (VPL) and expand from this input to reach the parietal-insula vestibular cortical (PIVC) area. The PIVC is located in the insula region and extends into the posterior parts of the insular lobe that receives, in mammals, mainly the ventral posterolateral thalamic nuclei, bilaterally (VPL, VPM, and VL; Dieterich and Brandt, 2020), while some fibers bypass directly to the PIVC. The ventral intraparietal area's (VIP) role in the cortical vestibular circuit, which shows somatosensory and visual optokinetic interactions, has been studied extensively (Dieterich and Brandt, 2020). In humans, it appears that certain stimuli project via ipsilateral connections to the PIVC. In contrast, other stimuli project bilaterally or even have a dominant input that crosses at least three different levels before reaching the cortex (Dieterich et al., 2017). Vestibular nuclei project indirect relays to the nucleus pre-positus hypoglossi (NPH) and supragenual nucleus (NPH/SGN), which projects to the head direction network (HDC), which in turn reaches the anterior vestibular-thalamic pathway (ADN) in the thalamus to reach the limbic and entorhinal cortex (Cullen et al., 2020; Taube and Yoder, 2020). More work is needed to establish the vestibular connections of the cortex, since data are primarily limited to tetrapods at the moment.

For the auditory system, at least three connections interact with the superior olivary complex and the lateral lemniscus bilaterally and expand to the inferior colliculi bilaterally, providing bilateral auditory tonotopic information (Kandler et al., 2020; Lohmann and Friauf, 1996). The medial geniculate body (MGB) also provides bilateral input that combines to project to the primary auditory cortex (Malone et al., 2020; Rauschecker, 2020), corresponding to the bilateral connection in vestibular cortical projections (Dieterich et al., 2017). What is somewhat unique is the level of the feedback system from the different levels of the auditory connections (Grothe, 2020). Little information exists in auditory, lateral line, and electro-reception and is limited to a few examples outside mammals (Grothe et al., 2004).

In summary, we compare the level of different sensory inputs that show convergence to the telencephalon. The number of afferent connections varies from a single connection (olfactory bulb to cortex) to at least two or more intermediate connections. Maps can reflect (a) local receptor density and activity (somatosensory), (b) convergence of distributed receptors (olfactory), (c) continuous one-dimensional (tonotopic auditory) or (d) 2D (retinotopic and somatotopic) maps, or (e) convergence and segregation of information gathered by distinct sensory organs (vestibular and orotopic) maps. What defines the different terminal fields is that each is characterized among its area that primarily receives distinct thalamic and olfactory bulb terminals.

10.6 INTERACTION AND INTEGRATION OF CORTICAL SENSORY PROJECTIONS

The frontal lobe from the telencephalon is absent in all chordates rostral to the neuropore except in vertebrates that expand bilaterally beyond the neuropore. Segregated central projections of the somatosensory, visual and olfactory system are now demonstrated from lamprey to mammals (Figure 10.7; Suryanarayana et al., 2020, 2021). Note that with the double projections of the olfactory cortex in lamprey, there is a close proximation to the distinct projections of olfactory, vision, and somatosensory cortical projections in humans. Additional cortical projections are known in humans (Figure 10.7) that are either unclear in the auditory system beyond the tetrapods or are nearly unknown for the vestibular, lateral line, electro-reception, and taste buds, for which we have limited information as to the central projections in non-mammalian vertebrates.

Second-order cortical projections expand and interact with tertiary projections to integrate distinct sensory inputs into a cohesive perception. Strong connections are demonstrated between the olfactory and gustatory systems and are highlighted in multiple levels of integration (Shepherd, 2006; Witt, 2020). Chemical senses converge, to provide a whole perception of taste and smell information onto single neurons of the caudal orbitofrontal cortex (De Araujo et al., 2003). Smell and taste combine for their 'pleasantness' in the medial anterior orbitofrontal cortex, whereas the anterior insula and the anterior operculum respond to unimodal taste and olfactory stimuli in humans. More details in rats show that the agranular insula cortex is more sensitive to olfactory signals, whereas the tasting process is predominant in the dysgranular insula cortex (Mizoguchi et al., 2020; Samuelsen and Fontanini, 2017). A direct response to taste is demonstrated in the posterior olfactory (piriform) cortex for taste stimuli that cause olfactory stimulation through the nasal epithelium (Maier et al., 2012), suggesting gustatory stimulation influences olfactory processing.

Further information suggests that odor identification relies heavily on visual input, implying that visual-olfactory congruence is perceived in gradual and distinct

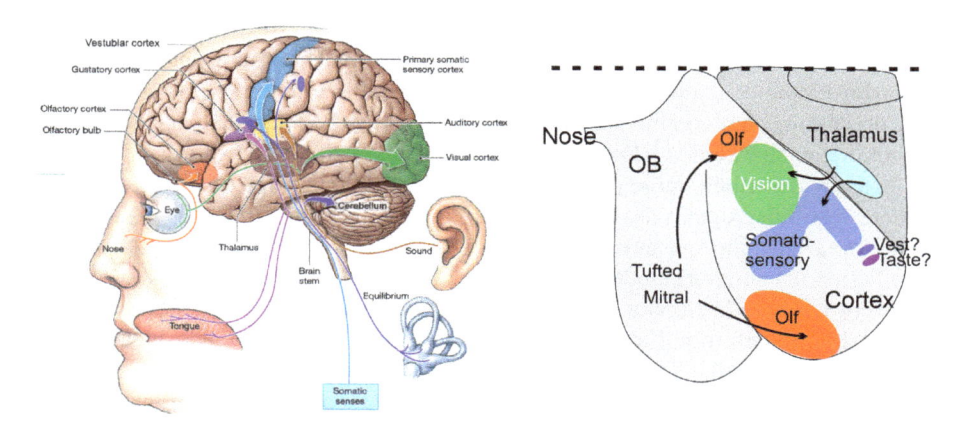

FIGURE 10.7 Comparisons of the human cortex (lateral view, left) with a lamprey (half brain, right). Both olfactory bulbs receive direct olfactory input that project to the olfactory cortex. In the lamprey, a unique projection exists from tufted receptors that project close to the visual sensory cortex (green), distinct from the other olfactory cortex that seems to be comparable to the piriform cortex (orange) innervated from mitral cells in human (left) and lamprey (right). Visual input is relayed to the thalamus to then reach the visual cortex (green) in both species. Somatosensory input projects via the thalamus to reach the somatosensory cortex (light blue) in both species. Auditory input is relayed through the thalamus to the auditory cortex (yellow) in humans (left) but is absent in lampreys. A gustatory cortex (lilac) receives connections from the sensory neurons carrying taste information in humans but is unclear in lampreys (Taste?). Vestibular input projects with multiple steps to reach the parietal-insula vestibular cortex (dark blue) in humans (left) but is unclear in lampreys (Vest?). Lateral line and electroreception are absent in tetrapods but are present in lampreys; however, the cortical regions of lampreys are unclear. OB, olfactory bulb; Olf, olfactory cortex; Vest?, vestibular cortex. (Modified from Suryanarayana et al., 2020, 2021.)

categories and is consistent with parallel processing of olfactory and visual stimuli that interact at various levels (Amsellem et al., 2018). Multisensory integration dependent on non-olfactory cues is challenging to record, such as auditory, visual, and somatosensory with olfactory matching tasks, and particularly visual-olfactory interactions (Meehan et al., 2017). Phase synchrony for integrating cross-modal odor processing with auditory stimulation highlights primary olfactory cortices (Zhou et al., 2019). These data suggest an integration beyond the traditional boundaries between different sensory systems, particularly olfaction and taste.

For vestibular connections, cortical structures involved include the parietal-insula vestibular cortex (PIVC), the visual temporal Sylvian area (VTS) in the retroinsula cortex, the superior temporal gyrus (STG), the inferior parietal lobule (IPL), in particular the ventral parietal area (VIP), the anterior cingulum, the hippocampus, and area 6a. All belong to a multisensory (vestibular) cortical circuit in which the neurons respond to not only one sensory modality. Functional imaging studies using vestibular, somatosensory, and visual optokinetic stimulation have suggested that such multisensory vestibular cortical areas are located and connected in corresponding regions in humans (Dieterich et al., 2017). The PIVC showed spatiotemporal characteristics in electrophysiological cortical registration that represented a gradual transformation of temporal responses, suggesting a hierarchy in cortical vestibular processing that combines vestibular sensation with somatosensory and visual optokinetic stimulation (Dieterich and Brandt, 2020). Further, visual signals effectively contribute more during quiet standing than more dynamic movements, where vestibular signals are relatively weighted (Peterka, 2018; Van Der Kooij and Peterka, 2011). Current evidence suggests that the cerebellum builds a dynamic prediction (an internal model) of the sensory consequences of self-motion during actively-generated behavior (Cullen et al., 2020). Spatial cells have been identified in the lateral entorhinal and postrhinal cortices and contain center-bearing and center-distance cells. These cells are considered egocentric that likely form the basis for the higher-order spatial representations or maps seen in the hippocampus. The extent to which the vestibular system contributes to these egocentric cell types is unknown (Taube and Yoder, 2020). The integration and broad projections are highlighted for integration with specific interactions to drive position and space information.

The auditory system is a recent formation from the basilar papilla/cochlea (Fritzsch and Elliott, 2017; Grothe et al., 2004) and develops in tetrapods (Fritzsch et al., 2013; Luo and Manley, 2020), to form a unique perception of a novel sense that is now comparable to vision in humans, forming maps for language and music (Rauschecker, 2020). A new model extends beyond speech processing and generally applies to both vision and audition (Rauschecker, 2018). Information from one sensory modality can influence another modality, implying the existence of polysensory neurons (Petro et al., 2017). Suggestions of direct, non-reciprocal projections from auditory areas to the visual cortex in primates suggest a shared characteristic of audiovisual integration (Majka et al., 2019). Importantly, conductive hearing loss can cause visual activity that demonstrates a direct auditory connection to vision (Teichert and Bolz, 2017). All evidence points out an influence by the auditory cortex on vision that is not reciprocated from the visual to the auditory, implying that the expansion of the auditory cortex is continuing to other cortical areas.

In summary, expanding from cephalochordates, bimodal inputs are provided by olfactory-taste interactions, auditory-vision interactions, and vestibular-vision interactions. Multimodal inputs are few but may show additional expansions with further investigations.

10.7 SUMMARY AND CONCLUSION

The eight sensory inputs to the brain from the sensory organs are the windows of the brain to the environment, permitting various behavioral interactions. A comparison of chordates shows the addition and diversification from existing precursors that lead to the formation of all sensory inputs in vertebrates. A notable additional bilateral expansion of the telencephalon is unique in vertebrates and is reflected as a primitive precursor in the neuropore of cephalochordates, the lancelets, and ascidians. A common theme is the multiplication of genes from a single set of genes in chordates that quadruple in vertebrates. Olfaction is unique in vertebrates with its addition to the anterior pole of the telencephalon, receiving sensory neurons that form unique olfactory bulbs and reach the cortex. The visual system started with simple neuro-sensory projections in chordates that show an individual multiplication of sensory cells and sensory neurons, including the precursors to the retina, protruding from the diencephalon. Trigeminal sensory information is added to central projections that are a substitute by the sensory neurons to reach the diencephalon on its way to the somatosensory cortex. Taste information requires the unique input of three distinct sensory neurons that reach the taste buds peripherally and end up in the solitary tract. From here, direct and indirect projections reach specific areas of the cortex to interact with olfactory information and integrate taste with olfaction. Vestibular input, lateral line, and electroreception, have a robust bilateral output that ends, with several additional connections, in the cortex after the auditory system had evolved in tetrapods. Interesting interactions are found between auditory and vestibular and auditory and vision in mammals. Moreover, there are connections between visual and vestibular input that integrates oculomotor output to end up as an integrated perspective of double and triple sensory systems. The secondary and tertiary connections are more complicated in demonstrating interactions, but these interactions may be detailed in the future with novel technologies.

ACKNOWLEDGMENTS

We thank all past and present researchers who helped with discussions. A specific thanks is given to Pin-Xian Xu that provided us images for *Eya1/2*.

REFERENCES

Abraira, V.E., and Ginty, D.D. (2013). The sensory neurons of touch. Neuron *79*, 618–639.
Alsina, B. (2020). Mechanisms of cell specification and differentiation in vertebrate cranial sensory systems. Current Opinion in Cell Biology *67*, 79–85.
Amsellem, S., Höchenberger, R., and Ohla, K. (2018). Visual–olfactory interactions: Bimodal facilitation and impact on the subjective experience. Chemical Senses *43*, 329–339.
Arendt, D., Musser, J.M., Baker, C.V., Bergman, A., Cepko, C., Erwin, D.H., Pavlicev, M., Schlosser, G., Widder, S., and Laubichler, M.D. (2016). The origin and evolution of cell types. Nature Reviews Genetics *17*, 744–757.
Arimura, N., Dewa, K.i., Okada, M., Yanagawa, Y., Taya, S.i., and Hoshino, M. (2019). Comprehensive and cell-type-based characterization of the dorsal midbrain during development. Genes to Cells *24*, 41–59.
Baker, C.V. H. (2019). The development and evolution of lateral line electroreceptors: Insights from comparative molecular approaches. In Electroreception: Fundamental Insights from Comparative Approaches. (Springer, Switzerland), pp. 25–62.
Bassham, S., and Postlethwait, J.H. (2005). The evolutionary history of placodes: A molecular genetic investigation of the larvacean urochordate *Oikopleura dioica*. Development *132*, 4259–4272.
Beckstead, R.M., Morse, J.R., and Norgren, R. (1980). The nucleus of the solitary tract in the monkey: Projections to the thalamus and brain stem nuclei. Journal of Comparative Neurology *190*, 259–282.
Bellono, N.W., Leitch, D.B., and Julius, D. (2018). Molecular tuning of electroreception in sharks and skates. Nature *558*, 122.

Bermingham, N.A., Hassan, B.A., Price, S.D., Vollrath, M.A., Ben-Arie, N., Eatock, R.A., Bellen, H.J., Lysakowski, A., and Zoghbi, H.Y. (1999). Math1: An essential gene for the generation of inner ear hair cells. Science *284*, 1837–1841.

Bermingham, N.A., Hassan, B.A., Wang, V.Y., Fernandez, M., Banfi, S., Bellen, H.J., Fritzsch, B., and Zoghbi, H.Y. (2001). Proprioceptor pathway development is dependent on Math1. Neuron *30*, 411–422.

Bertrand, N., Castro, D.S., and Guillemot, F. (2002). Proneural genes and the specification of neural cell types. Nature Reviews Neuroscience *3*, 517–530.

Birol, O., Ohyama, T., Edlund, R.K., Drakou, K., Georgiades, P., and Groves, A.K. (2016). The mouse Foxi3 transcription factor is necessary for the development of posterior placodes. Developmental Biology *409*, 139–151.

Bouchard, M., de Caprona, D., Busslinger, M., Xu, P., and Fritzsch, B. (2010). Pax2 and Pax8 cooperate in mouse inner ear morphogenesis and innervation. BMC Developmental Biology *10*, 89.

Bozzo, M., Lacalli, T.C., Obino, V., Caicci, F., Marcenaro, E., Bachetti, T., Manni, L., Pestarino, M., Schubert, M., and Candiani, S. (2021). Amphioxus neuroglia: Molecular characterization and evidence for early compartmentalization of the developing nerve cord. Glia *69*, 1654–1678.

Braun, K., Leubner, F., and Stach, T. (2020). Phylogenetic analysis of phenotypic characters of Tunicata supports basal Appendicularia and monophyletic Ascidiacea. Cladistics *36*, 259–300.

Briscoe, S.D., and Ragsdale, C.W. (2018). Homology, neocortex, and the evolution of developmental mechanisms. Science *362*, 190–193.

Briscoe, S.D., and Ragsdale, C.W. (2019). Evolution of the chordate telencephalon. Current Biology *29*, R647–R662.

Brzezinski, J.A., and Reh, T.A. (2015). Photoreceptor cell fate specification in vertebrates. Development *142*, 3263–3273.

Buck, L., and Axel, R. (1991). A novel multigene family may encode odorant receptors: A molecular basis for odor recognition. Cell *65*, 175–187.

Buzzi, A.L., Hintze, M.S., and Streit, A. (2019). Development of neurogenic placodes in vertebrates. eLS, 1–14.

Caicci, F., Gasparini, F., Rigon, F., Zaniolo, G., Burighel, P., and Manni, L. (2013). The oral sensory structures of Thaliacea (Tunicata) and consideration of the evolution of hair cells in Chordata. Journal of Comparative Neurology *521*, 2756–2771.

Campo-Paysaa, F., Marlétaz, F., Laudet, V., and Schubert, M. (2008). Retinoic acid signaling in development: tissue-specific functions and evolutionary origins. Genesis *46*, 640–656.

Catania, K.C., Jain, N., Franca, J.G., Volchan, E., and Kaas, J.H. (2000). The organization of somatosensory cortex in the short-tailed opossum (*Monodelphis domestica*). Somatosensory & Motor Research *17*, 39–51.

Chagnaud, B.P., Engelmann, J., Fritzsch, B., Glover, J.C., and Straka, H. (2017). Sensing external and self-motion with hair cells: A comparison of the lateral line and vestibular systems from a developmental and evolutionary perspective. Brain, Behavior and Evolution *90*, 98–116.

Chizhikov, V.V., Iskusnykh, I.Y., Fattakhov, N., and Fritzsch, B. (2021). Lmx1a and Lmx1b are redundantly required for the development of multiple components of the mammalian auditory system. Neuroscience *452*, 247–264.

Chon, U., LaFever, B.J., Nguyen, U., Kim, Y., and Imamura, F. (2020). Topographically distinct projection patterns of early-generated and late-generated projection neurons in the mouse olfactory bulb. Eneuro *7*(6).

Cleland, T.A., and Linster, C. (2003). Central olfactory structures. In Handbook of Olfaction and Gustation, R.L. Doty, ed. New York: Marcel Dekker, pp. 165–180.

Coste, B., Mathur, J., Schmidt, M., Earley, T.J., Ranade, S., Petrus, M.J., Dubin, A.E., and Patapoutian, A. (2010). Piezo1 and Piezo2 are essential components of distinct mechanically activated cation channels. Science *330*, 55–60.

Cullen, K.E., Zobeini, O.A., Woboonsaksakul, K.P., Wang, L., Stanley, O.R., Leavitt, O.M.E., and Chang, H.H.V. (2020). Self-motion. In The Senses, B. Fritzsch, ed. (Elsevier) pp. 483–495.

Cuthill, I.C., Partridge, J.C., Bennett, A.T., Church, S.C., Hart, N.S., and Hunt, S. (2000). Ultraviolet vision in birds. Advances in the Study of Behavior *29*, 159–214.

De Araujo, I.E., Rolls, E.T., Kringelbach, M.L., McGlone, F., and Phillips, N. (2003). Taste-olfactory convergence, and the representation of the pleasantness of flavour, in the human brain. European Journal of Neuroscience *18*, 2059–2068.

del Pilar Gomez, M., Angueyra, J.M., and Nasi, E. (2009). Light-transduction in melanopsin-expressing photoreceptors of Amphioxus. Proceedings of the National Academy of Sciences *106*, 9081–9086.

Dennis, D.J., Han, S., and Schuurmans, C. (2019). bHLH transcription factors in neural development, disease, and reprogramming. Brain Research *1705*, 48–65.

Dhande, O.S., and Huberman, A.D. (2014). Retinal ganglion cell maps in the brain: Implications for visual processing. Current Opinion in Neurobiology *24*, 133–142.

Di Bonito, M., Studer, M., and Puelles, L. (2017). Nuclear derivatives and axonal projections originating from rhombomere 4 in the mouse hindbrain. Brain Structure and Function *222*, 3509–3542.

Dieterich, M., and Brandt, T. (2020). Structural and functional imaging ofthe human bilateral vestibular network from the brainstem to the cortical hemispheres. In The Senses, B. Fritzsch, ed. (Elsevier), pp. 414–431.

Dieterich, M., Kirsch, V., and Brandt, T. (2017). Right-sided dominance of the bilateral vestibular system in the upper brainstem and thalamus. Journal of Neurology *264*, 55–62.

Dvorakova, M., Macova, I., Bohuslavova, R., Anderova, M., Fritzsch, B., and Pavlinkova, G. (2020). Early ear neuronal development, but not olfactory or lens development, can proceed without SOX2. Developmental Biology *457*, 43–56.

Elliott, K.L., Pavlínková, G., Chizhikov, V.V., Yamoah, E.N., and Fritzsch, B. (2021). Development in the mammalian auditory system depends on transcription factors. International Journal of Molecular Science *22*, 4189.

Erives, A., and Fritzsch, B. (2020). A Screen for gene paralogies delineating evolutionary branching order of early Metazoa. G3 (Bethesda) *10*, 811–826.

Erzurumlu, R.S., and Gaspar, P. (2020). How the barrel cortex became a working model for developmental plasticity: A historical perspective. Journal of Neuroscience *40*, 6460–6473.

Erzurumlu, R.S., Murakami, Y., and Rijli, F.M. (2010). Mapping the face in the somatosensory brainstem. Nature Reviews Neuroscience *11*, 252–263.

Finger, T.E. (2008). Sorting food from stones: The vagal taste system in Goldfish, Carassius auratus. Journal of Comparative Physiology A *194*, 135–143.

Fode, C., Gradwohl, G., Morin, X., Dierich, A., LeMeur, M., Goridis, C., and Guillemot, F. (1998). The bHLH protein NEUROGENIN 2 is a determination factor for epibranchial placode-derived sensory neurons. Neuron *20*, 483–494.

Fodor, A., Liu, J., Turner, L., and Swalla, B.J. (2020). Transitional chordates and vertebrate origins: Tunicates. Current Topics in Developmental Biology *141*, 149–171.

Fritzsch, B. (1996). Similarities and differences in lancelet and craniate nervous systems. Israel Journal of Zoology *42*, S147–S160.

Fritzsch, B., Beisel, K.W., Pauley, S., and Soukup, G. (2007). Molecular evolution of the vertebrate mechanosensory cell and ear. The International Journal of Developmental Biology *51*, 663.

Fritzsch, B., Eberl, D.F., and Beisel, K.W. (2010). The role of bHLH genes in ear development and evolution: Revisiting a 10-year-old hypothesis. Cellular and Molecular Life Sciences *67*, 3089–3099.

Fritzsch, B., and Elliott, K.L. (2017). Gene, cell, and organ multiplication drives inner ear evolution. Developmental Biology *431*, 3–15.

Fritzsch, B., Elliott, K.L., and Pavlinkova, G. (2019). Primary sensory map formations reflect unique needs and molecular cues specific to each sensory system. F1000 Research *8*.

Fritzsch, B., and Northcutt, G. (1993). Cranial and spinal nerve organization in amphioxus and lampreys: Evidence for an ancestral craniate pattern. Cells Tissues Organs *148*, 96–109.

Fritzsch, B., Pan, N., Jahan, I., Duncan, J.S., Kopecky, B.J., Elliott, K.L., Kersigo, J., and Yang, T. (2013). Evolution and development of the tetrapod auditory system: an organ of Corti-centric perspective. Evolution & Development *15*, 63–79.

Fritzsch, B., Pauley, S., Feng, F., Matei, V., and Nichols, D. (2006). The molecular and developmental basis of the evolution of the vertebrate auditory system. International Journal of Comparative Psychology *19*, 1–25.

Fujiyama, T., Yamada, M., Terao, M., Terashima, T., Hioki, H., Inoue, Y.U., Inoue, T., Masuyama, N., Obata, K., and Yanagawa, Y. (2009). Inhibitory and excitatory subtypes of cochlear nucleus neurons are defined by distinct bHLH transcription factors, Ptf1a and Atoh1. Development *136*, 2049–2058.

Gasparini, S., Resch, J.M., Gore, A.M., Peltekian, L., and Geerling, J.C. (2021). Pre-locus coeruleus neurons in rat and mouse. American Journal of Physiology-Regulatory Integrative and Comparative Physiology *320*, R342–r361.

Ghimire, S.R., and Deans, M.R. (2019). Frizzled3 and Frizzled6 cooperate with Vangl2 to direct cochlear innervation by type II spiral ganglion neurons. Journal of Neuroscience *39*, 8013–8023.

Glover, J.C., Elliott, K.L., Erives, A., Chizhikov, V.V., and Fritzsch, B. (2018). Wilhelm His' lasting insights into hindbrain and cranial ganglia development and evolution. Developmental Biology *444*, S14–S24.

Göhde, R., Naumann, B., Laundon, D., Imig, C., McDonald, K., Cooper, B.H., Varoqueaux, F., Fasshauer, D., and Burkhardt, P. (2021). Choanoflagellates and the ancestry of neurosecretory vesicles. Philosophical Transactions of the Royal Society B *376*, 20190759.

Gowan, K., Helms, A.W., Hunsaker, T.L., Collisson, T., Ebert, P.J., Odom, R., and Johnson, J.E. (2001). Crossinhibitory activities of Ngn1 and Math1 allow specification of distinct dorsal interneurons. Neuron *31*, 219–232.

Grothe, B. (2020). The auditory system function - An integrative perspective. In The Senses, B. Fritzsch, ed. (Elsevier), pp. 1–17.

Grothe, B., Carr, C.E., Casseday, J.H., Fritzsch, B., and Köppl, C. (2004). The evolution of central pathways and their neural processing patterns. In Evolution of the Vertebrate Auditory System (Springer, New York, NY), pp. 289–359.

Guido, W. (2018). Development, form, and function of the mouse visual thalamus. Journal of Neurophysiology *120*, 211–225.

Haberly, L.B. (2001). Parallel-distributed processing in olfactory cortex: New insights from morphological and physiological analysis of neuronal circuitry. Chemical Senses *26*, 551–576.

Haberly, L.B., and Price, J.L. (1978). Association and commissural fiber systems of the olfactory cortex of the rat. Journal of Comparative Neurology *178*, 711–740.

Herbert, H., Moga, M.M., and Saper, C.B. (1990). Connections of the parabrachial nucleus with the nucleus of the solitary tract and the medullary reticular formation in the rat. Journal of Comparative Neurology *293*, 540–580.

Hernandez-Miranda, L.R., Müller, T., and Birchmeier, C. (2017). The dorsal spinal cord and hindbrain: From developmental mechanisms to functional circuits. Developmental Biology *432*, 34–42.

Hertzano, R., Montcouquiol, M., Rashi-Elkeles, S., Elkon, R., Yücel, R., Frankel, W.N., Rechavi, G., Möröy, T., Friedman, T.B., and Kelley, M.W. (2004). Transcription profiling of inner ears from Pou4f3 ddl/ddl identifies Gfi1 as a target of the Pou4f3 deafness gene. Human Molecular Genetics *13*, 2143–2153.

Holland, L. (2020). Invertebrate origins of vertebrate nervous systems. In Evolutionary Neuroscience (Elsevier, Academic Press, London), pp. 51–73.

Holland, L.Z., and Daza, D.O. (2018). A new look at an old question: When did the second whole genome duplication occur in vertebrate evolution? Genome Biology *19*, 1–4.

Holland, N.D., and Somorjai, I.M. (2020). The sensory peripheral nervous system in the tail of a cephalochordate studied by serial blockface scanning electron microscopy. Journal of Comparative Neurology. *528*(15), 2569–2582

Hubel, D.H., and Wiesel, T.N. (1962). Receptive fields, binocular interaction and functional architecture in the cat's visual cortex. The Journal of Physiology *160*, 106.

Huberman, A.D., Feller, M.B., and Chapman, B. (2008). Mechanisms underlying development of visual maps and receptive fields. Annual Review of Neuroscience *31*, 479–509.

Imai, T. (2020). Odor coding in the olfactory bulb. In The Senses, B. Fritzsch, ed. (Elsevier, Academic Press, Amsterdam, London), pp. 640–649.

Imai, T., Yamazaki, T., Kobayakawa, R., Kobayakawa, K., Abe, T., Suzuki, M., and Sakano, H. (2009). Pre-target axon sorting establishes the neural map topography. Science *325*, 585–590.

Imayoshi, I., and Kageyama, R. (2014). bHLH factors in self-renewal, multipotency, and fate choice of neural progenitor cells. Neuron *82*, 9–23.

Iskusnykh, I.Y., Steshina, E.Y., and Chizhikov, V.V. (2016). Loss of Ptf1a leads to a widespread cell-fate misspecification in the brainstem, affecting the development of somatosensory and viscerosensory nuclei. Journal of Neuroscience *36*, 2691–2710.

Jahan, I., Kersigo, J., Elliott, K.L., and Fritzsch, B. (2021). Smoothened overexpression causes trochlear motoneurons to reroute and innervate ipsilateral eyes. Cell and Tissue Research, 1–14.

Jahan, I., Pan, N., Kersigo, J., and Fritzsch, B. (2010). Neurod1 suppresses hair cell differentiation in ear ganglia and regulates hair cell subtype development in the cochlea. PloS One *5*, e11661.

Kageyama, R., Shimojo, H., and Ohtsuka, T. (2019). Dynamic control of neural stem cells by bHLH factors. Neuroscience Research *138*, 12–18.

Kalish, B.T., Cheadle, L., Hrvatin, S., Nagy, M.A., Rivera, S., Crow, M., Gillis, J., Kirchner, R., and Greenberg, M.E. (2018). Single-cell transcriptomics of the developing lateral geniculate nucleus reveals insights into circuit assembly and refinement. Proceedings of the National Academy of Sciences *115*, E1051–E1060.

Kandler, K., Lee, J., and Pecka, M. (2020). The superior olivary complex. In The Senses, B. Fritzsch, ed. (Elsevier, Academic Press, Amsterdam, London), pp. 533–555.

Karten, H.J. (2013). Neocortical evolution: Neuronal circuits arise independently of lamination. Current Biology *23*, R12–R15.

Kersigo, J., D'Angelo, A., Gray, B.D., Soukup, G.A., and Fritzsch, B. (2011). The role of sensory organs and the forebrain for the development of the craniofacial shape as revealed by Foxg1-cre-mediated microRNA loss. Genesis *49*, 326–341.

Kim, W.-Y., Fritzsch, B., Serls, A., Bakel, L.A., Huang, E.J., Reichardt, L.F., Barth, D.S., and Lee, J.E. (2001). NeuroD-null mice are deaf due to a severe loss of the inner ear sensory neurons during development. Development *128*, 417–426.

Kim, K., Gibboney, S., Razy-Krajka, F., Lowe, E.K., Wang, W., and Stolfi, A. (2020). Regulation of neurogenesis by FGF signaling and Neurogenin in the invertebrate chordate Ciona. Frontiers in Cell and Developmental Biology *8*, 477.

Konno, A., Kaizu, M., Hotta, K., Horie, T., Sasakura, Y., Ikeo, K., and Inaba, K. (2010). Distribution and structural diversity of cilia in tadpole larvae of the ascidian Ciona intestinalis. Developmental Biology *337*, 42–62.

Korsching, S.I. (2020). Taste and smell in zebrafish. In The Senses, B. Fritzsch, ed. (Elsevier, Academic Press, Amsterdam, London), pp. 466–492.

Krolewski, R.C., Packard, A., Jang, W., Wildner, H., and Schwob, J.E. (2012). Ascl1 (Mash1) knockout perturbs differentiation of nonneuronal cells in olfactory epithelium. PloS One *7*, e51737.

Krüger, M., Ruschke, K., and Braun, T. (2004). NSCL-1 and NSCL-2 synergistically determine the fate of GnRH-1 neurons and control necdin gene expression. The EMBO Journal *23*, 4353–4364.

Krüger, M., Schmid, T., Krüger, S., Bober, E., and Braun, T. (2006). Functional redundancy of NSCL-1 and NeuroD during development of the petrosal and vestibulocochlear ganglia. European Journal of Neuroscience *24*, 1581–1590.

Lacalli, T.C., Holland, N., and West, J. (1994). Landmarks in the anterior central nervous system of amphioxus larvae. Philosophical Transactions of the Royal Society of London Series B: Biological Sciences *344*, 165–185.

Lai, H.C., Seal, R.P., and Johnson, J.E. (2016). Making sense out of spinal cord somatosensory development. Development *143*, 3434–3448.

Lamb, T.D. (2013). Evolution of phototransduction, vertebrate photoreceptors and retina. Progress in Retinal and Eye Research *36*, 52–119.

Lamb, T.D., Patel, H., Chuah, A., Natoli, R.C., Davies, W.I., Hart, N.S., Collin, S.P., and Hunt, D.M. (2016). Evolution of vertebrate phototransduction: Cascade activation. Molecular Biology and Evolution *33*, 2064–2087.

Lean, G.A., Liu, Y.J., and Lyon, D.C. (2019). Cell type specific tracing of the subcortical input to primary visual cortex from the basal forebrain. Journal of Comparative Neurology *527*, 589–599.

Liao, C.-C., Qi, H.-X., Reed, J.L., and Kaas, J.H. (2020). The somatosensory system of primates. In The Senses, B. Fritzsch, ed. (Elsevier, Academic Press, Amsterdam, London), pp. 180–197.

Liu, B., and Satou, Y. (2019). Foxg specifies sensory neurons in the anterior neural plate border of the ascidian embryo. Nature Communications *10*, 1–10.

Liu, B., and Satou, Y. (2020). The genetic program to specify ectodermal cells in ascidian embryos. Development, Growth & Differentiation *62*, 301–310.

Li, R., Wu, F., Ruonala, R., Sapkota, D., Hu, Z., and Mu, X. (2014). Isl1 and Pou4f2 form a complex to regulate target genes in developing retinal ganglion cells. PloS One *9*, e92105.

Li, J., Zhang, T., Ramakrishnan, A., Fritzsch, B., Xu, J., Wong, E.Y., Loh, Y.-H.E., Ding, J., Shen, L., and Xu, P.-X. (2020). Dynamic changes in cis-regulatory occupancy by Six1 and its cooperative interactions with distinct cofactors drive lineage-specific gene expression programs during progressive differentiation of the auditory sensory epithelium. Nucleic Acids Research *48*, 2880–2896.

Lohmann, C., and Friauf, E. (1996). Distribution of the calcium-binding proteins parvalbumin and calretinin in the auditory brainstem of adult and developing rats. Journal of Comparative Neurology *367*, 90–109.

Lorenzen, S.M., Duggan, A., Osipovich, A.B., Magnuson, M.A., and García-Añoveros, J. (2015). Insm1 promotes neurogenic proliferation in delaminated otic progenitors. Mechanisms of Development *138 Pt 3*, 233–245.

Lowenstein, E.D., Rusanova, A., Stelzer, J., Hernaiz-Llorens, M., Schroer, A.E., Epifanova, E., Bladt, F., Isik, E.G., Buchert, S., Jia, S., *et al.* (2021). Olig3 regulates early cerebellar development. eLife *10*.

Lunde, A., Okaty, B.W., Dymecki, S.M., and Glover, J.C. (2019). Molecular profiling defines evolutionarily conserved transcription factor signatures of major vestibulospinal neuron groups. eNeuro *6*.

Lundy Jr, R.F., and Norgren, R. (2015). Gustatory system. In The Rat Nervous System (Elsevier, Academic Press, Amsterdam), pp. 733–760.

Luo, Z.-X., and Manley, G.A. (2020). Origins and early evolution of mammalian ears and hearing function. In The Senses, B. Fritzsch, ed. (Elsevier, Academic Press, Amsterdam, London), pp. 207–252.

Ma, Q., Chen, Z., del Barco Barrantes, I., De La Pompa, J.L., and Anderson, D.J. (1998). neurogenin1 is essential for the determination of neuronal precursors for proximal cranial sensory ganglia. Neuron *20*, 469–482.

Ma, Q., Fode, C., Guillemot, F., and Anderson, D.J. (1999). Neurogenin1 and neurogenin2 control two distinct waves of neurogenesis in developing dorsal root ganglia. Genes & Development *13*, 1717–1728.

Macova, I., Pysanenko, K., Chumak, T., Dvorakova, M., Bohuslavova, R., Syka, J., Fritzsch, B., and Pavlinkova, G. (2019). Neurod1 is essential for the primary tonotopic organization and related auditory information processing in the midbrain. Journal of Neuroscience *39*, 984–1004.

Maier, J.X., Wachowiak, M., and Katz, D.B. (2012). Chemosensory convergence on primary olfactory cortex. Journal of Neuroscience *32*, 17037–17047.

Majka, P., Rosa, M.G., Bai, S., Chan, J.M., Huo, B.-X., Jermakow, N., Lin, M.K., Takahashi, Y.S., Wolkowicz, I.H., and Worthy, K.H. (2019). Unidirectional monosynaptic connections from auditory areas to the primary visual cortex in the marmoset monkey. Brain Structure and Function *224*, 111–131.

Malone, B.J., Hasenstaub, A.R., and Schreiner, C.E. (2020). Primary auditory cortex II. Some functional considerations. In The Senses, B. Fritzsch, ed. (Elsevier, Academic Press, Amsterdam, London), pp. 657–680.

Manni, L., Anselmi, C., Burighel, P., Martini, M., and Gasparini, F. (2018). Differentiation and induced sensorial alteration of the coronal organ in the asexual life of a tunicate. Integrative and Comparative Biology *58*, 317–328.

Maricich, S.M., Wellnitz, S.A., Nelson, A.M., Lesniak, D.R., Gerling, G.J., Lumpkin, E.A., and Zoghbi, H.Y. (2009). Merkel cells are essential for light-touch responses. Science *324*, 1580–1582.

Marlétaz, F., Firbas, P.N., Maeso, I., Tena, J.J., Bogdanovic, O., Perry, M., Wyatt, C.D., de la Calle-Mustienes, E., Bertrand, S., and Burguera, D. (2018). Amphioxus functional genomics and the origins of vertebrate gene regulation. Nature *564*, 64–70.

Matei, V., Pauley, S., Kaing, S., Rowitch, D., Beisel, K.W., Morris, K., Feng, F., Jones, K., Lee, J., and Fritzsch, B. (2005). Smaller inner ear sensory epithelia in Neurog 1 null mice are related to earlier hair cell cycle exit. Devlopmental Dynamics *234*, 633–650.

Maurer, K.A., Riesenberg, A.N., and Brown, N.L. (2014). Notch signaling differentially regulates Atoh7 and Neurog2 in the distal mouse retina. Development *141*, 3243–3254.

McLaughlin, D., Shapley, R., Shelley, M., and Wielaard, D.J. (2000). A neuronal network model of macaque primary visual cortex (V1): Orientation selectivity and dynamics in the input layer 4Cα. Proceedings of the National Academy of Sciences *97*, 8087–8092.

Meehan, T.P., Bressler, S.L., Tang, W., Astafiev, S.V., Sylvester, C.M., Shulman, G.L., and Corbetta, M. (2017). Top-down cortical interactions in visuospatial attention. Brain Structure and Function *222*, 3127–3145.

Meinhardt, H. (2015). Models for patterning primary embryonic body axes: the role of space and time. Seminars in Cell & Developmental Biology *42*, 103–117.

Miesfeld, J.B., Ghiasvand, N.M., Marsh-Armstrong, B., Marsh-Armstrong, N., Miller, E.B., Zhang, P., Manna, S.K., Zawadzki, R.J., Brown, N.L., and Glaser, T. (2020). The Atoh7 remote enhancer provides transcriptional robustness during retinal ganglion cell development. Proceedings of the National Academy of Sciences *117*, 21690–21700.

Mizoguchi, N., Muramoto, K., and Kobayashi, M. (2020). Olfactory signals from the main olfactory bulb converge with taste information from the chorda tympani nerve in the agranular insular cortex of rats. Pflügers Archiv-European Journal of Physiology *472*, 721–732.

Mombaerts, P., Wang, F., Dulac, C., Chao, S.K., Nemes, A., Mendelsohn, M., Edmondson, J., and Axel, R. (1996). Visualizing an olfactory sensory map. Cell *87*, 675–686.

Montcouquiol, M., Rachel, R.A., Lanford, P.J., Copeland, N.G., Jenkins, N.A., and Kelley, M.W. (2003). Identification of Vangl2 and Scrb1 as planar polarity genes in mammals. Nature *423*, 173–177.

Moody, S.A., and LaMantia, A.-S. (2015). Transcriptional regulation of cranial sensory placode development. In Current Topics in Developmental Biology *11*, 301–350.

Muthusamy, K., Hanna, C., Johnson, D.R., Cramer, C.H., Tebben, P.J., Libi, S.E., Poling, G.L., Lanpher, B.C., Morava, E., and Schimmenti, L.A. (2021). Growth hormone deficiency in a child with branchio-oto-renal spectrum disorder: Clinical evidence of EYA1 in pituitary development and a recommendation for pituitary function surveillance. American Journal of Medical Genetics Part A *185*, 261–266.

Nakano, Y., Wiechert, S., Fritzsch, B., and Bánfi, B. (2020). Inhibition of a transcriptional repressor rescues hearing in a splicing factor–deficient mouse. Life Science Alliance *3*, e202000841.

Negrón-Piñeiro, L.J., Wu, Y., and Di Gregorio, A. (2020). Transcription factors of the bHLH family delineate vertebrate landmarks in the nervous system of a simple chordate. Genes *11*, 1262.

Neville, K.R., and Haberly, L. (2004). Olfactory cortex. In The Synaptic Organization of the Brain, G.M. Shepherd, ed. (New York: Oxford University Press), pp. 415–454.

Nichols, D.H., and Bruce, L.L. (2006). Migratory routes and fates of cells transcribing the Wnt-1 gene in the murine hindbrain. Developmental dynamics: An official publication of the American Association of Anatomists *235*, 285–300.

Nieuwenhuys, R., and Puelles, L. (2015). Towards a New Neuromorphology. (Springer, Heidelberg).

Niimura, Y., Ihara, S., and Touhara, K. (2020). Mammalian olfactory and vomeronasal receptor families. In The Senses, B. Fritzsch, ed. (Elsevier, Academic Press, Amsterdam, London), pp. 516–535.

Ochocinska, M.J., Muñoz, E.M., Veleri, S., Weller, J.L., Coon, S.L., Pozdeyev, N., Michael Iuvone, P., Goebbels, S., Furukawa, T., and Klein, D.C. (2012). NeuroD1 is required for survival of photoreceptors but not pinealocytes: Results from targeted gene deletion studies. Journal of Neurochemistry *123*, 44–59.

Okubo, T., Pevny, L.H., and Hogan, B.L. (2006). Sox2 is required for development of taste bud sensory cells. Genes and Devlopment *20*, 2654–2659.

Osório, J., Mueller, T., Rétaux, S., Vernier, P., and Wullimann, M.F. (2010). Phylotypic expression of the bHLH genes Neurogenin2, Neurod, and Mash1 in the mouse embryonic forebrain. Journal of Comparative Neurology *518*, 851–871.

Panaliappan, T.K., Wittmann, W., Jidigam, V.K., Mercurio, S., Bertolini, J.A., Sghari, S., Bose, R., Patthey, C., Nicolis, S.K., and Gunhaga, L. (2018). Sox2 is required for olfactory pit formation and olfactory neurogenesis through BMP restriction and Hes5 upregulation. Development *145*, dev153791.

Pan, B., Askew, C., Galvin, A., Heman-Ackah, S., Asai, Y., Indzhykulian, A.A., Jodelka, F.M., Hastings, M.L., Lentz, J.J., Vandenberghe, L.H., *et al.* (2017). Gene therapy restores auditory and vestibular function in a mouse model of Usher syndrome type 1c. Nature Biotechnology *35*, 264–272.

Pan, N., Jahan, I., Lee, J.E., and Fritzsch, B. (2009). Defects in the cerebella of conditional Neurod1 null mice correlate with effective Tg (Atoh1-cre) recombination and granule cell requirements for Neurod1 for differentiation. Cell and Tissue Research *337*, 407–428.

Pani, A.M., Mullarkey, E.E., Aronowicz, J., Assimacopoulos, S., Grove, E.A., and Lowe, C.J. (2012). Ancient deuterostome origins of vertebrate brain signalling centres. Nature *483*, 289–294.

Penfield, W., and Boldrey, E. (1937). Somatic motor and sensory representation in the cerebral cortex of man as studied by electrical stimulation. Brain *60*, 389–443.

Pergner, J., and Kozmik, Z. (2017). Amphioxus photoreceptors-insights into the evolution of vertebrate opsins, vision and circadian rhythmicity. International Journal of Developmental Biology *61*, 665–681.

Peterka, R. J. (2018). Sensory integration for human balance control. Handbook of Clinical Neurology, *159*, pp. 27–42.

Petro, L., Paton, A., and Muckli, L. (2017). Contextual modulation of primary visual cortex by auditory signals. Philosophical Transactions of the Royal Society B: Biological Sciences *372*, 20160104.

Pisani, D., Rota-Stabelli, O., and Feuda, R. (2020). Sensory neuroscience: A taste for light and the origin of animal vision. Current Biology *30*, R773–r775.

Qi, H.-X., Liao, C.-C., Reed, J.L., and Kaas, J.H. (2020). Cortical and subcortical plasticity after sensory loss in the somatosensory system of primates. In The Senses, B. Fritzsch, ed. (Elsevier, Academic Press, Amsterdam, London), pp. 399–418.

Qian, Y., Fritzsch, B., Shirasawa, S., Chen, C.-L., Choi, Y., and Ma, Q. (2001). Formation of brainstem (nor) adrenergic centers and first-order relay visceral sensory neurons is dependent on homeodomain protein Rnx/Tlx3. Genes & Development *15*, 2533–2545.

Qiu, X., and Müller, U. (2018). Mechanically gated ion channels in mammalian hair cells. Frontiers in Cellular Neuroscience *12*, 100.

Rauschecker, J.P. (2018). Where, when, and how: are they all sensorimotor? Towards a unified view of the dorsal pathway in vision and audition. Cortex 98, 262–268.

Rauschecker, J.P. (2020). The auditory cortex of primates including man with reference to speech. In The Senses, B. Fritzsch, ed. (Elsevier, Academic Press, Amsterdam, London), pp. 791–811.

Reiprich, S., and Wegner, M. (2015). From CNS stem cells to neurons and glia: Sox for everyone. Cell and Tissue Research *359*, 111–124.

Riddiford, N., and Schlosser, G. (2016). Dissecting the pre-placodal transcriptome to reveal presumptive direct targets of Six1 and Eya1 in cranial placodes. eLife *5*, e17666.

Roper, S.D., and Chaudhari, N. (2017). Taste buds: Cells, signals and synapses. Nature Reviews Neuroscience *18*, 485–497.

Ryan, K., Lu, Z., and Meinertzhagen, I.A. (2018). The peripheral nervous system of the ascidian tadpole larva: Types of neurons and their synaptic networks. Journal of Comparative Neurology *526*, 583–608.

Samuelsen, C.L., and Fontanini, A. (2017). Processing of intraoral olfactory and gustatory signals in the gustatory cortex of awake rats. Journal of Neuroscience *37*, 244–257.

Saper, C., and Loewy, A. (1980). Efferent connections of the parabrachial nucleus in the rat. Brain Research *197*, 291–317.

Schlosser, G. (2015). Vertebrate cranial placodes as evolutionary innovations—the ancestor's tale. Current Topics in Developmental Biology *111*, 235–300.

Schlosser, G. (2018). A short history of nearly every sense—the evolutionary history of vertebrate sensory cell types. Integrative and Comparative Biology 58, 301–316.

Schwob, J.E., Jang, W., Holbrook, E.H., Lin, B., Herrick, D.B., Peterson, J.N., and Hewitt Coleman, J. (2017). Stem and progenitor cells of the mammalian olfactory epithelium: Taking poietic license. Journal of Comparative Neurology 525, 1034–1054.

Shaker, T., Dennis, D., Kurrasch, D.M., and Schuurmans, C. (2012). Neurog1 and Neurog2 coordinately regulate development of the olfactory system. Neural Development 7, 1–23.

Shepherd, G.M. (2006). Smell images and the flavour system in the human brain. Nature 444, 316–321.

Shibata, S.B., Ranum, P.T., Moteki, H., Pan, B., Goodwin, A.T., Goodman, S.S., Abbas, P.J., Holt, J.R., and Smith, R.J. (2016). RNA interference prevents autosomal-dominant hearing loss. The American Journal of Human Genetics 98, 1101–1113.

Simões-Costa, M., and Bronner, M.E. (2015). Establishing neural crest identity: A gene regulatory recipe. Development 142, 242–257.

Sokpor, G., Abbas, E., Rosenbusch, J., Staiger, J.F., and Tuoc, T. (2018). Transcriptional and epigenetic control of mammalian olfactory epithelium development. Molecular Neurobiology 55, 8306–8327.

Somorjai, I.M., Martí-Solans, J., Diaz-Gracia, M., Nishida, H., Imai, K.S., Escrivà, H., Cañestro, C., and Albalat, R. (2018). Wnt evolution and function shuffling in liberal and conservative chordate genomes. Genome Biology 19, 98.

Stellwagen, D., and Shatz, C. (2002). An instructive role for retinal waves in the development of retinogeniculate connectivity. Neuron 33, 357–367.

Steventon, B., Mayor, R., and Streit, A. (2012). Mutual repression between Gbx2 and Otx2 in sensory placodes reveals a general mechanism for ectodermal patterning. Developmental Biology 367, 55–65.

Stolfi, A., Ryan, K., Meinertzhagen, I.A., and Christiaen, L. (2015). Migratory neuronal progenitors arise from the neural plate borders in tunicates. Nature 527, 371–374.

Striedter, G.F., and Northcutt, R.G. (2019). Brains through Time: A Natural History of Vertebrates (Oxford University Press, New York, NY).

Suryanarayana, S.M., Pérez-Fernández, J., Robertson, B., and Grillner, S. (2020). The evolutionary origin of visual and somatosensory representation in the vertebrate pallium. Nature Ecology & Evolution 4, 639–651.

Suryanarayana, S.M., Pérez-Fernández, J., Robertson, B., and Grillner, S. (2021). Olfaction in Lamprey Pallium Revisited—Dual Projections of Mitral and Tufted Cells. Cell Reports 34, 108596.

Suzuki, D.G., and Grillner, S. (2018). The stepwise development of the lamprey visual system and its evolutionary implications. Biological Reviews 93, 1461–1477.

Tang, W.J., Chen, J.S., and Zeller, R.W. (2013). Transcriptional regulation of the peripheral nervous system in Ciona intestinalis. Developmental Biology 378, 183–193.

Tarchini, B., and Lu, X. (2019). New insights into regulation and function of planar polarity in the inner ear. Neuroscience Letters 709, 134373.

Taroc, E.Z.M., Naik, A.S., Lin, J.M., Peterson, N.B., Keefe, D.L., Genis, E., Fuchs, G., Balasubramanian, R., and Forni, P.E. (2020). Gli3 regulates vomeronasal neurogenesis, olfactory ensheathing cell formation, and GnRH-1 neuronal migration. Journal of Neuroscience 40, 311–326.

Tassinari, G., Bentivoglio, M., Chen, S., and Campara, D. (1997). Overlapping ipsilateral and contralateral retinal projections to the lateral geniculate nucleus and superior colliculus in the cat: A retrograde triple labelling study. Brain Research Bulletin 43, 127–139.

Taube, J.S., and Yoder, R.M. (2020). The impact of vestibular signals on cells responsible for orietation and navigation. In The Senses, B. Fritzsch, ed. (Elsevier, Academic Press, Amsterdam, London), pp. 496–511.

Teichert, M., and Bolz, J. (2017). Simultaneous intrinsic signal imaging of auditory and visual cortex reveals profound effects of acute hearing loss on visual processing. Neuroimage *159*, 459–472.

Thiery, A., Buzzi, A.L., and Streit, A. (2020). Cell fate decisions during the development of the peripheral nervous system in the vertebrate head. Current Topics in Developmental Biology *139*, 127–167.

Tosches, M.A., and Laurent, G. (2019). Evolution of neuronal identity in the cerebral cortex. Current Opinion in Neurobiology *56*, 199–208.

Touhara, K., Niimura, Y., and Ihara, S. (2016). Vertebrate Odorant Receptors. In Chemosensory Transduction (Elsevier, Academic Press, Amsterdam), pp. 49–66.

Triplett, J.W., and Feldheim, D.A. (2012). Eph and ephrin signaling in the formation of topographic maps. Seminars in Cell & Developmental Biology *23*, 7–15.

Urness, L.D., Bleyl, S.B., Wright, T.J., Moon, A.M., and Mansour, S.L. (2011). Redundant and dosage sensitive requirements for Fgf3 and Fgf10 in cardiovascular development. Developmental Biology *356*, 383–397.

van der Heijden, M.E., and Zoghbi, H.Y. (2020). Development of the brainstem respiratory circuit. Wiley Interdisciplinary Reviews: Developmental Biology *9*, e366.

Van Der Kooij, H., & Peterka, R.J. (2011). Non-linear stimulus-response behavior of the human stance control system is predicted by optimization of a system with sensory and motor noise. Journal of Computational Neuroscience 30(3), 759–778.

Veeman, M.T., Newman-Smith, E., El-Nachef, D., and Smith, W.C. (2010). The ascidian mouth opening is derived from the anterior neuropore: Reassessing the mouth/neural tube relationship in chordate evolution. Developmental Biology *344*, 138–149.

Veeman, M., and Reeves, W. (2015). Quantitative and in toto imaging in ascidians: Working toward an image-centric systems biology of chordate morphogenesis. Genesis *53*, 143–159.

Wang, V.Y., Rose, M.F., and Zoghbi, H.Y. (2005). Math1 expression redefines the rhombic lip derivatives and reveals novel lineages within the brainstem and cerebellum. Neuron *48*, 31–43.

Watson, C., Shimogori, T., and Puelles, L. (2017). Mouse Fgf8-Cre-LacZ lineage analysis defines the territory of the postnatal mammalian isthmus. Journal of Comparative Neurology *525*, 2782–2799.

Wilson, D.A., and East, B.S. (2020). Function of the olfactory cortex. In The Senses, B. Fritzsch, ed. (Elsevier, Academic Press, Amsterdam, London), pp. 661–674.

Winkley, K.M., Kourakis, M.J., DeTomaso, A.W., Veeman, M.T., and Smith, W.C. (2020). Tunicate gastrulation. In Current Topics in Developmental Biology (Elsevier, Academic Press, Cambridge, MA), pp. 219–242.

Witt, M. (2020). Anatomy and development of the human gustatory and olfactory systems. In The Senses, B. Fritzsch, ed. (Elsevier, Academic Press, Amsterdam, London), pp. 85–118.

Wright, T.J., and Mansour, S.L. (2003). Fgf3 and Fgf10 are required for mouse otic placode induction. Development *130*, 3379–3390.

Wu, F., Bard, J.E., Kann, J., Yergeau, D., Sapkota, D., Ge, Y., Hu, Z., Wang, J., Liu, T., and Mu, X. (2021). Single cell transcriptomics reveals lineage trajectory of retinal ganglion cells in wild-type and Atoh7-null retinas. Nature Communications *12*, 1–20.

Wullimann, M.F., and Grothe, B. (2013). The central nervous organization of the lateral line system. In The Lateral Line System (Springer, New York, NY), pp. 195–251.

Xiang, M., Maklad, A., Pirvola, U., and Fritzsch, B. (2003). Brn3c null mutant mice show long-term, incomplete retention of some afferent inner ear innervation. BMC Neuroscience *4*, 2.

Xu, P.X., Adams, J., Peters, H., Brown, M.C., Heaney, S., and Maas, R. (1999). Eya1-deficient mice lack ears and kidneys and show abnormal apoptosis of organ primordia. Nature Genetics *23*, 113–117.

Xu, J., Li, J., Zhang, T., Jiang, H., Ramakrishnan, A., Fritzsch, B., Shen, L., and Xu, P.X. (2021). Chromatin remodelers and lineage-specific factors interact to target enhancers to establish proneurosensory fate within otic ectoderm. Proceedings of the National Academy of Sciences of the United States of America *118*, e2025196118.

Xu, P.-X., Woo, I., Her, H., Beier, D.R., and Maas, R.L. (1997). Mouse Eya homologues of the Drosophila eyes absent gene require Pax6 for expression in lens and nasal placode. Development *124*, 219–231.

Yamoah, E.N., Li, M., Shah, A., Elliott, K.L., Cheah, K., Xu, P.-X., Phillips, S., Young Jr, S.M., Eberl, D.F., and Fritzsch, B. (2020). Using Sox2 to alleviate the hallmarks of age-related hearing loss. Ageing Research Reviews, 101042.

Zhang, T., Xu, J., and Xu, P.X. (2021). Eya2 expression during mouse embryonic development revealed by Eya2(lacZ) knockin reporter and homozygous mice show mild hearing loss. Developmental dynamics: An official publication of the American Association of Anatomists.

Zhou, G., Lane, G., Noto, T., Arabkheradmand, G., Gottfried, J.A., Schuele, S.U., Rosenow, J.M., Olofsson, J.K., Wilson, D.A., and Zelano, C. (2019). Human olfactory-auditory integration requires phase synchrony between sensory cortices. Nature Communications *10*, 1–12.

Zou, D., Silvius, D., Fritzsch, B., and Xu, P.-X. (2004). Eya1 and Six1 are essential for early steps of sensory neurogenesis in mammalian cranial placodes. Development *131*, 5561–5572.

Index

Note: Locators in *italics* represent figures and **bold** indicate tables in the text.

A

Accessory olfactory bulb (AOB), 37
Accessory optic system (AOS), 73–74
Acid tastes, *see* Sour taste
AIS, *see* Axon initial segment plasticity
ALL, *see* Anterior lateral line
ALLN, *see* Anterior lateral line nerve
Amphibians
    amphibian papilla, 175–176, *176*
    bilateral projection, 77
    inner ear neuronal proliferation, 178
    lateral line afferents, 207
    mechano- and electroreceptors, 244
    proliferation of hair cells, 183
    vestibular information processing, 162
Amphioxus (lancelet), 62–64, *63, 70*
    expression of *Wnt1* and *Wnt3,* 258
    formation of glial cells, 258
    'frontal eye,' 256, 258
    gene expression at MHB, 258, *259*
    lack of *Atoh* and *Neurog* expression, 257–258
    neural crest cells and placodes, 256, *257*
    'olfactorans' (urochordates, vertebrates), 256
Ampullae of Lorenzini, 227, 229; *see also* Electroreception
Ampullary organs (AO), 224, *226–227,* 227, 229–231, *232, 238, 243*
Antagonism, 32
Anterior lateral line (aLL), 225, 227
Anterior lateral line nerve (aLLN), 206
Anterior olfactory nucleus (AONpE), 34
Anterior vestibular-thalamic pathway (ADN), 275
Anteroventral cochlear nucleus (AVCN), 184
Anurans, 175, 211, 213, 224, 262
AO, *see* Ampullary organs
AOB, *see* Accessory olfactory bulb
AOS, *see* Accessory optic system
Ascidians; *see also* Tunicates
    'bipolar tail neurons,' 261
    expression of *Atoh8,* 260
    Lmx genes, 261
    Neurog and *Atoh,* 260–262
    opsin-1 gene is, 260
    vision receptors, 260
Auditory system
    cochlear nucleus, 184–186
    cortex projections, 189

development of neurons and peripheral connections, 178–181
evolution, 175, *176*
hair cells and heterogeneity, 181–184
higher-order projections, 187–189
in *Latimeria,* 175–176, *176*
Neurod1, 178, *179*
Neurog1, 178, 181, 182
neurotrophic factors, 180
phylogenetic differences, 175–178
SGN neurites, 179–180
spiral ganglion central projections, *185*
AVCN, *see* Anteroventral cochlear nucleus
Axon initial segment (AIS) plasticity, 48

B

Basal optic tract (BOT), 72, 73–74, 77
Basic helix-loop-helix (bHLH) genes
    downstream of *Sox2* and *Eya1,* 263–264, 267
    Eya2 expression, 263
    Eya1/Six1 expression, 268
    Foxi3 or *Fgf3/10,* 264
    genes involved in neurosensory development, 264, **265–266**
    NeuroD family, 267
    Neurog1 and *Neurog2,* 267
    Pou4f1 and *Pou4f2,* 267
    transcription factors, 238
BHLH, *see* Basic helix-loop-helix genes
Bioelectricity, 223–224, 230
Bitter taste, 113, 119, 120, 121; *see also* Taste buds
BMPs, *see* Bone morphogenetic proteins
Body patterning, 4
    Chordin and *BMP,* 4
    Nodal, Fgfs, and *Otx,* 4
    Noggin, Chordin, and *Follistatin,* 4
    Sog and *Dpp,* 4
Bone morphogenetic proteins (BMPs), 238
Bony fish
    electroreception in, 227, 229
    gene duplication, 233
    non-neopterygian, 235
    olfactory system, 37
BOR, *see* Branchio-oto-renal spectrum disorder
BOT, *see* Basal optic tract
Brain-derived neurotrophic factor (BDNF), 233
Branchio-oto-renal spectrum disorder (BOR), 263